D1210056

# X-Ray Diffraction

# X-Ray Diffraction

## In Crystals, Imperfect Crystals, and Amorphous Bodies

A. GUINIER

UNIVERSITÉ PARIS-SUD (ORSAY)

TRANSLATED BY

Paul Lorrain

AND

Dorothée Sainte-Marie Lorrain

DOVER PUBLICATIONS, INC.

NEW YORK

Published in Canada by General Publishing Company, Ltd., 30 Lesmill Road, Don Mills, Toronto, Ontario.

*Bibliographical Note*

This Dover edition, first published in 1994, is an unabridged and unaltered republication of the English edition of the work first published by W. H. Freeman and Company, San Francisco, in 1963. It was translated from the second French edition published by Dunod, Paris, in 1956

*Library of Congress Cataloging-in-Publication Data*

Guinier, André.
[Radiocristallographie. English]
X-ray diffraction in crystals, imperfect crystals, and amorphous bodies / A. Guinier ; translated by Paul Lorrain and Dorothée Sainte-Marie Lorrain.
p. cm.
Previously published: San Francisco : W.H. Freeman, 1963.
Includes index.
ISBN 0-486-68011-8
1. X-ray crystallography. I. Title.
QD945.G8513 1994
548′.83—dc20 94-7410
CIP

Manufactured in the United States of America
Dover Publications, Inc., 31 East 2nd Street, Mineola, N.Y. 11501

# PREFACE

This book is a revised version of the last part of *Théorie et Technique de la Radiocristallographie* (Dunod, Paris, 1956). Since there now exist a number of excellent texts on X-ray crystallography in the English language, it was judged unnecessary to translate Parts I to IV, which deal with the experimental methods, with descriptions of apparatus, and with the elementary results of X-ray diffraction by crystals. We have added an introductory chapter so as to obtain a logical sequence, and recent developments were integrated throughout the text.

We have chosen to start with the general theory of diffraction in the concise and elegant form rendered possible by the use of Fourier transforms. We then apply the general results to various atomic structures, to amorphous bodies, to crystals, and to imperfect crystals, and the elementary laws of X-ray diffraction on crystals follow as a special case. It would of course have been simpler to establish these laws directly, but the method we have used has the advantage of illustrating clearly the meaning of the general equations which are essential for the study of more complex cases. We have assumed that the reader is familiar with the elements of crystallography and X-ray diffraction as discussed, for example, in the elementary books listed at the end of Chapter 1.

We have completely ignored the problem of determining crystal structures, but we have discussed at length the cases of imperfect crystals, amorphous bodies, and liquids. This book should therefore be especially useful for solid-state physicists, metallographers, chemists, and even biologists. More generally, it should be useful to all those who require a thorough understanding of diffraction theory so as to interpret properly the information provided by modern X-ray diffraction instruments on line profiles, line intensities, diffuse scattering, and other phenomena.

In our modern scientific world it is well known that authors who do not publish in the English language are restricted to a much smaller audience than their Anglo-Saxon colleagues. To judge from the trend of the past few years, it appears that this situation will not change in the near future. Thus, for example, there are still relatively few American scientists who as a matter of routine utilize books written in foreign languages; on the other hand, nearly all European scientists utilize American books without difficulty. The exchange of scientific information between America and Europe is thus basically asymmetrical. Translation is therefore, and will long remain, a necessity for the dissemination of European books in America.

I wish to express my deep gratitude to Professor Paul Lorrain and Mrs. Dorothée Lorrain, who have, in fact, rewritten the book in English, while remaining faithful to the original text.

January, 1963

A. Guinier

# CONTENTS

CHAPTER

# 1

# FUNDAMENTALS OF X-RAY DIFFRACTION THEORY

## 1.1. General Properties of X-Rays

### 1.1.1. The Nature of X-Rays

X-rays are transverse electromagnetic radiations, like visible light, but of much shorter wavelength. The range of wavelengths corresponding to X-rays is ill-defined, but it extends from radiations which are identical to ultraviolet light to others which are identical to the gamma rays emitted by radioactive substances. We shall consider here only the range of wavelengths which is commonly used for X-ray crystallography, namely 0.5 to 2.5Å.

Electromagnetic radiation has two complementary aspects: some experiments require a wave interpretation, while others can be understood only in terms of photons. In this book we shall be concerned mainly with the first of these aspects; however, it is in some cases useful to consider photons of energy $h\nu = hc/\lambda$, where $h$ is Planck's constant, $6.62 \times 10^{-34}$ joule-second, and $c$ is the velocity of light, $2.998 \times 10^{8}$ m/sec.

### 1.1.2. Geometrical Optics of X-Rays

The geometrical optics of X-rays is simple compared to that of visible light. Whatever be the medium in which they propagate, X-rays have approximately the same velocity as light in a vacuum, the difference being always smaller than one part in $10^{4}$. *They are therefore practically not deviated by refraction.* The index of refraction $n$ is slightly smaller than unity; for a pure substance, the theoretical formula is

$$n = 1 - \delta = 1 - \frac{1}{2\pi} r_e N \lambda^2 , \qquad (1.1)$$

where $r_e$ is the classical radius of the electron, which is numerically equal to $2.818 \times 10^{-15}$ m, $N$ is the number of electrons per cubic meter, and $\lambda$ is the wavelength in meters [1].

Numerically, if $\rho$ is the density, if $Z$ is the number of electrons in the atomic group (which can be either an isolated atom, a molecule, or a unit cell in a crystal lattice) of atomic mass $M$, and if $\lambda$ is expressed in Ångströms,

$$\delta = 2.72 \times 10^{-6} \frac{Z}{M} \rho \lambda^2 .$$

Since $M/Z$ is always approximately equal to 2,

$$\delta \cong 1.3 \times 10^{-6} \rho \lambda^2 . \qquad (1.2)$$

Even for the densest substances and for $\lambda = 2$Å, $\delta$ is still of the order of $10^{-4}$. In most cases, $\delta$ is of the order of $10^{-5}$.

Since X-rays always propagate along straight lines, it follows that they cannot be focused with lenses. On the other hand, there do exist mirrors for X-rays. When a beam propagating in air meets the surface of a solid medium whose index $n$ is less than unity, total reflection phenomena can be expected for grazing incidence. The general formula for the critical angle of incidence, $\theta_c$, which is the angle between the incident ray and the reflecting surface, is

$$\sin\left(\frac{\pi}{2} - \theta_c\right) = n . \qquad (1.3)$$

However, since $n$ is close to unity, it follows from Eqs. (*1.2*) and (*1.3*) that

$$\theta_c = (2\delta)^{1/2} = 1.6 \times 10^{-3} \rho^{1/2} \lambda . \qquad (1.4)$$

For usual values of $\rho$ and $\lambda$, $\theta_c$ is of the order of 10–30′ [2]. When the angle of incidence is larger than $\theta_c$, the energy which is reflected according to the laws of optics is negligible, as can be shown from the Fresnel formulas [3].

### 1.1.3. Polarization of X-Rays

X-rays, like visible light, can be linearly polarized, either partially or totally, and the polarization factor will be used in many of our formulas. This polarization was demonstrated by the experiments of Barkla [4]. Polarized beams of X-rays can be obtained by scattering from a solid body and, for a scattering angle of 90°, polarization is complete, the electric vector being normal to the plane of the incident and scattered rays.

### 1.1.4. Definition of the Intensity of a Beam of X-Rays

The intensity of an approximately parallel beam of X-rays is the flux of energy which crosses a unit surface normal to the average ray of the beam per second. For a monochromatic plane wave, it is well known that the intensity is proportional to the square of the amplitude of the vibration.

The intensity of radiation emitted by a point source, or by a quasi-point source, in a given direction, is the energy emitted per second by the source per unit solid angle in the direction considered. For absolute intensity measurements, the simplest method is to determine the number of photons emitted or received per second, either per unit area or per unit solid angle.

### 1.1.5. Absorption of X-Rays

A beam of X-rays is attenuated in passing through matter. The interaction between the radiation and matter is complex and the photons which do not appear in the transmitted beam may have been subjected to various transformations.

(a) Some may remain photons but be deviated from their course, without loss of energy: this is the *scattered radiation*; it has the same wavelength as the incident beam. Others may also be scattered with a slight loss of energy, i.e., with a slight change of wavelength; this is *Compton scattering* or incoherent scattering.

(b) Photons can be absorbed by atoms by the *photoelectric effect*. After absorbing a photon, the atom is excited and an electron is ejected. The atom can then return to its ground state, emitting either another electron (*Auger effect*), or X-photons which are *fluorescence* X-rays whose wavelength is characteristic of the excited atom.

Total absorption or energy loss by passage through matter is due both to the photoelectric effect and to scattering, but the latter produces a relatively negligible effect except for short wavelength radiation in substances of low atomic weight.

The coefficient of absorption $\mu$ of a shield is defined as follows. We consider a beam of monochromatic X-rays of intensity $I$ incident perpendicularly on a thin sheet whose mass per unit area is $dp$. The intensity of the transmitted beam is then $I + dI$, with $dI < 0$. It is found experimentally that $dI$ is proportional both to $I$ and to $dp$. Thus

$$dI = -\mu I \, dp \, , \tag{1.5}$$

where $\mu$ is the mass absorption coefficient of the substance. This coefficient is usually expressed in c.g.s. units. For a shield of finite thickness, it is found by integrating Eq. (*1.5*) that the ratio of the transmitted intensity $I$ to the incident intensity $I_0$ is related to the mass $p$ per square centimeter according to the formula

$$\frac{I}{I_0} = \exp(-\mu p) \, . \tag{1.6}$$

We can also express the absorption as a function of the thickness of the shield $x$ (cm) and its density $\rho$:

$$\frac{I}{I_0} = \exp(-\mu\rho x) \, ; \tag{1.7}$$

the coefficient $\mu\rho$ is called the *linear absorption coefficient*. We shall use mostly the mass coefficient, which does not depend on the particular physical state of the material.

For a given element the coefficient $\mu$ increases, in general, with the wavelength, but this increase is not continuous. There are *absorption discontinuities* for certain wavelengths called, in increasing order, $\lambda_K$, $\lambda_L$, $\lambda_M$, $\cdots$.

On crossing a discontinuity the mechanism of absorption changes and the new electronic shell $K, L, M, \cdots$ is ionized by radiations of wavelengths shorter than the corresponding critical wavelength. At the same time X-rays of the $K, L, M, \cdots$ series characteristic of the excited atom are emitted by fluorescence.

## 1.2. X-Ray Diffraction Calculations

The diffraction of X-rays by matter results from the combination of two different phenomena: (a) scattering by each individual atom, and (b) interference between the waves scattered by these atoms. This interference occurs because the waves scattered by the individual atoms are coherent with the incident wave, and therefore between themselves. All of our scattering calculations will be based on these two elementary phenomena, whose principal characteristics will be recalled here.

### 1.2.1. Interference Calculations

Let us consider a monochromatic parallel beam of X-rays and a scattering center which can be either an atom or an electron. If we set $A_0 \cos 2\pi\nu t$ to be the amplitude of the incident vibration at the point $O$ (Figure 1.1), the amplitude of the scattered wave at the distance $r$ from

$O$ is given by

$$A = fA_0 \cos\left[2\pi\nu\left(t - \frac{r}{c}\right) - \psi\right],$$

where $f$, which is called the *scattering factor*, is the ratio between the incident and scattered amplitudes. The quantity $f$ is, in general, a function of the angle $2\theta$ between the incident and scattered rays, and $\psi$ is the scattering phase shift.

Using complex notation,

$$\begin{aligned} A &= fA_0 \exp\left[2\pi i\nu\left(t - \frac{r}{c}\right) - i\psi\right] \\ &= \left[f \exp\left(-i\psi\right)\right] A_0 \exp\left[2\pi i\nu\left(t - \frac{r}{c}\right)\right]. \end{aligned} \tag{1.8}$$

This introduces a complex scattering factor:

$$f \exp(-i\psi).$$

When the scattering center is a free electron, $\psi$ is equal to $\pi$, and this is generally so for an atom; the scattered wave then has a phase opposite to that of the incident wave. Later on we shall meet cases where the factor $f$ is itself a complex number.

1.2.1.1. THE **s** VECTOR AND RECIPROCAL SPACE. Let us now consider two scattering centers $O$ and $M$ and let us calculate the resultant scattered radiation in a given direction (Figure 1.1). This radiation results from the interference of the waves scattered by $O$ and $M$. Let us assume that the phase shift $\psi$ due to scattering is the same for the two atoms. This will evidently be the case if the atoms are identical; except for a few rare cases, this is correct even for different atoms.

The phase difference between the emitted waves depends on the respective positions of the two scattering centers $O$ and $M$. The wave fronts of the incident and diffracted waves passing through $O$ are $(\pi_0)$ and $(\pi)$. The path length for the ray going through $M$ is greater by $\delta = mM + Mn$, $m$ and $n$ being the projections of $O$ on the rays through $M$. We shall define the directions of the incident and scattered rays by the unit vectors $\mathbf{S}_0$ and $\mathbf{S}$. Vectorially,

$$\begin{aligned} mM &= \mathbf{S}_0 \cdot \mathbf{OM}, \\ Mn &= -\mathbf{S} \cdot \mathbf{OM}, \\ \delta &= -\mathbf{OM} \cdot (\mathbf{S} - \mathbf{S}_0). \end{aligned} \tag{1.9}$$

The phase difference is therefore

$$\varphi = \frac{2\pi\delta}{\lambda} = -2\pi\mathbf{OM} \cdot \frac{\mathbf{S} - \mathbf{S}_0}{\lambda}. \tag{1.10}$$

**FIGURE 1.1.** *Interference between the waves originating at two scattering centers.*

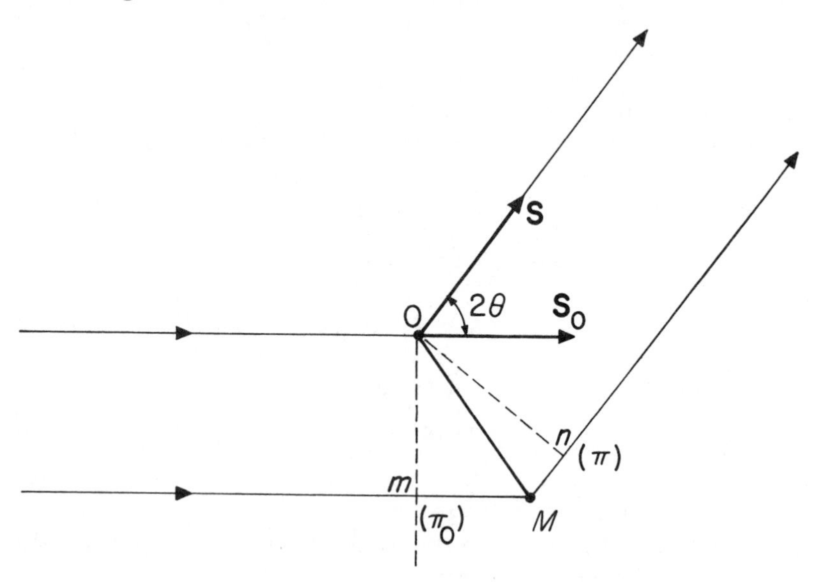


We shall call $\boldsymbol{s}$ the vector $(S - S_0)/\lambda$. This vector plays a fundamental role in scattering theory and will be used throughout all our scattering calculations. It is therefore important to understand its properties thoroughly. As shown in Figure 1.2, its direction is that of $ON$, which bisects the angle formed between $S$ and $-S_0$. Its length is equal to

**FIGURE 1.2.** *Definition of the vector* $\boldsymbol{s} = (S - S_0)/\lambda$.

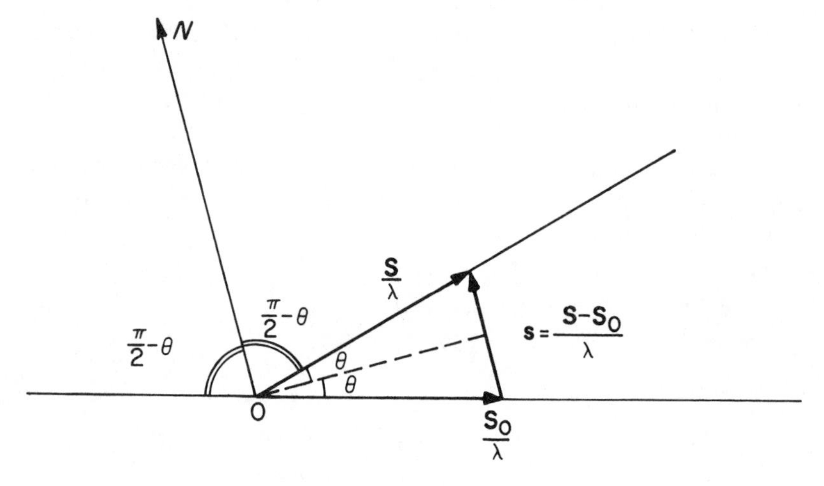

$s = (2/\lambda) \sin [(\boldsymbol{S}, \boldsymbol{S}_0)/2]$. For simplicity we shall set the scattering angle $(\boldsymbol{S}, \boldsymbol{S}_0)$ equal to $2\theta$. We therefore have the relation

$$s = \frac{2 \sin \theta}{\lambda}, \tag{1.11}$$

which will be used throughout this book.

Since the phase differences depend only on the vector $\boldsymbol{s}$, interference calculations do not depend explicitly on the three parameters $\boldsymbol{S}$, $\boldsymbol{S}_0$, and $\lambda$, but only on the combination $(\boldsymbol{S} - \boldsymbol{S}_0)/\lambda = \boldsymbol{s}$.

All scattering measurements on a given object can be reduced to the determination of the value of the scattered intensity $I$, for all possible values of $\boldsymbol{s}$. We shall represent the function $I(\boldsymbol{s})$ in the so-called *reciprocal space*. For each point in this space there corresponds a vector $\boldsymbol{s}$ which is equal to the vector between the origin and that point.

To each measurement there corresponds one point in reciprocal space but, inversely, for each point of this space, that is for a given vector $\boldsymbol{s}$, there correspond all of the measurements for which the vector $\boldsymbol{s}$ bisects the angle between $\boldsymbol{S}$ and $-\boldsymbol{S}_0$, and for which the diffraction angle is related to the wavelength according to Eq. (*1.11*). All measurements so defined are equivalent from the point of view of the diffraction conditions.

Let us consider in particular the group of diffraction experiments performed on a given object with a given parallel monochromatic beam of X-rays. In such a case $\boldsymbol{S}_0$ and $\lambda$ are fixed. We investigate the dependence of the diffracted intensity on the direction of the vector $\boldsymbol{S}$. From a point $S_0$, chosen as the origin of the reciprocal space (Figure 1.3), let us draw the vector $\boldsymbol{S_0O} = -\boldsymbol{S}_0/\lambda$ and, using $O$ as center, let us draw the sphere of radius $1/\lambda$. For any given direction of observation, the point $S$ defined by the vector $\boldsymbol{OS} = \boldsymbol{S}/\lambda$ is located on the sphere defined above and $S_0S$ represents the vector $\boldsymbol{s} = (\boldsymbol{S} - \boldsymbol{S}_0)/\lambda$.

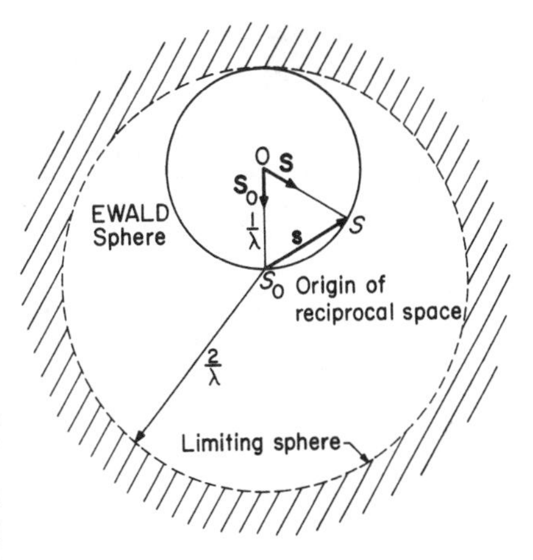


**FIGURE 1.3.** *The reciprocal space explored by successive sections through the Ewald sphere.*

This is the *Ewald sphere*. The investigation of the diffraction in all space by a fixed object and for a given incident ray involves effectively the determination of the values of the diffracted intensity $I(\boldsymbol{s})$ at the surface of the Ewald sphere. In other words, one effectively takes a section of reciprocal space through the Ewald sphere.

If, with the object fixed in position, we change the direction of the incident rays, $\lambda$ being kept constant, and if we perform a series of experiments to determine the diffraction intensity as a function of the direction of diffraction, we effectively determine $I(\boldsymbol{s})$ at the surface of another Ewald sphere, rotated with respect to the first one around the origin $\boldsymbol{S}_0$ of reciprocal space. We therefore see that if we utilize successively all possible orientations of $\boldsymbol{S}_0$ we can explore completely that portion of reciprocal space which is contained within the sphere of radius $2/\lambda$ centered on the origin (Figure 1.3). With radiations of shorter and shorter wavelengths, it is theoretically possible to explore as large a volume as desired in reciprocal space.

1.2.1.2. CALCULATION OF THE RESULTANT AMPLITUDE. We now have to calculate the amplitude resulting from the addition of the two waves originating at $O$ and at $M$. This is a classical problem in optics and we first consider the geometrical method of Fresnel. We use a diagram where the amplitude is represented by a vector whose length is proportional to its modulus and which forms with an arbitrary direction an angle equal to the phase angle. The amplitude resulting from the interference of several waves will be represented by the vector sum of the component amplitudes.

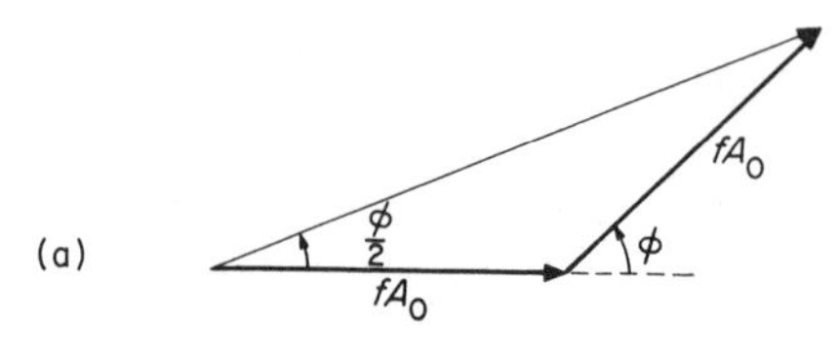


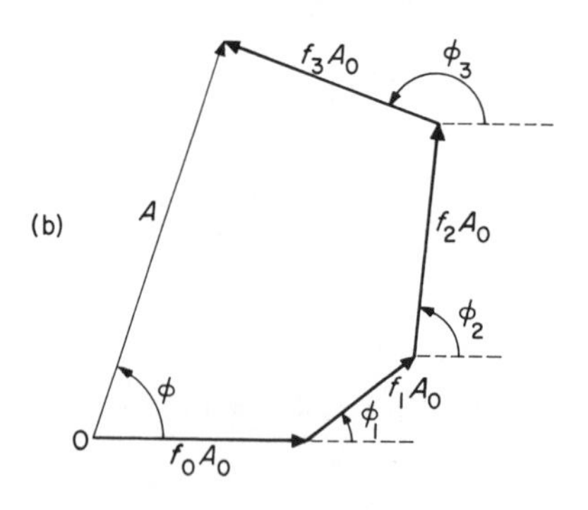


**FIGURE 1.4.** *Fresnel construction.* (a) *For two equal amplitudes.* (b) *For the general case.*

Let us consider two waves scattered by two identical atoms with scattering factor $f$ and for which the phase difference [Eq. (*1.10*)] is $\varphi$, as in Figure 1.4a. The length of $A_1$ and $A_2$ is $fA_0$, so that

$$A = 2fA_0 \cos \frac{\varphi}{2} . \qquad (1.12)$$

The maximum absolute value is obtained when the phase difference $\varphi$ is zero; then $A = 2fA_0$. But the value is zero for $\varphi = \pi$, the two waves cancelling completely by interference.

The Fresnel method is easily generalized for any number of scattering centers with scattering factors $f_0, f_1, f_2 \cdots f_n$ and corresponding phase differences $0, \varphi_1, \varphi_2 \cdots \varphi_n$. Figure 1.4b shows that

$$A = A_0[(\Sigma_0^n f_i \cos \varphi_i)^2 + (\Sigma_0^n f_i \sin \varphi_i)^2]^{1/2} \qquad (1.13)$$

In general we shall use an analytic method which can be deduced from the geometrical construction of Fresnel, using complex notation for all the operations on sinusoidal quantities.

The complex amplitude is equal to $A \exp(i\varphi)$, $A$ being the modulus and $\varphi$ the phase with respect to an arbitrary origin. The algebraic sum of component imaginary amplitudes is given by

$$A = A_0 \Sigma_0^n f_i \exp(i\varphi_i) . \qquad (1.14)$$

This is a complex number whose modulus is that of the amplitude and whose argument is the phase with respect to the common origin. We can deduce from this the amplitude as given in Eq. (*1.13*):

$$|\Sigma_0^n f_i \exp i\varphi_i| = |\Sigma_0^n f_i \cos\varphi_i + i\,\Sigma_0^n f_i \sin\varphi_i|$$
$$= [(\Sigma_0^n f_i \cos\varphi_i)^2 + (\Sigma_0^n f_i \sin\varphi_i)^2]^{1/2} .$$

The intensity is given by the square of the modulus of the amplitude. It is also the product of the complex amplitude by its complex conjugate,

$$I = |A|^2 = AA^* . \qquad (1.15)$$

### 1.2.2. Scattering by a Free Electron

We now have to evaluate the scattering factor. The simplest type of scattering center is a free electron.

Let us consider a free electron in a parallel beam of X-rays of intensity $I_0$. Let us first consider a plane-polarized wave propagating along $OX$, with an electric vector $\boldsymbol{E}_0$, incident on an electron at rest at the origin but completely free from any restraining force (Figure 1.5). This electron oscillates because it is subjected to an alternating acceleration whose amplitude $\gamma = \boldsymbol{E}_0 e/m$. According to classical electromagnetic theory [5], an accelerated electron emits electromagnetic radiation whose electric vector at the point $P$ is given by

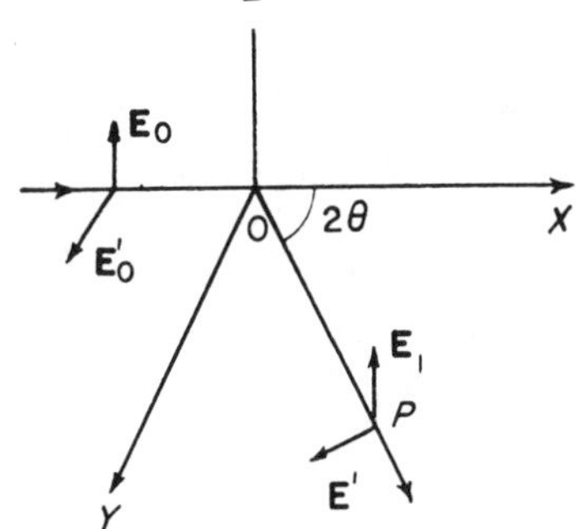


**FIGURE 1.5.** *Scattering of a wave by a free electron.*

$$\boldsymbol{E} = \left(\frac{\mu_0}{4\pi}\frac{e}{r}\sin\varphi\right)\gamma$$

in rationalized MKSA units; $r$ is the distance $OP$ and $\varphi$ is the angle between $\mathbf{OP}$ and the direction of acceleration of the electron; $\mu_0 = 4\pi \times 10^{-7}$; $e = 1.6\times10^{-19}$ coulomb; $r$ is expressed in meters and $E$ in volts per meter. The electric vector is situated in the plane containing $\mathbf{OP}$ and $\boldsymbol{\gamma}$. There is therefore incident on $P$ a radiation of the same frequency as that of the incident wave and whose amplitude is

$$E = \frac{\mu_0}{4\pi} \frac{e^2}{m} E_0 \frac{\sin\varphi}{r} . \tag{1.16}$$

On the other hand, scattered wave is $\pi$ radians out of phase with the incident wave. Let the $OXY$ plane be that containing the vectors $\mathbf{OP}$ and $\mathbf{OX}$, and let $2\theta$ be the angle of scattering. Let us first assume that $\mathbf{E}_0$ is directed along $\mathbf{OZ}$. We then have

$$E_\perp = \frac{\mu_0}{4\pi} \frac{e^2}{m} \frac{E_0}{r} . \tag{1.17}$$

The ratio of the incident intensity at $O$ to the scattered intensity at $P$ is equal to the ratio of the squares of the amplitudes of the electric field. Therefore the flux of energy crossing a unit surface per second at $P$ is

$$I_0 \frac{e^4}{m^2 r^2} \left(\frac{\mu_0}{4\pi}\right)^2 . \tag{1.18}$$

This surface subtends at $O$ a solid angle of $1/r^2$ and the energy scattered in the direction $OP$ per second per unit solid angle, i.e. the intensity of scattering, is therefore

$$I_\perp = I_0 \left(\frac{\mu_0}{4\pi}\right)^2 \frac{e^4}{m^2} = r_e^2 I_0 , \tag{1.19}$$

where $(\mu_0/4\pi)(e^2/m)$ is a quantity which has the dimension of a length and which is called the classical radius of of the electron, $r_e$. Its numerical value is $2.818\times10^{-15}$ m.

If we now assume that the electric vector of the incident wave is in the $XOY$ plane, the scattered intensity becomes, according to Eqs. (*1.16*) and (*1.19*),

$$I_\parallel = r_e^2 I_0 \cos^2 2\theta . \tag{1.20}$$

For a beam of unspecified polarization the calculation is made by decomposing the beam into two separate beams whose electric vectors are respectively perpendicular and parallel to the plane of the incident and scattered rays, and in the proportions $k_\perp$ and $k_\parallel$. The scattered intensity is then

$$I_e = r_e^2 I_0 (k_\perp + k_\parallel \cos^2 2\theta) . \tag{1.21}$$

If, in particular, the incident beam is not polarized,

$$k_\perp = k_\parallel = \tfrac{1}{2},$$

and Eq. (*1.21*) becomes

$$I_e = r_e^2 I_0 \frac{1 + \cos^2 2\theta}{2}. \tag{1.22}$$

This is the *Thomson formula.* In MKSA units,

$$I_e = 7.90 \times 10^{-30} \left( \frac{1 + \cos^2 2\theta}{2} \right) I_0 . \tag{1.23}$$

We recall here that *$I_e$ is the energy scattered per unit solid angle per second by an electron situated in a beam carrying a power of $I_0$ per square meter of cross-section.* If, as is usually the case, we measure the intensity from the energy flux per square centimeter, the numerical coefficient of Eq. (*1.23*) becomes $7.90 \times 10^{-26}$. The scattered beam at the distance $r$ from the scattering electron has an intensity of $I_e/r^2$.

Equation (*1.19*) shows that only electrons scatter X-rays to an appreciable extent. The reason for this is that the mass of the scattering particle appears as $m^2$ in the denominator. In the case of the proton, whose charge is equal to that of the electron but whose mass is 1836 times larger, the scattered intensity is $(1836)^2$ times smaller than that scattered by an electron, and therefore relatively negligible.

The Thomson formula plays an essential role in all scattering calculations which involve absolute values. It is indeed most convenient to express the intensity scattered by a given sample in terms of the intensity scattered by an isolated electron substituted for the sample, all other experimental conditions remaining the same. This ratio is called *the scattering power of the sample,* which we shall calculate theoretically. To obtain the observed experimental intensity from this scattering power, it is sufficient to multiply it by the intensity $I_e$ given by the Thomson formula corresponding to the measurement.

It is unfortunately impossible to verify the Thomson formula experimentally because it is physically impossible to obtain a scatterer which is formed exclusively of free electrons. One might think that the best scatterer for this purpose would be a light element in which the binding energy of the electrons is low. However, such experiments with light elements have led to the discovery of an effect which is completely different from the predictions of classical theory. It is called the *Compton effect.*

### 1.2.3. Incoherent or Compton Scattering

From the point of view of its spectrum, the radiation scattered by an object is not simple. Only a part of it has the same wavelength as the incident radiation; the rest has a slightly longer wavelength, the difference depending on the angle of scattering.

This second type of radiation was discovered by A. H. Compton (1926) and is easily interpreted with the corpuscular model of radiation. We consider the scattering as the result of the impact of an incident photon $h\nu$ on an electron, and we apply the usual laws of mechanics (Figure 1.6).

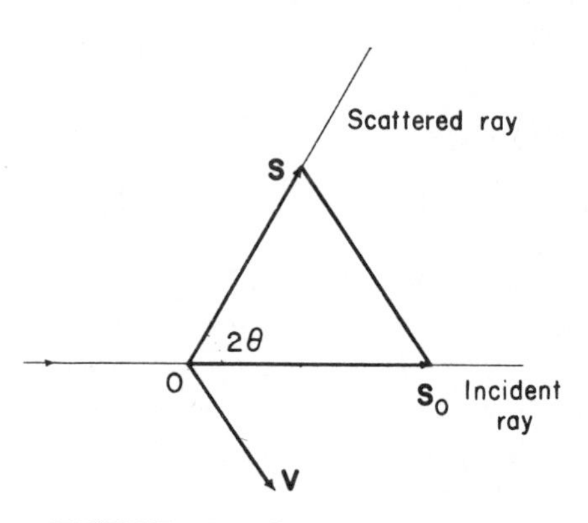


**FIGURE 1.6.** *Compton scattering.*

Let $S_0$ and $S$ be the unit vectors along the incident and scattered rays, $2\theta$ the scattering angle, $h\nu'$ the scattered photon, and $v$ the recoil velocity of the electron after collision. The electron is assumed to be at rest before the collision.

If relativistic effects are neglected, we have the following two relations.

Conservation of energy: $$h\nu = h\nu' + (1/2)\,mv^2 \qquad (1.24)$$

Conservation of momentum: $$(h\nu/c)\mathbf{S}_0 = (h\nu'/c)\mathbf{S} + m\mathbf{v}\,. \qquad (1.25)$$

If we assume that $\nu'$ is approximately equal to $\nu$ and set $\lambda' = \lambda + d\lambda$, we find from Eqs. (*1.24*) and (*1.25*) that

$$\begin{aligned} d\lambda(\text{Å}) &= \frac{2h}{mc}\sin^2\theta = 0.048\sin^2\theta \\ &= 0.024\,(1 - \cos 2\theta)\,. \end{aligned} \qquad (1.26)$$

The change in wavelength is independent of $\lambda$; it is relatively small for medium wavelengths but not negligible. Thus, for a scattering angle of 180°, in the case of Cu K$\alpha$, it is 10 times larger than the separation of the doublet $K\alpha_1 - K\alpha_2$.

According to quantum theory, *scattering from a free electron occurs only through the Compton effect.* However, the scattered intensity is correctly given by the classical formula of Thomson with a correction factor of $(\nu'/\nu)^3 \sim 1 - (3d\lambda/\lambda)$ which is equal to unity within a few percent in the case of medium wavelengths [6]. Since the incident and the scattered radiations do not have the same wavelength, there is no definite phase relation between them and they are therefore *incoherent.* In other words, the waves scattered by Compton effect by the various electrons in the scatterer never interfere and their intensities simply add.

### 1.2.4. Scattering by an Atomic Electron

We now have to relate classical scattering theory and the theory of the Compton effect to explain the experimental fact that the two types of scattering occur simultaneously. It is indeed observed that the wavelength of the incident radiation and that to be expected from Compton scattering are both found in the scattered radiation.

Let us consider the simplest case-the hydrogen atom, which contains a single electron. According to wave mechanics this electron can occupy various states, corresponding to a discontinuous set of negative energies and to a continuous band of positive energies.

If the electron occupies the same state after the collision of the photon on the atom, its energy is unchanged and the photon has therefore lost no energy; the photon has been scattered without its wavelength being changed. But if the electron changes state, the photon energy must also change. If the transitions occurred between quantized states, one would observe discrete values for the wavelengths re-emitted, the change in frequency $\Delta E/h$ being very small compared to the frequency of the X-rays. Such discrete lines in the scattered radiation, which would be analogous to Raman lines in optics, seem to have been observed (Das Gupta [17]). Whenever scattering is accompanied by an energy loss, the electron is ejected from the atom with a positive energy. This is the Compton effect for a bound electron, and it is very similar to that described above for a free electron initially at rest. There is, however, a difference which comes from the motion of the scattering electron: for a given scattering angle the wavelength of the scattered radiation is not exactly that given by Eq. (*1.26*) but rather covers a band centered on this average value. Compton radiation from a bound electron therefore has a definite width.

The following essential facts about scattering intensity have been deduced from quantum mechanical calculations. It is assumed that the frequency of the radiation is not very high ($\lambda \sim 1$Å) and that it is far removed from the frequencies corresponding to the absorption discontinuities.

(a) There is both coherent and Compton scattering.

(b) The total intensity scattered per electron (coherent + Compton) is given by the classical formula of Thomson.

(c) The intensity of the coherent scattering is calculated as follows. It is known from wave mechanics that the point electron of classical theory must be replaced by a smooth distribution of electric charge. Using the charge of the electron as a unit, the density $\rho(\boldsymbol{x})$ of the cloud at the extremity of the vector $\boldsymbol{x}$ near the nucleus is $\rho(\boldsymbol{x}) = |\psi(\boldsymbol{x})|^2$, where $\psi(\boldsymbol{x})$ is the wave function satisfying the Schroedinger equation. As in classical theory, the volume $dv$ carrying a charge $\rho dv$ scatters a wave whose amplitude is that scattered by a single electron, multiplied by $\rho dv$. The elementary waves which are scattered by the electronic cloud interfere, since they are coherent, and the total effect can be calculated by using the interference formulas of Section 1.2.1.

At the point determined by the vector $\boldsymbol{x}$, whose origin is at the center of the atom, the charge $\rho dv$ scatters a wave whose phase, with respect to that at the origin, is

$$\varphi = -2\pi i \boldsymbol{x} \cdot \left(\frac{\boldsymbol{S} - \boldsymbol{S}_0}{\lambda}\right) = -2\pi i \boldsymbol{x} \cdot \boldsymbol{s}\,, \tag{1.27}$$

from Eq. (*1.10*). Using as a unit the amplitude scattered by an electron at the origin, the total amplitude is

$$f = \int \rho(\boldsymbol{x}) \exp(-2\pi i \boldsymbol{x} \cdot \boldsymbol{s}) dv\,. \tag{1.28}$$

The integral must be extended to the whole volume of the atom where the electron density is different from zero. The quantity $f$ is called the *scattering factor of the electron.*

If the electronic wave function has spherical symmetry, the density $\rho$ depends only on the length of the vector $\boldsymbol{x}$. The integral of Eq. (*1.28*) can be calculated and the result, which will be justified in Section 2.5, is

$$f = \int_0^\infty 4\pi x^2 \rho(x) \frac{\sin 2\pi s x}{2\pi s x}\, dx\,, \tag{1.29}$$

where, according to Eq. (*1.11*), $s = (2 \sin \theta)/\lambda$. The scattering power of the electron is $f^2$, i.e. the coherent scattered intensity is

$$I_{\text{coh}} = I_e f^2\,, \tag{1.30}$$

$I_e$ being the intensity as given by the Thomson formula. The scattering intensity of the electron depends solely on $s = (2 \sin \theta)/\lambda$ only in the case of an electron with spherical symmetry, but in other cases it would be a function of $\boldsymbol{s} = (\boldsymbol{S} - \boldsymbol{S}_0)/\lambda$.

(d) Since the total intensity scattered by the electron is $I_e$ according to the Thomson formula, the intensity of the Compton radiation is

$$I_{\text{Compton}} = I_e - I_{\text{coh}} = I_e(1 - f^2)\,. \tag{1.31}$$

The scattering factor $f$ is always equal to unity for $\theta = 0$; it decreases with an increase in $\theta$, the decrease being more rapid if the electron cloud is more extended. Inversely, Compton scattering is always zero for $\theta = 0$. It increases with $\theta$ and, for small angles, is proportional to $\sin^2 \theta$. Thus, when the electron is strongly bound, i.e. when the electron cloud is very small, coherent scattering remains important up to relatively large scattering angles. However, for weakly bound electrons coherent scattering occurs mostly close to the direction of the primary ray, and for larger scattering angles there remains only the incoherent scattering. At the limit, for the free electron, only Compton scattering is observed.

### 1.2.5. Scattering by an Atom with Several Electrons

One of the approximate methods utilized in wave mechanics for the case of a real atom is to look for wave functions for individual electrons and to assume that the total electron density is the sum of the densities corresponding to the different electrons. Under these conditions, for coherent scattering the amplitude of the waves scattered by the $Z$ electrons simply add, and the total intensity is

$$I_{\text{coh}} = f^2 I_e = (\Sigma_1^Z f_j)^2 I_e \,, \qquad (1.32)$$

where $f_j$, the scattering factor for the $j$ electron, is

$$f_j = \int \rho_j(\boldsymbol{x}) \exp(-2\pi i \boldsymbol{s} \cdot \boldsymbol{x}) dv \,.$$

The quantity $f$ is called the *atomic scattering factor*. It is sometimes called the *atomic structure factor*, but it is preferable to reserve this term for the coefficient defined in the next section.

In the case of incoherent radiation there is no interference between the waves scattered by the electrons, and the total intensity is the sum of the individual intensities:

$$I_{\text{incoh}} = \Sigma_1^Z (1 - f_j^2) I_e \,.$$

One should really subtract a correction term which results from the Pauli exclusion principle, according to which only those transitions are permitted which bring the electron either to an unoccupied state or to one which is occupied by an electron of opposite spin.

This term, which is generally of little importance, is the following:

$$f_{jk} = \int \psi_j^* \psi_k \exp(-2\pi i \boldsymbol{s} \cdot \boldsymbol{x}) dv \,.$$

Thus, finally,

$$I_{\text{incoh}} = I_e (Z - \Sigma f_j^2 - \Sigma_j \Sigma_k f_{jk}) \,. \qquad (1.33)$$

The sums are evaluated over all the $Z$ electrons of the atom.

### 1.2.6. Structure Factor for a Group of Atoms ((Molecule or unit cell of a crystal)

Let us consider a group of atoms such as the atoms of a molecule or those of a unit cell in a crystal. The atoms $A_1, A_2, \cdots, A_r$, have scattering factors $f_1, f_2, \cdots, f_r$, and their positions are defined with respect to an arbitrary origin by means of the vectors $\boldsymbol{x}_1, \boldsymbol{x}_2, \cdots, \boldsymbol{x}_r$. If we select as a unit the amplitude scattered by an electron at the origin, Eq.

(*1.14*) gives a total amplitude.

$$F(\boldsymbol{s}) = \sum_1^r f_i \exp(-2\pi i \boldsymbol{s} \cdot \boldsymbol{x}_i) . \qquad (1.34)$$

$F(\boldsymbol{s})$ is called the *structure factor* of the group of atoms. It can be expressed as a function of the electron density in the group of atoms $\rho(\boldsymbol{x})$ by an equation similar to Eq. (*1.28*):

$$F(\boldsymbol{s}) = \int \rho(\boldsymbol{x}) \exp(-2\pi i \boldsymbol{s} \cdot \boldsymbol{x}) dv . \qquad (1.35)$$

The diffracted intensity is

$$\begin{aligned} I(\boldsymbol{s}) &= |F(\boldsymbol{s})|^2 = |\sum_1^r f_i \exp(-2\pi i \boldsymbol{s} \cdot \boldsymbol{x}_i)|^2 \\ &= (\sum_1^r f_i \cos 2\pi \boldsymbol{s} \cdot \boldsymbol{x}_i)^2 + (\sum_1^r f_i \sin 2\pi \boldsymbol{s} \cdot \boldsymbol{x}_i)^2 . \end{aligned} \qquad (1.36)$$

The amplitude $F(\boldsymbol{s})$, i.e. the structure factor, depends on the position of the origin since it contains the coordinates of the vector $\boldsymbol{x}_i$, but its modulus, and therefore the scattered intensity, does not. Therefore we can write

$$\begin{aligned} I(\boldsymbol{s}) &= |F(\boldsymbol{s})|^2 = F(\boldsymbol{s}) \cdot F^*(\boldsymbol{s}) \\ &= \sum_i \sum_j f_i f_j \exp[-2\pi i \boldsymbol{s} \cdot (\boldsymbol{x}_i - \boldsymbol{x}_j)] . \end{aligned} \qquad (1.37)$$

The total intensity therefore depends only on the distances between the atoms in the group.

The structure factor is maximum for $\boldsymbol{s} = 0$, or, experimentally, for very small scattering angles. This maximum value is simply equal to the sum of the scattering factors of the individual atoms:

$$F(0) = \sum_1^r f_i .$$

Since the atomic scattering factor for $\boldsymbol{s} = 0$ is equal to the atomic number, $F(0)$ is equal to the total number of electrons in the group of atoms considered.

### 1.2.7. Theoretical Calculation of the Atomic Scattering Factor

Equations (*1.28*), (*1.32*) and (*1.33*) can be integrated numerically to obtain the scattering factor if the electronic wave functions are known. The Hartree or self-consistent field method gives the best results for the wave function. James and Brindley were the first to use it to determine the scattering factors of atoms up to the atomic number 25. Interpolations were used in certain cases. More refined calculations are now available [7] and the best values are given in the *International Tables of Crystallography*[1].

Figure 1.7a gives the electron density distributions in the $K^+$ ion for

[1] *International Tables of Crystallography*, Vol. 111, The Kynoch Press, Birmingham (1962), Chapter 3.3.

**FIGURE 1.7.** (a) *Radial electron densities for the various shells of the $K^+$ ion.* (b) *Scattering factors for the individual electrons and total scattering factor for the $K^+$ ion (James [1].*

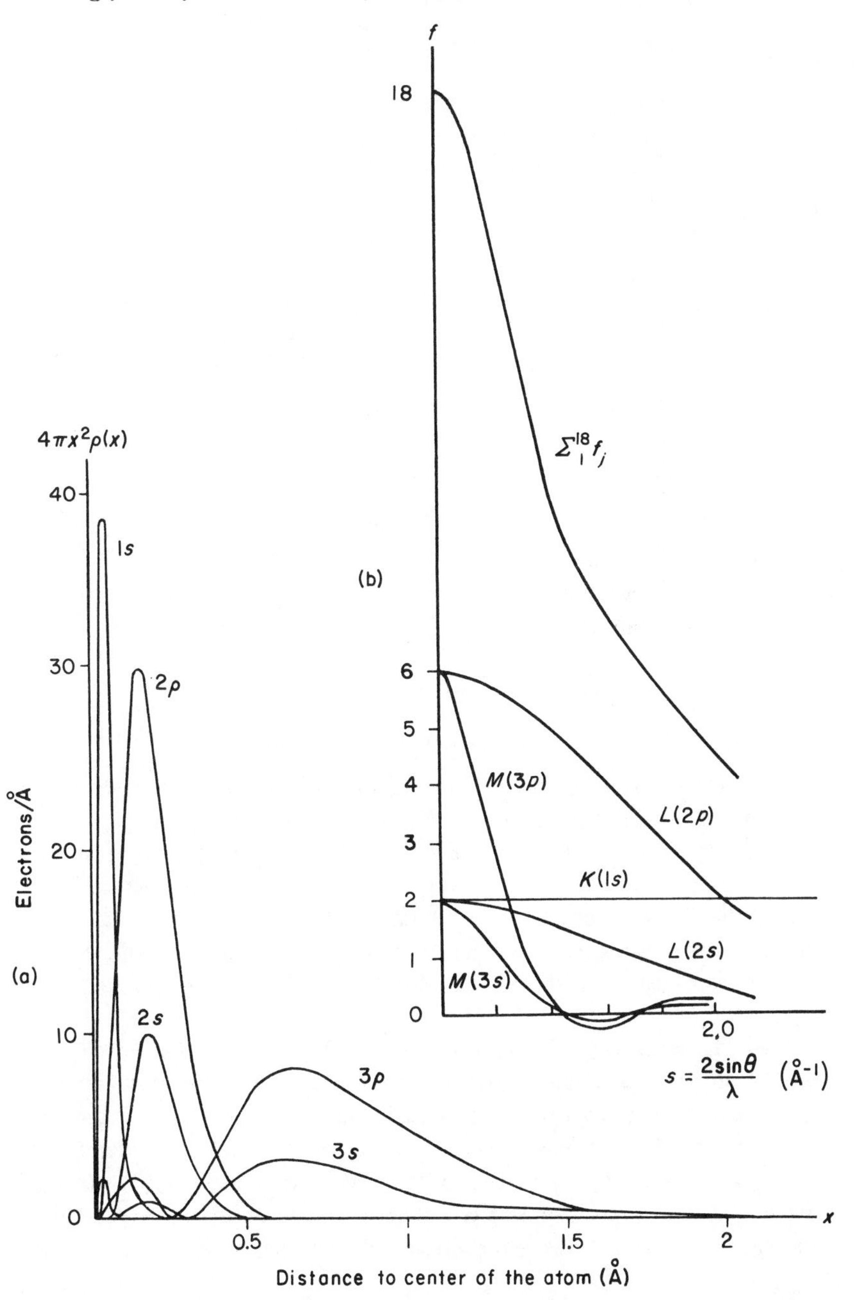

**FIGURE 1.8.** *Radial electron density for the neon atom as calculated with the Thomas-Fermi approximation* (*Compton and Allison* [*6*]).

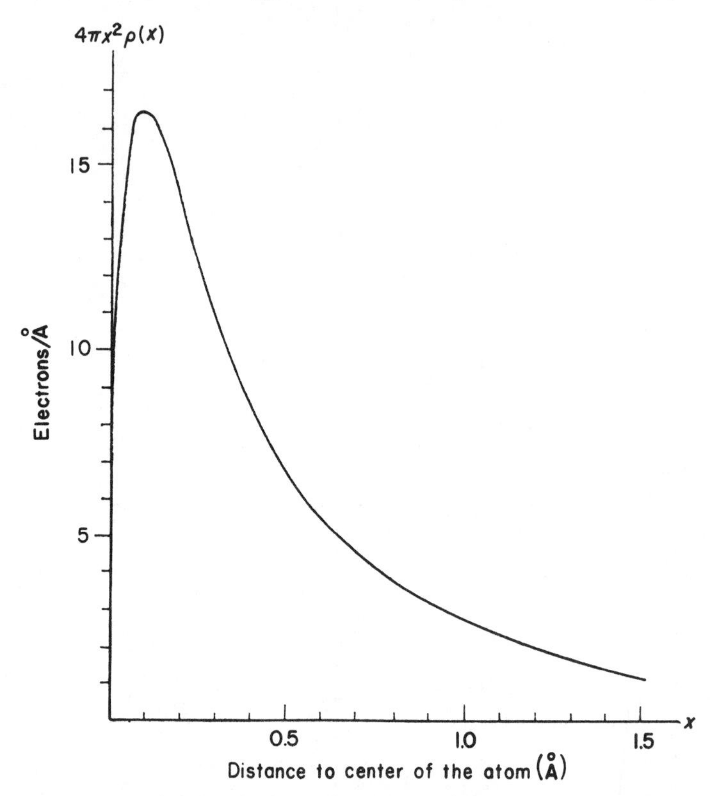


the two electrons of the K (1s) shell, as well as for the 8 electrons of the L (2s and 2p) shell and for the 8 electrons of the M (3s and 3p) shell. Figure 1.7b gives the values of the $f$ factor corresponding to these various electron shells and the total scattering factor. In the case of the heavy atoms the Hartree solution is not known and we use the Thomas-Fermi approximation based on a continuous distribution which is analogous for all atoms. Figure 1.8 shows this distribution in the case of the neon atom.

There is never any interference between the Compton waves scattered by the atoms of a scatterer, whatever its structure. The total intensity of the Compton radiation is therefore simply the sum of the intensities scattered by the various atoms. The Compton scattering factors are given in the *International Tables of Crystallography* (see also [8]) but they are strictly valid only for isolated atoms. It has been suggested that in the case of crystals, where the ejected electron is not entirely free, the

Compton intensity may be smaller, especially for small angles (Laval [9]). But recent measurements of the Compton scattering are in good agreement with the more precise theoretical calculations [10].

It is interesting to note that the scattering factor is not very sensitive to inaccuracies in the calculation of the electron density and that the different theories give quite similar results. These have been well confirmed by experiment [11], and the values of $f$ which are given in the tables are usually sufficient for crystallographic calculations. However, for recent measurements of high precision, it has been found necessary to find very accurate values of $f$, especially for the atoms which are found in organic compounds. Equation (*1.29*) is based on the assumption that the atom has symmetry of revolution, which is not correct for incomplete shells. For example, the distribution of the $2p$ electrons of oxygen has only one axis of symmetry. The general Eq. (*1.28*) shows that the factor $f_{2p}$ depends on the angle between this axis and the vector $\mathbf{s}$. MacWeeny [12] has calculated the difference between the scattering factors for two extreme positions of the atom—$\mathbf{s}$ parallel and perpendicular to the axis of the atom—as well as the difference with the usual factor found in the tables of James and Brindley (Figure 1.9). He has also calculated the influence of distortions in the electron clouds which occur when the atom is chemically bound. The error here is always quite small.

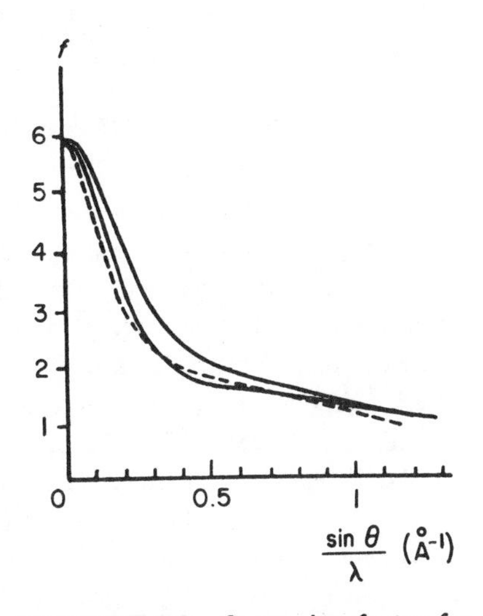


**FIGURE 1.9.** *Scattering factor for the carbon atom according to MacWeeny* [*12*] *for two perpendicular directions of* $\mathbf{s}$ (*solid lines*), *and according to James and Brindley* [*6*] (*dotted line*).

Figure 1.7 shows that the electrons which are most important for coherent scattering are those of the inner shell. Other electrons, which are represented by a cloud of much large extent, give very little coherent scattering for large values of $(\sin\theta)/\lambda$. It is only for very small values of $(\sin\theta)/\lambda$, and therefore of the scattering angle $\theta$, that they have any effect. As will be seen from our study of scattering from crystals, it is not possible to measure $f$ in the case of most crystals for $(\sin\theta)/\lambda$ smaller than 0.1, for example. It is therefore practically impossible to determine with X-rays whether the nodes of an aluminum crystal contain Al atoms with 13 electrons or $Al^{+++}$ ions with 10 electrons.

For a given atom the intensity of the Compton radiation is of the same order of magnitude as the coherent scattering. However, if the waves scattered by N atoms are in phase, the intensity is $N^2$ times larger for the coherent radiation, while it is only N times larger for the incoherent scattering. Since N is very large, in practice, it follows that in the usual scattering phenomena Compton radiation is negligible. It is, however, necessary to take it into account in certain special cases, such as the scattering by amorphous bodies or by imperfect crystals, which we shall discuss at length; hence our relatively extensive treatment of the Compton effect.

### 1.2.8. Anomalies in the Atomic Scattering Factor: The Dispersion Effect

The above calculations are valid only if the incident radiation has a frequency which is large compared to that of an absorption discontinuity for the scattering atom. In fact, the scattered radiation arises from the excitation of electronic vibrations by the incident radiation. From the classical point of view, resonance effects are to be expected when the applied frequency approaches a critical value, and this should modify the scattered radiation. The strongly bound electrons, and especially those of the K shell, are most important from this point of view and the dispersion anomaly in the scattering factor is important when the incident wavelength is close to the K discontinuity of the scatterer. It is also observed in all of the region where $\lambda$ is larger than $\lambda_K$ for, in the case of these radiations, the K electron vibrations are weakly excited and they have little effect on the scattering coefficient. The true factor for long wavelength is thus smaller than the theoretical value shown in the tables, which is valid only for very short wavelengths.

A quantum mechanical theory of the phenomenon has been elaborated by Hönl [13], and experiments have been attempted to check its results. These experiments, however, require measurements of the scattered intensity and their accuracy is not high. There is a wide divergence between the results of the various experiments, but there is nevertheless no contradiction with theory.

In the absence of reliable experimental results, it is best to use the theoretical results in the calculations. We shall summarize here the useful data without proof. A thorough discussion will be found in James [14].

The scattering factor is complex:

$$f = f_0 + \Delta f' + i\Delta f'' ,$$

where $f_0$ is the normal coefficient corresponding to the value of $s = (2\sin\theta)/\lambda$. The modulus of $f$ which determines the amplitude of the scattered wave is

$$|f| = [(f_0 + \Delta f')^2 + \Delta f''^2]^{1/2} . \tag{1.38}$$

In the case where $\Delta f''$ is sufficiently small, $|f|$ can be calculated by the approximate formula

$$|f| = f_0 + \Delta f' + \frac{1}{2}\frac{\Delta f''^2}{f_0 + \Delta f'} . \tag{1.39}$$

In the normal case the scattered wave is shifted in phase by $\pi$ radians

**FIGURE 1.10.** *The correction factors $\Delta f'$ and $\Delta f''$ near an absorption discontinuity $\lambda_K$.*

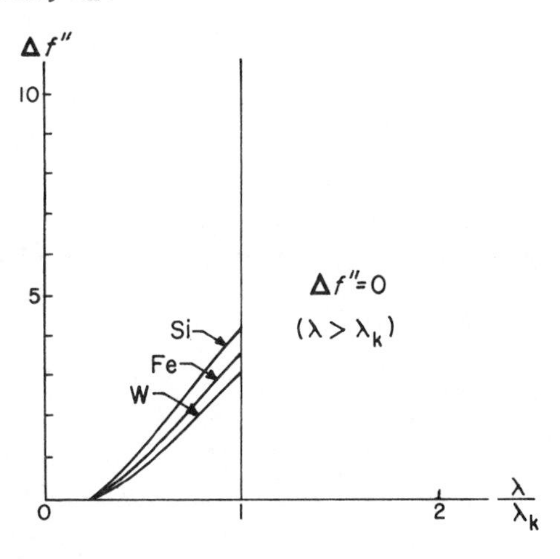


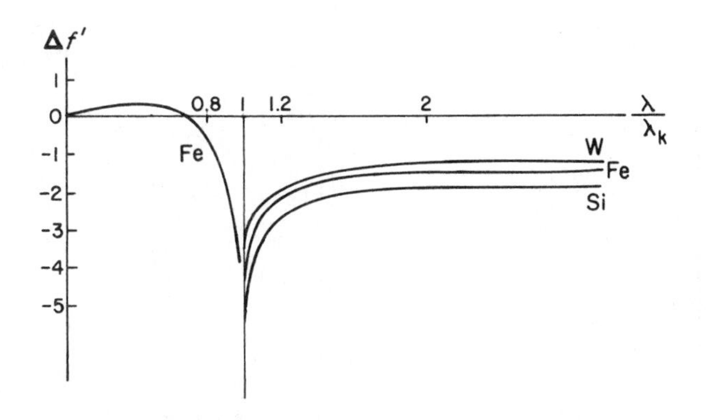

with respect to the incident wave but, near an anomaly, the phase shift is reduced to $\pi - \psi$, with

$$\tan \psi = \frac{\Delta f''}{f_0 + \Delta f'} . \tag{1.40}$$

The correction terms $\Delta f'$ and $\Delta f''$ are independent of the scattering angle, since the electrons producing the scattering are concentrated in the immediate neighborhood of the nucleus. These terms depend solely on $\lambda$ and on the atomic number of the scatterer. If we consider only

**FIGURE 1.11.** *The correction factors $\Delta f'$ and $\Delta f''$ for anomalous dispersion as functions of the atomic number Z for three commonly used wavelengths.*

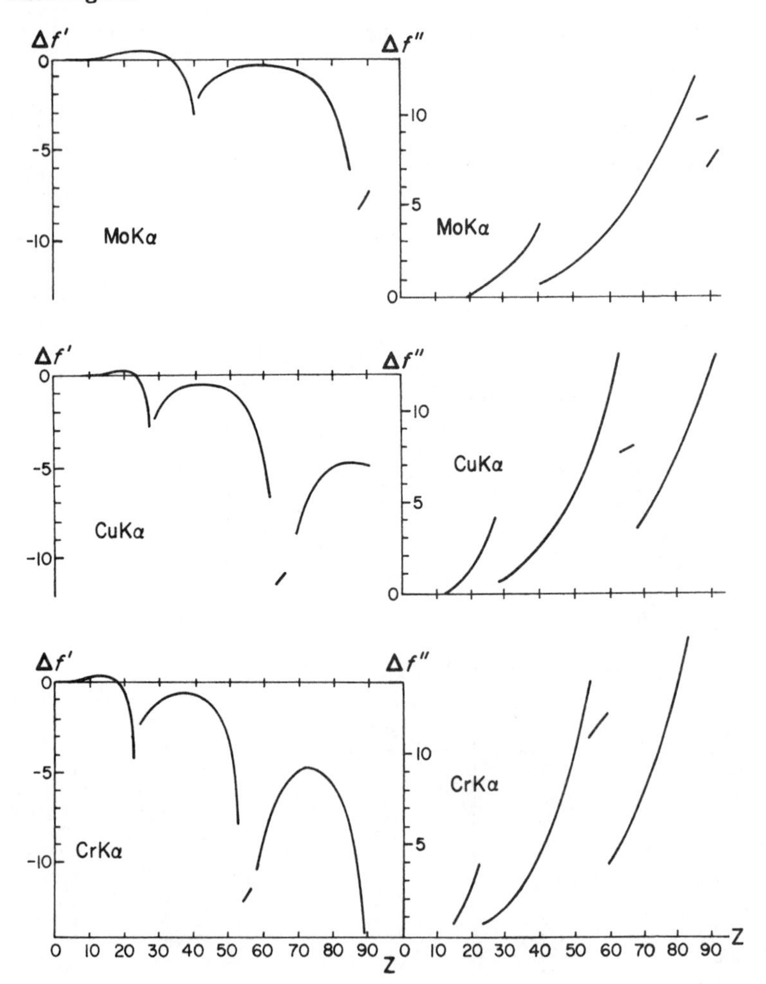

the effects of the K electrons, the correction depends mostly on the ratio $\lambda/\lambda_K$ and varies little from one element to the next. This will be evident from the curves of Figure 1.10, which were calculated by James. It will be observed that $f_0 - |f|$ is of the order of 1 to 1.5 for long wavelengths.

If we take into account the effect not only of the K but also of the L and M shells, the results are more complicated. The calculations are also rather unreliable because the oscillator strengths which are required are not well known. Figure 1.11 shows the results of calculations by Dauben and Templeton [15], following the method of Parratt and Hempstead for three commonly used wavelengths (Cr K$\alpha$, Cu K$\alpha$, and Mo K$\alpha$) as functions of the atomic number of the scatterer. These curves are more accurate than those of Figure 1.10, but they apply to only three wavelengths and it would be difficult to interpolate to obtain the values at other wavelengths. It is important to note that *the correction for anomalous dispersion can be important—as high as 7 or 8 units for heavy atoms.*

One interesting application of the anomalous atomic scattering factor is in the case of a crystal composed of two elements of neighbouring atomic numbers, for example Cu and Zn. Here the scattering factors are nearly equal and X-rays cannot, in general, distinguish between the two types of atoms. Thus, for $(\sin\theta)/\lambda = 0.3$, the tables give

$$f_{0\text{Cu}} = 17.8\,,$$
$$f_{0\text{Zn}} = 18.2\,.$$

However, if we use Cu K$\beta$ radiation ($\lambda = 1.392$Å), the ratios $\lambda/\lambda_K$ for Cu and for Zn are respectively 1.0085 and 1.08. Then, from the curves of Figure 1.10,

$$\Delta f'_{\text{Cu}} = -5.2\,, \qquad \Delta f''_{\text{Cu}} = 0\,,$$
$$\Delta f'_{\text{Zn}} = -2.8\,, \qquad \Delta f''_{\text{Zn}} = 0\,,$$

whence

$$f_{\text{Cu}} = 12.6\,,$$
$$f_{\text{Zn}} = 15.4\,.$$

The difference in response between the two atoms is therefore much enhanced. In fact, the quantity which appears in the calculations is

$$\left(\frac{f_{\text{Cu}} - f_{\text{Zn}}}{f_{\text{Cu}} + f_{\text{Zn}}}\right)^2 .$$

If we use the Cu K$\beta$ line instead of some other line—Mo K$\alpha$, for example—the contrast is increased from 0.05% to 4%. This method can be especially important for the study of alloys composed of neighboring elements.

The phase shift associated with anomalous dispersion is also used in crystallography because of the following property. Let us consider a pair of atoms, $A$ and $B$, separated by a distance $\boldsymbol{u}$. The structure factor of the pair is

$$F(\boldsymbol{s}) = f_A + f_B \exp(-2\pi i \boldsymbol{s} \cdot \boldsymbol{u}) \quad (1.41)$$

and, for $-\boldsymbol{s}$,

$$F(-\boldsymbol{s}) = f_A + f_B \exp(2\pi i \boldsymbol{s} \cdot \boldsymbol{u})\,. \quad (1.42)$$

Let us first suppose that $f_A$ and $f_B$ are real. Then $F(\boldsymbol{s})$ and $F(-\boldsymbol{s})$ are complex conjugates and their moduli are therefore equal. The measured intensities in the two cases are therefore also equal. However, if the wavelength is chosen so that the phase shift is appreciable for atom $B$, then $F(\boldsymbol{s})$ and $F(-\boldsymbol{s})$ are not complex conjugates any more, $|F(\boldsymbol{s})|^2 \neq |F(-\boldsymbol{s})|^2$, and the function $I(\boldsymbol{s})$ is no more centrosymmetrical as required by Friedel's law, which applies to normal cases. This consequence of anomalous dispersion is the basis of an ingenious technique used in crystal structure determinations [16].

SOME ELEMENTARY BOOKS ON X-RAY CRYSTALLOGRAPHY

C. S. BARRETT, *Structure of Metals*, McGraw-Hill, New York (1952).
M. J. BUERGER, *Elementary Crystallography*, Wiley, New York (1956).
G. L. CLARK, *Applied X-Rays*, McGraw-Hill, New York (1953).
B. D. CULLITY, *Elements of X-Ray Diffraction* Addison Wesley, Reading (1956).
R. GLOCKER, *Material Prüfung mit Röntgenstrahlen*, (3rd Ed.) Springer Verlag, Berlin (1949).
A. GUINIER, *X-Ray Technology*, Hilger and Watts, London (1952).
H. KLUG and L. ALEXANDER, *X-Ray Diffraction Procedures*, Wiley, NewYork (1954).
K. LONSDALE, *Crystals and X-Rays*, Bell, London (1948).
F. C. PHLLIPS, *An Introduction to Crystallography*, Longmans Green, London, (1946).
A. TAYLOR, *An Introduction to X-Ray Metallography*, Chapman and Hall, London (1954).

REFERENCES

1. R. W. JAMES, *Optical Principles of X-Ray Diffraction*, Bell, London (1948), 53.
2. N. WAINFAN, N. J. SCOTT and L. G. PARRATT, *J. Appl. Phys.*, **30** (1959), 1604.
3. J. STRONG, *Concepts of Classical Optics*, Freeman, San Francisco (1958).
4. C. G. BARKLA, *Proc. Roy. Soc.* London A, **77** (1906), 247.
5. W. PANOFSKY and M. PHILLIPS, *Classical Electricity and Magnetism*, Addison-Wesley, Reading, p. 325.
6. A. H. COMPTON and S. K. ALLISON, *X-Rays in Theory and Experiment*, MacMillan, London (1935), p. 38.
7. A. L. VEENENDAAL, C. H. MACGILLAVRY, B. STAM, M. L. POTTERS and M. J. H. ROMGEN, *Acta Cryst.* **12** (1959), 242. H. C. FREEMAN and J. E. W. SMITH. *Acta Cryst.*, **11** (1958), 819. B. DAWSON, *Acta Cryst.*, **14** (1961), 1117, 1271.

8. A. J. Freeman, *Acta Cryst.*, **12** (1959), 261, 271, 274, 929; **13** (1960), 190, 618.
9. J. Laval, *C. R. Acad. Sci. Paris.*, **215** (1942), 279, 359.
10. J. Corbeau, *C. R. Acad. Sci. Paris.*, **253** (1961), 1553. C. B. Walker. *Phys. Rev.* **103** (1956), 547.
11. J. Ibers, *Acta. Cryst.*, **12** (1959), 347. R. B. Roof. *J. Appl. Phys.*, **30** (1959), 1599.
12. R. MacWeeny, *Acta. Cryst.*, **4** (1951), 513; **5** (1952), 463; **6** (1953), 631; **7** (1954), 180.
13. H. Hönl, *Ann. Phys.*, **18** (1933), 625.
14. R. W. James, *Optical Principles of X-Rays Diffraction, Bell,* London (1948), Chapter IV.
15. C. H. Dauben and D. H. Templeton, *Acta. Cryst.*, **8** (1955), 841.
16. J. M. Bijvoet, *Endeavour,* **14** (1955) 71. R. Kern, A. Rimsky and J. C. Monier, *Bull. Soc. Franc. Minér. Cryst.*, **181** (1958), 103.
17. K. Das Gupta, *Nature,* **156** (1950), 563; **167** (1951), 313; *Phy. Rev. Letters,* **3** (1959), 38.

CHAPTER

# 2

# GENERAL THEORY OF X-RAY DIFFRACTION FOR AN ARBITRARY STRUCTURE

## 2.1. Calculation of the Diffracted Amplitude

Let us consider a group of atoms arranged in some arbitrary manner and placed in a parallel monochromatic beam of X-rays. Each atom emits a scattered wave which is coherent with the incident radiation, and these waves interfere with each other, so that we have a diffracted radiation whose intensity depends on the direction of observation. This dependence is related to the atomic structure of the object. We shall now establish the laws governing this dependence in the case of a general structure, later applying these laws to the particular cases of crystals and of amorphous substances. We shall then consider the case of imperfect crystals, which is important for solid state physics. The well-known Bragg laws for X-ray diffraction by crystals will follow as a particular case.

Let us consider $N$ atoms situated at the points $\boldsymbol{x}_1, \boldsymbol{x}_2, \cdots, \boldsymbol{x}_n$, the origin being chosen arbitrarily. Let $f_1, f_2, \cdots, f_n$ be the scattering factors for the $N$ atoms (Section 1.2.5), the amplitude of the scattered wave of the $n^{\text{th}}$ atom being $f_n$ times larger than if it were replaced by an isolated electron. It is assumed that the object is small enough that absorption can be neglected. The incident beam is then of the same intensity for all the atoms.

Let $\boldsymbol{S}_0$ and $\boldsymbol{S}$ be unit vectors directed respectively along the incident beam and in the direction of observation. According to Eqs. (*1.10*) and

(*1.14*), the amplitude of the wave resulting from the addition of $N$ wavelets is given by

$$A = \sum_1^N f_n \exp\left(-2\pi i \frac{\boldsymbol{S} - \boldsymbol{S}_0}{\lambda} \cdot \boldsymbol{x}_n\right). \qquad (2.1)$$

The phases are given with respect to a wavelet which would be produced by an isolated electron at the origin of the object space.

The vector $\boldsymbol{s} = (\boldsymbol{S} - \boldsymbol{S}_0)/\lambda$, discussed in Section 1.2.1.1, defines a point in reciprocal space, and in this space the diffracted amplitude is given by

$$A(\boldsymbol{s}) = \sum_1^N f_n \exp(-2\pi i \boldsymbol{s} \cdot \boldsymbol{x}_n). \qquad (2.2)$$

We can now generalize our definition of the diffracting object by replacing the individual atoms by a function representing the *electron density*, $\rho(\boldsymbol{x})$. This function is the number of electrons per unit volume in a small element of volume at $\boldsymbol{x}$. Equation (*2.1*) for the amplitude $A(\boldsymbol{s})$ of the diffracted wave is now replaced by

$$A(\boldsymbol{s}) = \int \rho(\boldsymbol{x}) \exp(-2\pi i \boldsymbol{s} \cdot \boldsymbol{x}) dv_x, \qquad (2.3)$$

the integration being extended to all of the object space.

Equations (*2.2*) and (*2.3*) can be shown to correspond if the centers of the individual atoms are considered to be at $\boldsymbol{x}_1, \boldsymbol{x}_2, \cdots, \boldsymbol{x}_n$. Since the atoms are impenetrable, each point where $\rho(\boldsymbol{x})$ is not zero is in the electron cloud of one of the $N$ atoms. Then, if $\rho_n(\boldsymbol{x} - \boldsymbol{x}_n)$ is the density of the electron cloud of the $n^{\text{th}}$ atom whose center is situated at $\boldsymbol{x}_n$,

$$\rho(\boldsymbol{x}) = \sum_1^N \rho_n(\boldsymbol{x} - \boldsymbol{x}_n)$$

The integral of Eq. (*2.3*) can thus be written as a sum:

$$\begin{aligned} A(\boldsymbol{s}) &= \int \sum_1^N \rho_n(\boldsymbol{x} - \boldsymbol{x}_n) \exp(-2\pi i \boldsymbol{s} \cdot \boldsymbol{x}) dv_x \\ &= \sum_1^N \exp(-2\pi i \boldsymbol{s} \cdot \boldsymbol{x}_n) \left[ \int \rho_n(\boldsymbol{x} - \boldsymbol{x}_n) \exp[-2\pi i \boldsymbol{s} \cdot (\boldsymbol{x} - \boldsymbol{x}_n)] \, dv_x \right]. \end{aligned}$$

However, according to Section 1.2.5, the second integral is equal to $f_n$ and we are thus led back to Eq. (*2.2*).

Instead of the classical definition which we have given for the electron density, it is preferable to consider $\rho(\boldsymbol{x})$ as representing the sum of the squares of the absolute values of the wave functions for all the electrons in the scattering object.

Equation (*2.3*) is easily interpreted in terms of *Fourier transforms*, which have now become essential both to the theory of X-ray diffraction and to its applications. *The necessary mathematical background is given in Appendix A. The reader who is not already familiar with this field of mathematics should study this appendix before proceeding further.*

The function $A(\mathbf{s})$ of Eq. (2.3) is the Fourier transform of $f(\mathbf{x})$.

The basic property of this transformation is that $\rho(\mathbf{x})$ can be deduced inversely from $A(\mathbf{s})$ by an analogous equation:

$$\rho(\mathbf{x}) = \int A(\mathbf{s}) \exp(2\pi i \mathbf{s}\cdot\mathbf{x}) dv_s \, . \tag{2.4}$$

The only change is in the sign of the exponent, and the integral of Eq. (*2.4*) is extended to all of the reciprocal space, just as the integral of Eq. (*2.3*) was extended to all of the object space.

## 2.2. Calculation of the Diffracted Intensity

Experimentally, the diffracted amplitude is not interesting since we have no way of determining the relative phases of the waves which are diffracted in various directions of space; we can only measure their intensities. The only part of Eqs. (*2.2*) and (*2.3*) which has any experimental significance is the modulus of the complex number, the square of this modulus being the intensity.

In the theoretical calculations and in their applications, it is useful to utilize the *scattering power* of the object under consideration. Let us consider an object which is small enough that absorption is negligible, and then let us replace the object by a single free electron, all other experimental conditions remaining unchanged. The scattering power of the object, $I_N(\mathbf{s})$, is the ratio of the measured values for the radiation scattered by the object and by the isolated electron. It is also the number of free electrons which, *scattering independently*, would produce the same effect as the object, under the given experimental conditions. The unit scattering power,

$$I(\mathbf{s}) = \frac{I_N(\mathbf{s})}{N} \, , \tag{2.5}$$

is the scattering power per atom if the object is composed of $N$ atoms or, more generally, per atomic group if the object is composed of $N$ such groups (molecules, or unit cells in the case of a crystal).

If $F(\mathbf{s})$ is the structure factor of the atomic group, i.e. the scattering factor for the atom, or the structure factor of the group of atoms under consideration (Section 1.2.6), we set

$$\mathscr{I}(\mathbf{s}) = \frac{I(\mathbf{s})}{F^2} = \frac{I_N(\mathbf{s})}{NF^2} \, , \tag{2.5a}$$

where $\mathscr{I}(s)$ is the *interference function*. Its value is determined by the interference between the scattered waves, and it would be equal to unity for any vector $\mathbf{s}$ if the atomic groups scattered incoherently.

The purpose of the theoretical calculations is to determine $I(\mathbf{s})$ as a function of $\mathbf{s}$. In a given experiment, we find the scattered intensity by multiplying the unit scattering power by the number of atomic groups and by the intensity scattered by a single electron under the given experimental conditions. This intensity is given by the Thomson eq. (*1.21*) or (*1.23*). In the case of an extended object for which the absorption is not negligible, the observed intensity is reduced and a correction is necessary. The reader will find an example of such a calculation in Section 4.7.

Theoretically, Eq. (*2.3*) gives the value of $A(\mathbf{s})$ when the structure of the object is known. Therefore

$$I_N(\mathbf{s}) = |\, A(\mathbf{s}) \,|^2 \,. \tag{2.6}$$

However, it is *impossible to use Eq.* (*2.4*) *directly* because it involves not only the amplitude but also the phase of the wave, and this is not observable. This is an essential difficulty in the determination of atomic structures. If the amplitude function could be determined in all of reciprocal space, X-ray diffraction could be used like a true microscope with very high resolution to determine a true picture of the structure of the object. Unfortunately, diffraction patterns cannot be used for this purpose. We shall see just what information can be obtained from them in the most general case.

Let us first substitute Eq. (*2.2*) in (*2.6*):

$$\begin{aligned} I_N(\mathbf{s}) &= |\, A(\mathbf{s}) \,|^2 = A(\mathbf{s})A^*(\mathbf{s}) \\ &= \textstyle\sum_1^N f_n \exp(-2\pi i \mathbf{s}\cdot \mathbf{x}_n) \times \sum_1^N f_{n'} \exp(2\pi i \mathbf{s}\cdot \mathbf{x}_{n'}) \\ &= \textstyle\sum_1^N \sum_1^N f_n f_{n'} \exp[-2\pi i \mathbf{s}\cdot(\mathbf{x}_n - \mathbf{x}_{n'})] \,. \end{aligned} \tag{2.7}$$

The intensity must be a real quantity; this can be seen by regrouping the terms of the double summation, as follows. When $n = n'$, there are $N$ terms whose sum is $\sum f_n^2$. If $n \neq n'$, there are $N(N-1)$ terms, of which two correspond to a given pair of numbers $n$ and $n'$. One is the conjugate complex of the other and their sum is

$$f_n f_{n'}(\cos[2\pi \mathbf{s}\cdot(\mathbf{x}_n - \mathbf{x}_{n'})] + \cos[2\pi \mathbf{s}\cdot(\mathbf{x}_{n'} - \mathbf{x}_n)]) \,.$$

Setting $\mathbf{x}_{nn'}$ as the vector from atom $n$ to atom $n'$, the intensity becomes

$$I_N(\mathbf{s}) = \textstyle\sum_1^N f_n^2 + \sum\sum_{n\neq n'} f_n f_{n'} \cos(2\pi \mathbf{s}\cdot \mathbf{x}_{nn'}) \,. \tag{2.8}$$

For identical atoms or atomic groups with structure factor $F$, the unit scattering power is

$$I(\mathbf{s}) = \frac{1}{N} I_N(\mathbf{s}) = F^2 + \frac{F^2}{N} \sum\sum_{n\neq n'} \cos(2\pi \mathbf{s}\cdot \mathbf{x}_{nn'}) \,, \tag{2.9}$$

and the interference function is

$$\mathscr{I}(\boldsymbol{s}) = \frac{I_N(\boldsymbol{s})}{NF^2} = 1 + \frac{1}{N} \sum_{n \neq n'}\sum \cos(2\pi\boldsymbol{s}\cdot\boldsymbol{x}_{nn'}) . \tag{2.10}$$

Let us now find the value of the intensity corresponding to the general expression for the amplitude as a function of electron density in object space. We shall denote this density as $\rho(\boldsymbol{u})$, changing the variable $\boldsymbol{x}$ to $\boldsymbol{u}$ for convenience. We shall proceed as previously:

$$\begin{aligned} I_N(\boldsymbol{s}) &= A^*(\boldsymbol{s})A(\boldsymbol{s}) \\ &= \int \rho(\boldsymbol{u}) \exp(2\pi i\boldsymbol{s}\cdot\boldsymbol{u})\, dv_u \int \rho(\boldsymbol{u}') \exp(-2\pi i\boldsymbol{s}\cdot\boldsymbol{u}')\, dv_{u'} \\ &= \iint \rho(\boldsymbol{u})\rho(\boldsymbol{u}') \exp[-2\pi i\boldsymbol{s}\cdot(\boldsymbol{u}' - \boldsymbol{u})]\, dv_u dv_{u'} . \end{aligned}$$

In terms of the new variables $\boldsymbol{u}$ and $\boldsymbol{x} = \boldsymbol{u}' - \boldsymbol{u}$,

$$\begin{aligned} I_N(\boldsymbol{s}) &= \iint \rho(\boldsymbol{u})\rho(\boldsymbol{x} + \boldsymbol{u}) \exp(-2\pi i\boldsymbol{s}\cdot\boldsymbol{x})\, dv_x dv_u \\ &= \int \mathscr{P}(\boldsymbol{x}) \exp(-2\pi i\boldsymbol{s}\cdot\boldsymbol{x})\, dv_x , \end{aligned} \tag{2.11}$$

where

$$\mathscr{P}(\boldsymbol{x}) = \int \rho(\boldsymbol{u})\rho(\boldsymbol{x} + \boldsymbol{u})\, dv_u . \tag{2.12}$$

It will be useful to express these equations in another form for the case where the object is composed of $N$ identical atoms or atomic groups. Setting $F$ as their common scattering or structure factor and $\rho_a$ as the local *atomic* density instead of the electron density $\rho$, we have

$$A(\boldsymbol{s}) = F(\boldsymbol{s}) \int \rho_a(\boldsymbol{u}) \exp(-2\pi i\boldsymbol{s}\cdot\boldsymbol{u})\, dv_u .$$

If $V$ is the volume of the object and $v_1$ is the average volume available for each atom or atomic group,

$$N = \frac{V}{v_1}$$

The value of $I_N(\boldsymbol{s})$ is calculated as previously, whence the value of the interference function:

$$\mathscr{I}(s) = \frac{I_N(s)}{N|F|^2} = v_1 \int \frac{\mathscr{P}_a(\boldsymbol{x})}{V} \exp(-2\pi i\boldsymbol{s}\cdot\boldsymbol{x})\, dv_x , \tag{2.13}$$

where

$$\mathscr{P}_a(\boldsymbol{x}) = \int \rho_a(\boldsymbol{u})\rho_a(\boldsymbol{u} + \boldsymbol{x})\, dv_u . \tag{2.14}$$

These integrals must be extended to the whole of the object space. The above functions $\mathscr{P}(\boldsymbol{x})$ and $\mathscr{P}_a(\boldsymbol{x})$, which are defined with respect to the electrons of the object [Eq. (*2.12*)] or with respect to the atoms [Eq. (*2.14*)], are called the *Patterson functions.* They are also called *autocorrelation functions* of the electron or atomic density.

According to Eq. (*2.11*), *the function $I_N(\boldsymbol{s})$ in reciprocal space is the transform of the function $\mathscr{P}(\boldsymbol{x})$ defined in object space.* Inversely, inverting Eq. (*2.11*), we can write

$$\mathscr{P}(\boldsymbol{x}) = \int I_N(\boldsymbol{s}) \exp(2\pi i \boldsymbol{s}\cdot\boldsymbol{x})\, dv\,. \tag{2.15}$$

Theoretically, it is not impossible to determine $I_N(\boldsymbol{s})$ in all of reciprocal space. It is therefore possible to determine $\mathscr{P}(\boldsymbol{x})$ experimentally, and *all ideal diffraction experiments reduce to such a determination.*

Let us consider this point more closely. The ratio of the integral $\mathscr{P}(\boldsymbol{x})$ to $V$ is the *average value of the product of the electronic densities at two points separated by the vector* $\boldsymbol{x}$, the origin of $\boldsymbol{x}$ being inside the volume $V$. Similarly, the ratio $\mathscr{P}_a(\boldsymbol{x})/V$ is the analogous expression for atomic densities. X-rays therefore reveal only a highly complex function of the distribution of matter in the object, and it is often difficult to interpret this function directly in terms of a particular structure. Equations (*2.11*) are highly important, because all the different expressions which we shall find for the diffracted intensities are simply consequences of the general formula, when certain particular hypotheses are made concerning the electron density $\rho(\boldsymbol{x})$.

It may seem odd that we should have attempted to replace expression (*2.6*) for the intensity by (*2.11*), which is obviously much more complex, but the latter is often the only one which can be used. We often have to deal with objects in which the atoms or molecules are in motion, so that we know only their *statistical distribution.* This is the case in gases and liquids. The motion is extremely slow compared to the frequencies of X-rays, but it is extremely rapid compared to the duration of any experimental measurement. If $\boldsymbol{x}_1, \boldsymbol{x}_2, \cdots, \boldsymbol{x}_n$ are the positions of the atoms at a given moment, Eq. (*2.2*) gives the correct value for the instantaneous diffracted amplitude. The observations, however, only give the average value of the diffraction intensity corresponding to the successive configurations of the object. Equation (*2.11*) for the intensity permits such an averaging, while the expression for the amplitude cannot, in general, be used. There is no relation between the average amplitude and the average intensity. For example, if the atoms move as a group in a straight line, the average amplitude is zero in all directions, but the intensity remains constant because, according to Eq. (*2.7*), it depends only on the relative positions of the atoms. In practice, the amplitude can be used only when its modulus is easily calculated. Thus, if it can be shown that the modulus of the amplitude is zero, the average intensity is, of course, also zero.

## 2.3. Average Value of the Function $I_N(s)$ in Reciprocal Space

The atomic scattering factors $f_n$ of Eq. (2.7) depend on the absolute value of $\mathbf{s}$. It is, however, possible to use this equation by considering the electrons themselves as diffracting centers, each one having a scattering power of unity.

Let us set $\mathcal{N}$ to be the total number of electrons. Then, for the $N$ atoms,

$$\mathcal{N} = \Sigma_1^N f_n(0) ,$$

since the atomic scattering factor is equal to the atomic number for $(\sin\theta)/\lambda = 0$. Setting $\mathbf{y}_{pp'}$ as the vector between the electrons $p$ and $p'$, Eq. (2.8) becomes

$$I_N(\mathbf{s}) = \mathcal{N} + \sum_{p \neq p'}\sum \exp(-2\pi i\mathbf{s}\cdot\mathbf{y}_{pp'}) , \tag{2.16}$$

where we have kept an imaginary part in each term. The intensity is therefore given by a term which is independent of $\mathbf{s}$, plus a set of terms which oscillate symmetrically about zero. If we therefore calculate the average value of $I_N$ in a volume of reciprocal space centered on the origin and large enough to contain a sufficiently large number of oscillations of each one of the terms in the double summation, all of these will have an average value of zero and the average value of $I_N$ in all of reciprocal space will be $\mathcal{N}$.

Let us perform this operation mathematically. We integrate Eq. (2.16) in reciprocal space:

$$\int I_N(\mathbf{s})\, dv_s = \int \mathcal{N}\, dv_s + \sum_{pp'}\sum \int \exp(-2\pi i\mathbf{s}\cdot\mathbf{y}_{pp'})\, dv_s .$$

We define the vectors $\mathbf{y}_{pp'}$ and $\mathbf{s}$ respectively in object space and in reciprocal space by their rectangular coordinates $\mathbf{s}_i$ and $(\mathbf{y}_{pp'})_i$, where $i = 1$, 2, or 3. Integrating over a parallelepiped such that $-L_i < s_i < L_i$, each integral has the following form

$$\int \exp[-2\pi i[\Sigma\, s_i(y_{pp'})i]]\, dv_s = \int_{-L_1}^{+L_1} \exp[-2\pi i s_1(y_{pp'})_1]\, ds_1$$

$$\times \int_{-L_2}^{+L_2} \exp[-2\pi i s_2(y_{pp'})_2]\, ds_2 \int_{-L_3}^{+L_3} \exp[-2\pi i s_3(y_{pp'})_3]\, ds_3 . \tag{2.17}$$

Nwo

$$\int_{-L_i}^{+L_i} \exp[-2\pi i s_i (y_{pp'})_i]\, ds_i = 2L_i \frac{\sin[2\pi L_i (y_{pp'})_i]}{2\pi L_i (y_{pp'})_i} .$$

The average value of $I_N(\mathbf{s})$ in the volume $V = 8\, L_1 L_2 L_3$ is therefore

$$\frac{\int I_N(s)\, dv_s}{V} = \overline{I_N(s)}$$

$$= \mathcal{N} + \frac{\sin[2\pi L_1 (y_{pp'})_1]}{2\pi L_1 (y_{pp'})_1} \frac{\sin[2\pi L_2 (y_{pp'})_2]}{2\pi L_2 (y_{pp'})_2} \frac{[\sin 2\pi L_3 (y_{pp'})]}{2\pi L_3 (y_{pp'})_3} .$$

When $L_1$, $L_2$, $L_3$ tend to infinity, the functions $(\sin Q)/Q$ all tend to zero and

$$\overline{I_N(s)} \rightarrow \mathcal{N} . \tag{2.18}$$

It will be observed that the condition which is required to obtain the correct result for the average value over a certain volume is that at least one of the products $L_i(y_{pp'})_i$ for all the pairs of electrons be large compared to unity. The size of this volume of integration must therefore be large compared to the inverse of the distance between the two electrons which are closest in the object.

The unit scattering power in all of reciprocal space has an average value of

$$\overline{I(s)} = \frac{\overline{I_N(s)}}{N} = \frac{\mathcal{N}}{N} .$$

If the atomic group reduces to a single atom of atomic number $Z$, $\mathcal{N} = NZ$, and therefore $\overline{I(s)} = Z$.

Finally, according to Eq. (*2.10*), *the average of the interference function is equal to unity.* The interferences therefore produce variations in the spatial distribution of the diffracted radiation without changing the average intensity, which is independent of the positions of the atoms, and which is the same as if the atoms scattered independently. This theorem is reminiscent of the principle of the conservation of energy. It must be noted, however, that we have calculated the integral of the diffraction power in an abstract space, namely reciprocal space. It is *not* correct to infer that the total energy of the waves scattered by a given set of atoms in any direction is independent of the mutual positions of the atoms.

In many cases the interference function is periodic in reciprocal space, and then the average value in all of space is equal to the average value of the function in the unit cell of reciprocal space. *This average is equal to unity.*

## 2.4. Diffraction by a Statistically Homogeneous Object

Most of the objects which we have to investigate are composed of a very large number of atoms or molecules whose positions cannot be defined at any given instant. In these cases, Eq. (*2.8*) gives a quantity which cannot be observed—the diffraction by a particular configuration of the atoms. What is measured is the average intensity, which depends only on the macroscopic state of the homogeneous object, such as the composition, temperature, and pressure of a gas. All of these data determine the statistical distribution of the atoms. We must therefore express the observed intensity as a function of the statistical distribution of the component atoms. According to Eq. (*2.11*), the intensity $I_N(\mathbf{s})$ is the Fourier transform of the function $\mathscr{P}(\mathbf{x})$; therefore the required average intensity is the transform of the average value $\overline{\mathscr{P}(\mathbf{x})}$, which we shall now calculate.

### 2.4.1. Calculation of the Function $\overline{\mathscr{P}(x)}$ for an Infinite Homogeneous Object

Let us consider matter which is *homogeneous*, in which the distribution of the atoms with respect to one of them is statistically independent of the atom which is chosen to be at the origin. If the object is finite, the atoms which are close to the external surface have different surroundings than those near the center, but this is due to geometrical factors. We consider that the finite object is identical to a volume cut out from infinite homogeneous matter by an imaginary surface. In other words, we neglect all *physical* effects of the surface as could be produced, for example, by surface tension. Figure 2.1 illustrates this hypothesis: the

**FIGURE 2.1.** *Structure of a finite object.* (a) *Strictly homogeneous.* (b) *With a distorted superficial layer.*

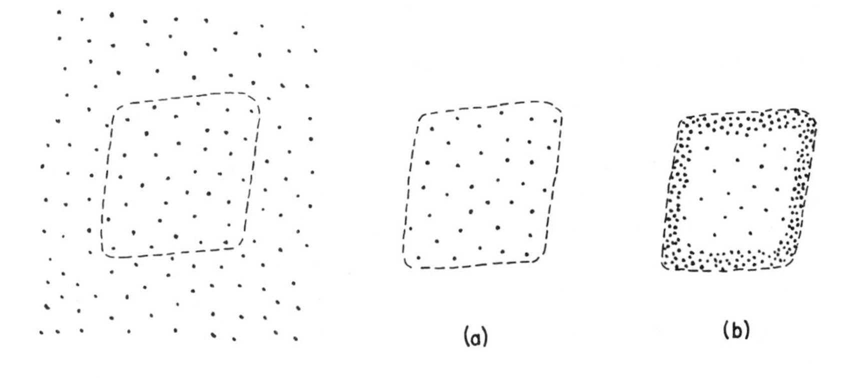

finite object is assumed to have a structure as in *a* and not as in *b*.

Under these conditions, it is possible to distinguish between the effects of finite size and those of the average atomic structure in the diffraction pattern, for they are very different.

Let us call $\rho_\infty(\boldsymbol{x})$ the function defining the distribution of electrons in infinite matter, and $\sigma(\boldsymbol{x})$ a function which is equal to unity inside the external surface of the real object and equal to zero outside. The function $\sigma(\boldsymbol{x})$ is called the *form factor of the object* [1]. The density of matter inside the object is

$$\rho(\boldsymbol{x}) = \rho_\infty(\boldsymbol{x})\sigma(\boldsymbol{x}) . \tag{2.19}$$

We shall calculate the required function $\overline{\mathscr{P}(\boldsymbol{x})}$ in two stages: first, for infinite matter; second, for the real object. We shall define its electron density by the product given in Eq. (*2.19*).

We shall assume that the matter under consideration is *composed of a single type of atom or group of atoms*, and we shall define its statistical distribution as follows. (For brevity, we shall mention only atoms, but it is to be understood that groups of atoms are also considered.) Let us consider an atom situated at $O$, or a group of atoms whose center is at $O$. We require the probability $dP(\boldsymbol{x})$ of finding the center of a second atom in a small volume $dv$ situated at the extremity of the vector $\boldsymbol{x}$ whose origin is at $O$. Let us suppose, that, in any sufficiently large volume $V$, the average number of atoms is $\bar{N}$. We shall set $v_1 = V/\bar{N}$, where $v_1$ is the average volume available for each atom. In the case of a crystal, $v_1$ is the volume $V_c$ of the unit cell. If the atomic distribution is completely random and independent of the fact that *we know* that there is an atom at the origin, the probability of finding a second atom in the volume $dv$ at the extremity of the vector $\boldsymbol{x}$ is $dv/v_1$. However, the positions of the atoms are generally not independent of each other and, for this reason, the probability $dP(\boldsymbol{x})$ is not $dv/v_1$ but rather

$$dP(\boldsymbol{x}) = P(\boldsymbol{x})\frac{dv}{v_1} \tag{2.20}$$

where $P(\boldsymbol{x})$ *is the distribution function which defines the statistical configuration of the atoms.* For example, if these are ordered in space as in a crystal, $P(\boldsymbol{x})$ is zero for all points in space except when $\boldsymbol{x}$ is one of the vectors of the crystal lattice. If the atomic distribution is random, one can at least say that, if the atoms are represented by spheres of radius $r$, $P(\boldsymbol{x})$ is zero for $|\boldsymbol{x}| < 2r$. In all cases, except in crystals where the atoms are distributed in orderly fashion over large distances, atomic forces do not act over more than a few atomic diameters and, at large distances from the atom situated at the origin, the atomic distribution becomes random and $P(\boldsymbol{x})$ tends toward unity for $|\boldsymbol{x}| \gg r$. Therefore $P(\boldsymbol{x})$ fluctuates about unity only for small values of $\boldsymbol{x}$. These fluctuations correspond to the *small-range order* which characterizes the distribution of atoms in matter. This orderly distribution at small distances is found in all forms of matter, but in different degrees for gases, liquids, amorphous solids, and crystals. It is only in crystals that there exists a long-range order.

Since there is necessarily an atom at the origin $\boldsymbol{x} = 0$, by assumption, we shall define the overall statistical distribution with respect to any given atom by

$$z(\boldsymbol{x}) = \delta(\boldsymbol{x}) + \frac{1}{v_1} P(\boldsymbol{x}) . \tag{2.21}$$

The delta or Dirac function, $\delta(\boldsymbol{x})$, has the following properties:

$$\delta(\boldsymbol{x}) = 0 \text{ when } \boldsymbol{x} \neq 0,$$
$$\delta(\boldsymbol{x}) = \infty \quad \text{when} \quad \boldsymbol{x} = 0,$$

but

$$\int_v \delta(\boldsymbol{x})\, dv_x = 1$$

for any volume $v$ around the origin.

If there is no long-range order, the distribution function $P(\boldsymbol{x})$ tends toward unity when $\boldsymbol{x}$ is not very small, and it is useful to write $z(\boldsymbol{x})$ in the form [2]

$$z(\boldsymbol{x}) = \delta(\boldsymbol{x}) + \frac{1}{v_1} + \frac{1}{v_1} [P(\boldsymbol{x}) - 1] , \tag{2.22}$$

where $[P(\boldsymbol{x}) - 1]$ is the oscillating part of $P(\boldsymbol{x})$.

We shall use the Fourier transform of $z(\boldsymbol{x})$, which we shall call $Z(\boldsymbol{s})$. This is the sum of the transforms of the three terms forming $z(\boldsymbol{x})$ of Eq. (*2.22*), two of which can be written explicitly. The transform of the delta function, $\delta(\boldsymbol{x})$, is unity (see Appendix A). Inversely, the transform of the constant $1/v_1$ is the delta function at the origin multiplied by the coefficient $1/v_1$, or $(1/v_1)\delta(\boldsymbol{s})$. We can therefore write that

$$Z(\boldsymbol{s}) = 1 + \frac{1}{v_1} \delta(\boldsymbol{s}) + \frac{1}{v_1} \text{transf}\, [P(\boldsymbol{x}) - 1] . \tag{2.23}$$

We shall now relate the function $\overline{\mathscr{P}_a(\boldsymbol{x})}/V$ of Eq. (*2.13*) to the statistical distribution $z(\boldsymbol{x})$. For *infinite, homogeneous matter*, the ratio $\overline{\mathscr{P}_a(\boldsymbol{x})}/V$ is independent of the volume of integration used in Eq. (*2.14*). We shall call $p_a(\boldsymbol{x})$ the ratio $\overline{\mathscr{P}_a(\boldsymbol{x})}/V$ which is the average value of the integral in Eq. (*2.14*) calculated over a unit volume chosen at random in the object. We now divide this unit volume into small elements $dv_u$. Those which do not contain atoms contribute nothing to the integral, while for those which do, $\rho_a(\boldsymbol{u})dv_u = 1$. The unit volume contains $1/v_1$ atoms, and the integral is therefore the sum of $1/v_1$ terms $\rho_a(\boldsymbol{u} + \boldsymbol{x})$. The average value of $\rho_a(\boldsymbol{u} + \boldsymbol{x})$ is simply $z(\boldsymbol{x})$, since in each term the vector $\boldsymbol{x}$ is such that its origin is at the position of an atom. We therefore have

$$\frac{\overline{\mathscr{P}_a(\boldsymbol{x})}}{V} = p_a(\boldsymbol{x}) = \frac{z(\boldsymbol{x})}{v_1} . \tag{2.24}$$

### 2.4.2. Diffraction by a Homogeneous Finite Object

Let us now introduce the form factor. The diffracting object is defined by its electron density [Eq. (*2.19*)],

$$\rho(\boldsymbol{x}) = \rho_\infty(\boldsymbol{x})\sigma(\boldsymbol{x}) .$$

We shall first calculate the diffracted amplitude, i.e. the Fourier transform of $\rho(\boldsymbol{x})$. This problem is easily solved because of a property of Fourier transforms which is stated below. (It is justified in Appendix A.5.).

Let us consider two point functions $f(\boldsymbol{x})$ and $g(\boldsymbol{x})$, whose transforms in reciprocal space are respectively $F(\boldsymbol{s})$ and $G(\boldsymbol{s})$. The transform $Y(\boldsymbol{s})$ of the product $y(\boldsymbol{x}) = f(\boldsymbol{x})g(\boldsymbol{x})$ is the *faltung of the transforms*, defined by

$$Y(\boldsymbol{s}) = F(\boldsymbol{s}) * G(\boldsymbol{s}) = \int F(\boldsymbol{u}) G(\boldsymbol{s} - \boldsymbol{u})\, dv_u = \int F(\boldsymbol{s} - \boldsymbol{u}) G(\boldsymbol{u})\, dv_u . \tag{2.25}$$

The integral is evaluated over all of reciprocal space.

We shall also use the *inverse* property. The transform of a faltung,

$$y(\boldsymbol{x}) = f(\boldsymbol{x}) * g(\boldsymbol{x}) = \int f(\boldsymbol{u}) g(\boldsymbol{x} - \boldsymbol{u})\, dv_u ,$$

is the product of the transforms of the two functions:

$$Y(\boldsymbol{s}) = F(\boldsymbol{s}) G(\boldsymbol{s}) . \tag{2.26}$$

According to Eq. (*2.3*), the amplitude is the transform of $\rho(\boldsymbol{x})$ and therefore of the product $\rho_\infty(\boldsymbol{x})\sigma(\boldsymbol{x})$:

$$A(\boldsymbol{s}) = \int \rho_\infty(\boldsymbol{x})\sigma(\boldsymbol{x}) \exp(-2\pi i \boldsymbol{s} \cdot \boldsymbol{x})\, dv_x .$$

We must first calculate the transform of $\rho_\infty(\boldsymbol{x})$; that is, $A_\infty(\boldsymbol{s})$.

We now introduce the transform of the form function $\sigma(\boldsymbol{x})$, which we shall call $\Sigma(\boldsymbol{s})$:

$$\Sigma(\boldsymbol{s}) = \int \sigma(\boldsymbol{x}) \exp(-2\pi i \boldsymbol{s} \cdot \boldsymbol{x})\, dv_x . \tag{2.27}$$

The faltung theorem gives us the required amplitude:

$$A(\boldsymbol{s}) = A_{\infty}(\boldsymbol{s}) * \Sigma(\boldsymbol{s}) = \int A_{\infty}(\boldsymbol{u})\Sigma(\boldsymbol{s}-\boldsymbol{u})\,dv_u\,. \tag{2.28}$$

We can interpret this formula as follows: the function $A(\boldsymbol{s})$ is the superposition of the transforms $\Sigma(\boldsymbol{s})$ centered on each point of reciprocal space, multiplied by a coefficient which is equal to the value at this point of the transform of the electron density of the infinite homogeneous object.

However, as we have seen above, the expression for the amplitude is not very useful. It is therefore important to express directly the effect of the finite size of the object on the diffracted intensity.

According to Eq. (*2.13*), the interference function for the object, $\mathscr{I}(\boldsymbol{s}) = [I_N(\boldsymbol{s})]/NF^2$. is equal, within a factor $v_1/V$, to the transform of $\mathscr{P}_a(\boldsymbol{x})$, or of the integral

$$\mathscr{P}_a(\boldsymbol{x}) = \int \rho_a(\boldsymbol{u})\rho_a(\boldsymbol{u}+\boldsymbol{x})\,dv_u\,.$$

Let us introduce the form factor. Then

$$\mathscr{P}_a(\boldsymbol{x}) = \int \rho_{a\infty}(\boldsymbol{u})\rho_{a\infty}(\boldsymbol{u}+\boldsymbol{x})\sigma(\boldsymbol{u})\sigma(\boldsymbol{u}+\boldsymbol{x})\,dv_u\,. \tag{2.29}$$

The product $\sigma(\boldsymbol{u})\sigma(\boldsymbol{u}+\boldsymbol{x})$ is equal to unity if the two points $\boldsymbol{u}$ and $\boldsymbol{u}+\boldsymbol{x}$ are *both* inside the object, and it is equal to zero otherwise. Let us consider the part of the object $\mathscr{V}(\boldsymbol{x})$ where the product $\sigma(\boldsymbol{u})\sigma(\boldsymbol{u}+\boldsymbol{x})$ is equal to unity (Figure 2.2). Equation (*2.29*) can be rewritten as

$$\mathscr{P}_a(\boldsymbol{x}) = \int_{\mathscr{V}(\boldsymbol{x})} \rho_{a\infty}(\boldsymbol{u})\rho_{a\infty}(\boldsymbol{u}+\boldsymbol{x})\,dv_u\,. \tag{2.30}$$

if we limit the volume of integration to the domain $\mathscr{V}(\boldsymbol{x})$. We shall call its volume $V\cdot V(\boldsymbol{x})$: it is given by the expression

$$V\cdot V(\boldsymbol{x}) = \int \sigma(\boldsymbol{u})\sigma(\boldsymbol{u}+\boldsymbol{x})\,dv_u\,. \tag{2.31}$$

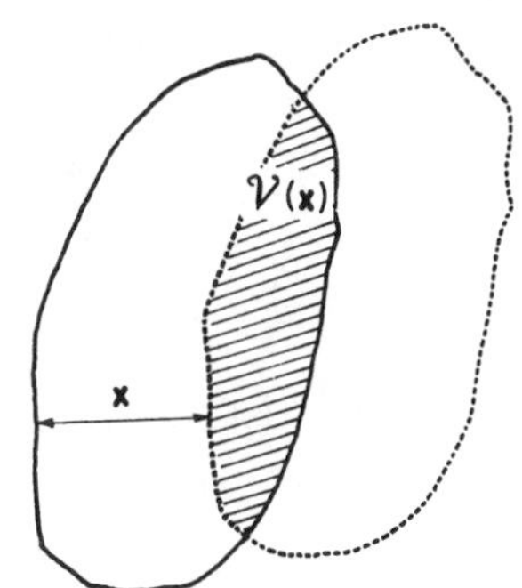


**FIGURE 2.2.** *Definition of the function V(x). The volume of the hatched body $\mathscr{V}(x)$ is $V\cdot V(x)$, V being the volume of the original object.*

Because of the homogeneity of the matter under consideration, we can state that $\mathscr{P}_a(\boldsymbol{x})$ is equal to the product of the average value of $\rho_{a\infty}(\boldsymbol{u})\rho_{a\infty}(\boldsymbol{u}+\boldsymbol{x})$ in the object, multiplied by the volume of integration $V\cdot V(\boldsymbol{x})$. This average value was calculated in Eq. (*2.24*) and it was called $p_a(\boldsymbol{x})$. Therefore,

$$\mathscr{P}_a(\boldsymbol{x}) = p_a(\boldsymbol{x})\,V\cdot V(\boldsymbol{x}) = \frac{V}{v_1}\,z(\boldsymbol{x})\,V(\boldsymbol{x})\,. \tag{2.32}$$

It is a simple matter to give a geometrical interpretation to $V(\boldsymbol{x})$: $[V \cdot V(\boldsymbol{x})]$ is the volume which is common to the object and to its ghost obtained by displacing the object by the distance $\boldsymbol{x}$, as in Figure 2.2. This construction shows that $V(\boldsymbol{x})$ has the following properties.

(a) $V(0)$ is equal to unity for, when $\boldsymbol{x} = 0$, the object and its ghost coincide.

(b) $V(\boldsymbol{x})$ is centrosymmetric since $V(\boldsymbol{x}) = V(-\boldsymbol{x})$.

(c) $V(\boldsymbol{x})$ decreases as $\boldsymbol{x}$ increases. It becomes zero when $\boldsymbol{x}$ is larger than the maximum diameter of the object in the direction of $\boldsymbol{x}$.

Substituting Eq. (*2.32*) in (*2.13*), we obtain for the interference function

$$\mathscr{I}(\boldsymbol{s}) = \int z(\boldsymbol{x})\, V(\boldsymbol{x}) \exp(-2\pi i \boldsymbol{s}\cdot\boldsymbol{x})\, dv_x\,. \qquad (2.33)$$

We therefore have to calculate the Fourier transform of the product $z(\boldsymbol{x})\,V(\boldsymbol{x})$. We have called $Z(\boldsymbol{s})$ the transform of $z(\boldsymbol{x})$ [Eq. (*2.23*)], and we now have to find the transform of $V(\boldsymbol{x})$ [Eq. (*2.31*)].

If we select $\boldsymbol{x} + \boldsymbol{u} = \boldsymbol{w}$ as a variable, $V(\boldsymbol{x})$ can be written

$$V(\boldsymbol{x}) = \frac{1}{V} \int \sigma(\boldsymbol{w})\sigma[-(\boldsymbol{x} - \boldsymbol{w})]\, dv_w\,,$$

and, according to Eq. (*2.25*), $[V \cdot V(\boldsymbol{x})]$ is the faltung of the functions $\sigma(\boldsymbol{x})$ and $\sigma(-\boldsymbol{x})$, $\Sigma(\boldsymbol{s})$ being the Fourier transform of $\sigma(\boldsymbol{x})$. As shown in Appendix A.3, the transform of $\sigma(-\boldsymbol{x})$ is $\Sigma^*(\boldsymbol{s})$.

According to Eq. (*2.26*), the transform of $V(\boldsymbol{x})$ is therefore

$$\frac{1}{V}\,\Sigma(\boldsymbol{s})\,\Sigma^*(\boldsymbol{s}) = \frac{|\,\Sigma(\boldsymbol{s})\,|^2}{V}\,,$$

so that

$$|\,\Sigma(\boldsymbol{s})\,|^2 = V \int V(\boldsymbol{x}) \exp(-2\pi i \boldsymbol{s}\cdot\boldsymbol{x})\, dv_x\,. \qquad (2.34)$$

To obtain the transform of the product $z(\boldsymbol{x})\,V(\boldsymbol{x})$, we now have to take the faltung of the transforms of both factors. The interference function is therefore

$$\mathscr{I}(\boldsymbol{s}) = \frac{1}{V}\, Z(\boldsymbol{s}) * |\,\Sigma(\boldsymbol{s})\,|^2 = \frac{1}{V} \int Z(\boldsymbol{u})\,|\,\Sigma(\boldsymbol{s} - \boldsymbol{u})\,|^2\, dv_u\,. \qquad (2.35)$$

The unit scattering power per atom of the object is

$$I(\boldsymbol{s}) = F^2 \mathscr{I}(\boldsymbol{s}) = \frac{F^2}{V}\, Z(\boldsymbol{s}) * |\,\Sigma(\boldsymbol{s})\,|^2\,, \qquad (2.35a)$$

and the total scattering power is

$$I_N(\boldsymbol{s}) = NF^2 \mathscr{I}(\boldsymbol{s}) = \frac{F^2}{v_1}\, Z(\boldsymbol{s}) * |\,\Sigma(\boldsymbol{s})\,|^2\,. \qquad (2.35b)$$

In the most general case where the structure is defined, not with respect to identical atoms or groups of atoms but with respect to the electron density $\rho(\boldsymbol{x})$, it can be shown from Eq. (*2.12*) that the scattering power for an object of volume $V$ is given by

$$I_N(\boldsymbol{s}) = V \int p(\boldsymbol{x}) V(\boldsymbol{x}) \exp(-2\pi i \boldsymbol{s} \cdot \boldsymbol{x})\, dv_x ,$$

where $p(\boldsymbol{x})$ is the average value of the integral

$$\mathscr{P}(\boldsymbol{x}) = \int \rho(\boldsymbol{u}) \rho(\boldsymbol{x} + \boldsymbol{u})\, dv_x ,$$

calculated over a unit volume of the substance under consideration. Therefore

$$I_N(\boldsymbol{s}) = [\text{transf } p(\boldsymbol{x})] * |\Sigma(\boldsymbol{s})|^2 . \qquad (2.36)$$

The following remarks will clarify the physical meaning of these expressions for the diffracted intensity.

(a) Let us first discuss a few properties of $\Sigma(\boldsymbol{s})$. We shall use two different definitions for this quantity: $\Sigma(\boldsymbol{s})$ is the transform of the form factor $\sigma(\boldsymbol{x})$, or

$$\Sigma(\boldsymbol{s}) = \int \sigma(\boldsymbol{x}) \exp(-2\pi i \boldsymbol{s} \cdot \boldsymbol{x})\, dv_x ,$$

and, according to Eq. (*2.34*),

$$|\Sigma(\boldsymbol{s})|^2 = V \int V(\boldsymbol{x}) \exp(-2\pi i \boldsymbol{s} \cdot \boldsymbol{x})\, dv_x .$$

Since $\sigma(\boldsymbol{x})$ is real, $\Sigma(\boldsymbol{s}) = \Sigma^*(-\boldsymbol{s})$, and $|\Sigma(\boldsymbol{s})|^2$ is therefore centrosymmetric. Its maximum value is $|\Sigma(0)|^2$, but

$$\Sigma(0) = \int \sigma(\boldsymbol{x})\, dv_x = V ,$$

and thus

$$|\Sigma(0)|^2 = V^2 . \qquad (2.37)$$

The function $|\Sigma(\boldsymbol{s})|^2$ is thus appreciably different from zero near the origin of reciprocal space, over a region which is small when the transform $V(\boldsymbol{x})$ extends over a large region of object space—that is, when the scattering object is large. If we invert the integral of Eq. (*2.34*), we obtain

$$V(\boldsymbol{x}) = \frac{1}{V} \int |\Sigma(\boldsymbol{s})|^2 \exp(2\pi i \boldsymbol{s} \cdot \boldsymbol{x})\, dv_s ,$$

and, since $V(0) = 1$,

$$\int |\Sigma(\boldsymbol{s})|^2 \, dv_s = V \,. \tag{2.38}$$

If we replace the decreasing function $|\Sigma(\boldsymbol{s})|^2$ by a constant which is equal to its maximum $V^2$ in a volume $w$ such that $V^2 w$ is equal to the integral

$$\int |\Sigma(\boldsymbol{s})|^2 \, dv_s = V \,,$$

we find that $w = 1/V$.

We obtain in this way an order of magnitude for the width of the function $|\Sigma(\boldsymbol{s})|^2$ in reciprocal space. The quantity $w$ is the volume of a cube measuring $2s_0$ on the side such that

$$(2s_0)^3 \cong \frac{1}{V} \,.$$

Let us now consider an object which is very small but not submicroscopic. For example, let us take $V = (10^{-2}\,\text{mm})^3$. The corresponding value for $s_0$ is $(1/2)10^{-5}\,\text{Å}^{-1}$, and $s_0$ is approximately $10^5$ times smaller than the unit cell of the reciprocal lattice of a common inorganic crystal.

Since $s = (2 \sin \theta)/\lambda$, where $2\theta$ is the scattering angle corresponding to a point in reciprocal space which is at a distance $s$ from the origin, $s_0 = (1/2)\,10^{-5}\,\text{Å}^{-1}$ corresponds for CuK$\alpha$ radiation to the angle $\theta_0 = [(1/2)10^{-4}]^\circ$. This is a very small angle indeed compared to those which are ordinarily used in the measurement of diffracted intensities.

(b) There is one particular case where the faltung of two functions, defined in Eq. (*2.25*), reduces to a simple operation, as will be shown in Appendix A.5: when one of the functions $f(\boldsymbol{x})$ reduces to a sharp peak of width $\boldsymbol{\varepsilon}$ near the origin, where $\boldsymbol{\varepsilon}$ is sufficiently small so that in the interval $(\boldsymbol{x}-\boldsymbol{\varepsilon}, \boldsymbol{x}+\boldsymbol{\varepsilon})$ the other function varies little from its value $g(\boldsymbol{x})$, the faltung of the two functions is simply the product of $g(\boldsymbol{x})$ and the area under the peak for $f(\boldsymbol{x})$.

Let us now consider an object with uniform electron density $\rho$. The function $p(\boldsymbol{x})$ of Eq. (*2.36*) is then equal to $\rho^2$ and the transform of $p(\boldsymbol{x})$ is $\rho^2\delta(\boldsymbol{s})$. The function $\delta(\boldsymbol{s})$ is infinitely narrow with respect to the peak of $|\Sigma(\boldsymbol{s})|^2$, even if it is extremely sharp. Therefore

$$I_N(\boldsymbol{s}) = \rho^2\delta(\boldsymbol{s}) * |\Sigma(\boldsymbol{s})|^2 = \rho^2 |\Sigma(\boldsymbol{s})|^2 \,.$$

This result can be deduced from the general equation [Eq. (*2.3*)], for in this case $\rho(\boldsymbol{x}) = \rho\sigma(\boldsymbol{x})$. The diffracted amplitude is the transform of $\rho(\boldsymbol{x})$, that is $\rho\Sigma(\boldsymbol{s})$, and the intensity is the square of the modulus of the amplitude, or $\rho^2 |\Sigma(\boldsymbol{s})|^2$.

2.4.2.1. DIFFRACTION BY A DISORDERED STRUCTURE. Let us return to the general expression for $Z(\boldsymbol{s})$ given in Eq. (2.23). If we use it in Eq. (2.35), we obtain two terms:

$$\mathscr{I}(\boldsymbol{s}) = \frac{1}{V}\left[\frac{1}{v_1}\,\delta(\boldsymbol{s}) * |\,\Sigma(\boldsymbol{s})\,|^2 + \Phi(\boldsymbol{s}) * |\,\Sigma(\boldsymbol{s})\,|^2\right], \tag{2.39}$$

where

$$\Phi(\boldsymbol{s}) = 1 + \frac{1}{v_1}\int [P(\boldsymbol{x}) - 1]\exp(-2\pi i\boldsymbol{s}\cdot\boldsymbol{x})\,dv_x\,. \tag{2.40}$$

The first faltung has already been calculated (in the last section):

$$\frac{1}{v_1}\,\delta(\boldsymbol{s}) * |\,\Sigma(\boldsymbol{s})\,|^2 = \frac{|\,\Sigma(\boldsymbol{s})\,|^2}{v_1}\,. \tag{2.41}$$

In this first term, the peak of the function $|\,\Sigma(\boldsymbol{s})\,|^2$ is considered to be very broad compared to that of the delta function. However, in the second product, we can in general consider that $|\,\Sigma(\boldsymbol{s})\,|^2$ has a very sharp peak since, in disordered structures, $\Phi(\boldsymbol{s})$ has no sharp maxima and often varies slowly with $\boldsymbol{s}$. We can therefore equate this second term to the product of $\Phi(\boldsymbol{s})$ and the integral of the peak for $|\,\Sigma(\boldsymbol{s})\,|^2$, i.e., $V$ from Eq. (2.38). The interference function can finally be written as

$$\mathscr{I}(\boldsymbol{s}) = \frac{1}{V v_1}\,|\,\Sigma(\boldsymbol{s})\,|^2 + \left[1 + \frac{1}{v_1}\int [P(\boldsymbol{x}) - 1]\exp(-2\pi i\boldsymbol{s}\cdot\boldsymbol{x})\,dv_x\right]. \tag{2.42}$$

The first term corresponds, within a multiplying factor, to the peak for $|\,\Sigma(\boldsymbol{s})\,|^2$ at the center of reciprocal space, or, experimentally, to *very small angles*. It depends on the shape and size of the object and not on its internal structure. In practice, whenever the dimensions of the object are larger than about one micron, it becomes impossible to observe this central diffraction peak which is masked by the undeflected beam. It is therefore generally possible to neglect this first term of the expression for the intensity, *except when the diffracting object is very small. This particular case will be studied in Chapters 5 and 10.*

In general, the intensity measured experimentally is represented exclusively by the second term of Eq. (2.42), which is the only one which we shall retain:

$$\mathscr{I}(\boldsymbol{s}) = 1 + \frac{1}{v_1}\int [P(\boldsymbol{x}) - 1]\exp(-2\pi i\boldsymbol{s}\cdot\boldsymbol{x})\,dv_x\,. \tag{2.43}$$

This term is independent of the volume and shape of the object. It depends exclusively on the *statistical distribution of the atoms in homogeneous and infinite matter*. If this distribution is statistically uniform, $P(\boldsymbol{x}) = 1$, and the interference function is constant and equal to unity in reciprocal space. The variations of this function about this average value show the irregularities in the atomic distribution.

Inversely, if it is possible to determine experimentally the interference function in all of reciprocal space, the inversion of Eq. (*2.43*) leads to the determination of the statistical distribution of the atoms in the object, as defined by the function $P(\boldsymbol{x})$:

$$P(\boldsymbol{x}) = 1 + v_1 \int [\mathscr{I}(\boldsymbol{s}) - 1] \exp(2\pi i \boldsymbol{s} \cdot \boldsymbol{x})\, dv_s \,. \tag{2.44}$$

This is the most complete result which X-ray diffraction can provide about the structure of an object formed of identical atoms or groups of atoms when it is statistically homogeneous.

Equation (*2.43*) involves an approximation which we shall now discuss. The exact equation for $\mathscr{I}(\boldsymbol{s})$ is given by the second term of Eq. (*2.39*):

$$\begin{aligned}\mathscr{I}(\boldsymbol{s}) &= \frac{1}{V}[\Phi(\boldsymbol{s}) * |\Sigma(\boldsymbol{s})|^2 \\ &= \left\{1 + \frac{1}{v_1}\text{transf}\,[P(\boldsymbol{x}) - 1]\right\} * \frac{|\Sigma(\boldsymbol{s})|^2}{V} \\ &= 1 + \frac{1}{v_1}\left\{\text{transf}\,[P(\boldsymbol{x}) - 1] * \frac{|\Sigma(\boldsymbol{s})|^2}{V}\right\}.\end{aligned} \tag{2.44a}$$

However, the transform of $|\Sigma(\boldsymbol{s})|^2$ is $V \cdot V(\boldsymbol{x})$, from Eq. (*2.34*), and hence Eq. (*2.44a*) can be rewritten as

$$\mathscr{I}(\boldsymbol{s}) - 1 = \frac{1}{v_1}\text{transf}\,[P(\boldsymbol{x}) - 1] * \text{transf}\,V(\boldsymbol{x})\,.$$

Applying now the theorem for the faltung of two functions,

$$\mathscr{I}(\boldsymbol{s}) - 1 = \frac{1}{v_1}\int V(\boldsymbol{x})[P(\boldsymbol{x}) - 1] \exp(-2\pi i \boldsymbol{s} \cdot \boldsymbol{x})\, dv_x \,. \tag{2.45}$$

The approximation of Eq. (*2.43*) therefore consists in replacing $V(\boldsymbol{x})$ by unity.

When there is disorder at large distances in the sample, $P(\boldsymbol{x})$ tends to unity when $|\boldsymbol{x}|$ increases. Let us call $\boldsymbol{x}_0$ a vector such that, for $|\boldsymbol{x}| > |\boldsymbol{x}_0|$, we can consider that $[P(\boldsymbol{x}) - 1]$ is zero. Since $V(\boldsymbol{x})$ decreases as $\boldsymbol{x}$ increases, we can find an upper limit for the error introduced by replacing $V(\boldsymbol{x})$ by unity in Eq. (*2.45*) if we replace $V(\boldsymbol{x})$ by $V(\boldsymbol{x}_0)$. We find that the true relative error is less than $1 - V(\boldsymbol{x}_0)$. To evaluate $V(\boldsymbol{x}_0)$ we consider a cylindical object parallel to $\boldsymbol{x}_0$, as in Figure 2.3. Then $V(\boldsymbol{x}_0)$ is equal to $(D - x_0)/D$ if $D$ is the diameter of the object parallel to $\boldsymbol{x}_0$. The error introduced in the interference function by neglecting the size of the object is of the order of $x_0/D$. For distances larger than $x_0$, the distribution of the atoms around one of them is practically uniform, and $D$ is the order of magnitude of the diameter of the particle.

It is therefore permissible to use Eq. (2.43) *if the region of influence of an atom is small compared to the volume of the object.* This applies to all amorphous bodies, even if the sample is very small, for the disorder is practically complete for distances larger than about ten atomic distances, i.e. larger than about 30 Å. This would not, however, be true for a crystal, whatever its size.

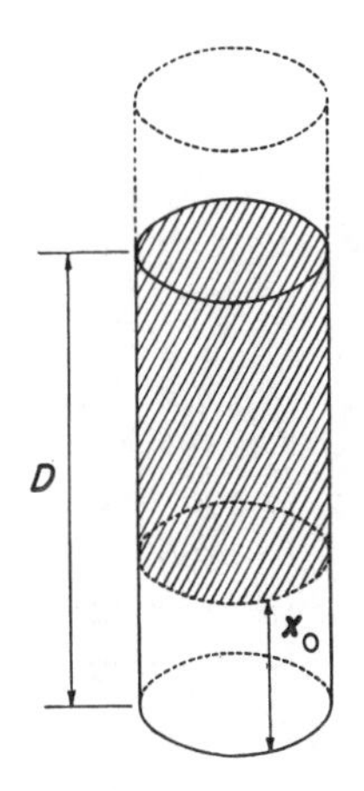


**FIGURE 2.3.** *Calculation of the error made when the effect of the size of the object is neglected.*

The above general discussion follows directly from some fundamental work by Ewald [1], Hosemann [2], Mering [3], and Wilson [4] which allows us to distinguish clearly between the effects of size and of structure in the object. Previous proofs of Eq. (2.43) often involved unnecessary hypotheses, such as those of Compton and Allison [5]. Also, the calculations given in many papers on the diffraction by random structures are obscured by the simultaneous consideration of size and structure effects. Certain authors have arrived at particular results which follow immediately from the above general theory.

## 2.5. The Particular Case of an Isotropic Object

We now introduce the additional hypothesis that the matter under consideration is isotropic. Then $P(\boldsymbol{x})$, and therefore $z(\boldsymbol{x})$, depend only on the modulus of $\boldsymbol{x}$ and not on its direction. These conditions are found in gases, in most liquids, in amorphous bodies, and also, on the average, in fine powders, even if the grains are anisotropic, so long as their orientations are random. The observed diffraction intensity as given by Eq. (2.43) is then only a function of $|\boldsymbol{s}| = s = (2\sin\theta)/\lambda$.

The calculation of the transform of $[P(\boldsymbol{x}) - 1]$ can be done by defining the modulus of $\boldsymbol{x}$, the angle $\alpha$ it forms with $\boldsymbol{s}$, and the azimuthal angle $\varphi$ formed by the plane containing $\boldsymbol{x}$ and $\boldsymbol{s}$ with an arbitrary plane. Then

$$\int [P(\boldsymbol{x}) - 1] \exp(-2\pi i \boldsymbol{s}\cdot\boldsymbol{x})\, dv_x$$

$$= \int_{x=0}^{x=\infty} \int_{\alpha=0}^{\alpha=\pi} \int_{\varphi=0}^{\varphi=2\pi} [P(x) - 1)] \exp(-2\pi i s\, x \cos\alpha)\, 2\pi x^2 \sin\alpha\, d\alpha\, dx\, \frac{d\varphi}{2\pi}$$

$$= 2\int_0^\infty [P(x) - 1]\, \frac{\sin(2\pi s x)}{s}\, x\, dx\,. \tag{2.46}$$

Then, from Eq. (2.43), the interference function is

$$\mathscr{I}(s) = 1 + \frac{2}{s v_1} \int_0^\infty [P(x) - 1] \sin(2\pi s x) x \, dx \, . \tag{2.47}$$

The distribution function $P(x)$ is found experimentally by inversion of the above equation:

$$x[P(x) - 1] = 2v_1 \int_0^\infty [\mathscr{I}(s) - 1] \sin(2\pi s x) s \, ds \, . \tag{2.48}$$

These equations can be written in other equivalent forms. Let us set $\rho_a(x)$ as the atomic density at a distance $x$ from the atom chosen to be at the origin. Then $\rho_a(x)$ tends toward the average density $\rho_0$ when $x$ becomes large, and $4\pi x^2 \rho_a(x) dx$ represents the number of atoms situated between $x$ and $x + dx$. From the definition of $P(x)$ given in Eq. (*2.20*), we find that $P(x) = \rho_a(x)/\rho_0$ since $\rho_0 = 1/v_1$. Equation (*2.47*) then becomes

$$\mathscr{I}(s) = 1 + \int_0^\infty 4\pi x^2 [\rho_a(x) - \rho_0] \frac{\sin(2\pi s x)}{2\pi s x} dx \, , \tag{2.49}$$

and, by Fourier inversion,

$$4\pi x^2 \rho_a(x) = 4\pi x^2 \rho_0 + 8\pi x \int_0^\infty [\mathscr{I}(s) - 1] \sin(2\pi s x) s \, ds \, . \tag{2.50}$$

X-ray diffraction by a homogeneous and isotropic body therefore gives the average number of neighboring atoms as a function of the distance to the atom chosen to be at the origin.

### 2.5.1. Limiting Value of the Diffracted Intensity at Small Angles

In the following chapters we shall study the dependence of the intensity on $s$, i.e. on the angle of diffraction, for different types of distribution functions $P(x)$ characteristic of amorphous substances. It is, however, possible to find a very general property of the limiting value of the intensity at small angles for isotropic bodies.

It is important to realize that we are dealing with the limit $s = 0$ of Eq. (*2.47*). This expression was obtained by eliminating from the more rigorous one given in Eq. (*2.42*) the central diffraction peak, which depends solely on the shape of the object and not on its structure. This peak is, in fact, not observed experimentally when the object is sufficiently large. The quantity $\mathscr{I}'_0$ which we shall calculate is therefore the limiting value of the scattering power as observed in practice with the usual instruments, when the object has a volume of, let us say, at least $(10^{-2}\,\text{mm})^3$.

Equation (*2.47*) can be rewritten as

$$\mathscr{I}(s) = 1 + \frac{1}{v_1} \int_0^\infty [P(x) - 1] \frac{\sin(2\pi s x)}{2\pi s x} 4\pi x^2 \, dx \, ,$$

and, for $s = 0$,

$$\mathscr{I}'(0) = 1 + \frac{1}{v_1}\int_0^\infty [P(x) - 1] 4\pi x^2 dx . \qquad (2.51)$$

The calculation of $\mathscr{I}'(0)$ is rather delicate, because one must avoid proceeding in a manner which, although it appears to be natural, is incorrect. It will be instructive to discuss this error. Since $P(x)dv/v_1$ represents the number of atoms in the volume $dv$ situated at the distance $x$ from the origin, it would appear that the integral of this quantity represents the number of neighbors of the atom at the origin, or $N-1$ in the volume $V$, if $V$ contains $N$ atoms. On the other hand, $\int dv/v_1 = N$. We could therefore write

$$\frac{1}{v_1}\int_0^\infty [P(x) - 1]\, 4\pi x^2 dx = \frac{1}{v_1}\int_V [P(x) - 1]\, dv_x = (N-1) - (N) = -1.$$

The two integrals are equivalent since $[P(x) - 1]$ is zero whenever $x$ is larger than a few atomic distances, and then we would conclude that

$$\mathscr{I}'(0) = 1 - 1 = 0.$$

Now, experimentally—in the case of gases, for example (Section 3.1)—we find a limiting value for the intensity which is different from zero and the above calculation is therefore incorrect.

Let us examine the problem more closely [6]. Consider a volume $V$ which is very large compared to the interatomic distances and which is defined by a virtual surface in homogeneous matter. Because of thermal agitation in gases and in liquids and because of irregularities in disordered structures, the number of atoms effectively contained in a volume $V$ is ***not fixed***, but can fluctuate about an average value $\bar{N} = V/v_1$. Let us try to calculate the average number of pairs of atoms contained in $V$. At a given moment, this number is $N(N-1)$ and the required average value is $\overline{N(N-1)} = \overline{N^2} - \bar{N}$.

Let us evaluate this same number of pairs of atoms using the function $P(x)$. If we return to the very definition of $P(x)$ given in Section 2.4.1, we find that we have set this number to be equal to $\bar{N}\int P(x)dv/v_1$. Therefore

$$\bar{N}\int P(x)\frac{dv}{v_1} = \overline{N(N-1)} .$$

If $N$ is a constant, we find that

$$\int \frac{P(x)\,dv}{v_1} = N - 1$$

and that $\mathscr{I}'(0) = 0$. However, if $N$ fluctuates, *the average of the product* $\overline{N(N-1)}$ *is not the product of the averages*, and the integral $\int P(x)dv/v_1$ is not equal to $N-1$ but to

$$\frac{\overline{N(N-1)}}{\bar{N}} = \frac{\overline{N^2}}{\bar{N}} - 1$$

and

$$\frac{1}{v_1}\int_V [P(x) - 1]\, dv_x = \frac{1}{v_1}\int_0^\infty [P(x) - 1]4\pi x^2\, dx = \frac{\overline{N^2}}{\bar{N}} - 1 - \bar{N}\,,$$

whence

$$\mathscr{I}'(0) = \frac{\overline{N^2} - (\bar{N})^2}{\bar{N}} = \frac{\overline{(N-\bar{N})^2}}{\bar{N}}\,.$$

The scattering power $\mathscr{I}'(0)$ is therefore a measure of the *fluctuations in the number of atoms contained in a given volume.*

Kinetic theory [7] relates these fluctuations to the isothermal compressibility

$$\beta = -\frac{1}{v}\left(\frac{\partial v}{\partial p}\right)_T$$

through the relation

$$\frac{kT\beta}{v_1} = \frac{\overline{(N-\bar{N})^2}}{\bar{N}}\,,$$

and therefore

$$\mathscr{I}'(0) = \frac{kT\beta}{v_1}\,. \tag{2.52}$$

The scattering power at zero angle will therefore be larger for highly compressible fluids like gases than for less compressible fluids like liquids. For a perfect gas, $\beta = 1/p$, $pv_1 = kT$, and therefore

$$\mathscr{I}'(0) = 1\,.$$

It will be shown in Section 3.1.1 that the interference function for a perfect gas is equal to unity at all points of reciprocal space. We therefore had to show that its limiting value for small angles was also unity.

We shall also find in the next chapter how the central peak, which cannot be observed despite the fact that it does exist in reality, can be eliminated in a most natural way by the calculations in the case of a perfect gas. This is not the case, however, for all diffraction calculations and, in some cases, we find a formula which is correct for all the angles where the diffraction can be observed but such that, if we set $s = 0$, we do not find the limiting value of $\mathscr{I}'(0)$ for small angles [8].

## 2.6. The Debye Formula for Powder Patterns

Let us consider a diffracting object in which the relative positions of the atoms are fixed. We assume that the object rotates in such a manner that all possible orientations with respect to the incident beam are equally probable. We also assume that this motion is sufficiently rapid so that one observes only the average diffracted intensity. We shall see below that, experimentally, this is equivalent to observing the diffraction by a collection of identical objects with random orientations and positions. We shall call such an object a perfect powder. This condition can be achieved not only with solid grains but also with polyatomic molecules in the gaseous state or in solution.

For a given position of the object, its scattering power is given by Eq. (*2.8*):

$$I_N(\boldsymbol{s}) = \Sigma_1^N f_n^2 + \sum_{n \neq n'}\sum f_n f_{n'} \cos(2\pi \boldsymbol{s} \cdot \boldsymbol{x}_{nn'}) .$$

The observed scattering power is the sum of the averages of each of the terms:

$$\overline{I_N(\boldsymbol{s})} = \Sigma_1^N f_n^2 + \sum_{n \neq n'}\sum f_n f_{n'} \overline{\cos(2\pi \boldsymbol{s} \cdot \boldsymbol{x}_{nn'})} . \qquad (2.53)$$

To calculate the average of the cosine term, we proceed as we did previously in arriving at Eq. (*2.46*):

$$\overline{\cos(2\pi \boldsymbol{s} \cdot \boldsymbol{x}_{nn'})} = \int_{\alpha=0}^{\alpha=\pi}\int_{\varphi=0}^{\varphi=2\pi} \cos(2\pi s x_{nn'} \cos\alpha) 2\pi \sin\alpha \, d\alpha \, \frac{d\varphi}{2\pi}$$
$$= \frac{\sin(2\pi s x_{nn'})}{2\pi s x_{nn'}} ,$$

where $x_{nn'}$ is the distance between the atoms $n$ and $n'$. The Debye formula for the intensity is

$$\overline{I_N(s)} = \Sigma_1^N \Sigma_1^N f_n f_{n'} \frac{\sin(2\pi s x_{nn'})}{2\pi s x_{nn'}} . \qquad (2.54)$$

Let us consider an object composed of atoms or of identical groups of atoms with structure factor $F$. We consider first in Eq. (*2.54*) the $N$ terms related to a given atom, and we notice that each pair is counted twice, the distance $x_{nn'}$ being equal to $x_{n'n}$. We shall call $x_m$ the interatomic distances in the object. There are $[N(N-1)]/2$ of these. According to the Debye formula the interference function is

$$\mathscr{I}(s) = \frac{\overline{I_N(s)}}{NF^2} = 1 + \frac{2}{N}\sum_m \frac{\sin(2\pi s x_m)}{2\pi s x_m}. \tag{2.55}$$

The powder diagram depends only on the lengths of all the interatomic vectors; *it does not depend on their mutual orientations.* Thus, for a group of four atoms situated at the summits of a regular tetrahedron of side $a$, the interference function is $1 + [3\sin(2\pi sa)]/(2\pi sa)$ and, for a pair of atoms separated by a distance $a$,

$$\mathscr{I}(s) = 1 + \frac{\sin(2\pi sa)}{2\pi sa}. \tag{2.55a}$$

The powder diagrams of these two atomic groups are very analogous: the maxima and minima are situated at exactly the same angles, as shown in Figure 2.4. This example shows that a powder diffraction pattern certainly does not provide an easy method for determining the structure of an object.

**FIGURE 2.4.** *Interference function.* (a) *Disoriented molecules with four atoms at the corners of a regular tetrahedron of edge a.* (b) *Disoriented diatomic molecules of length a.*

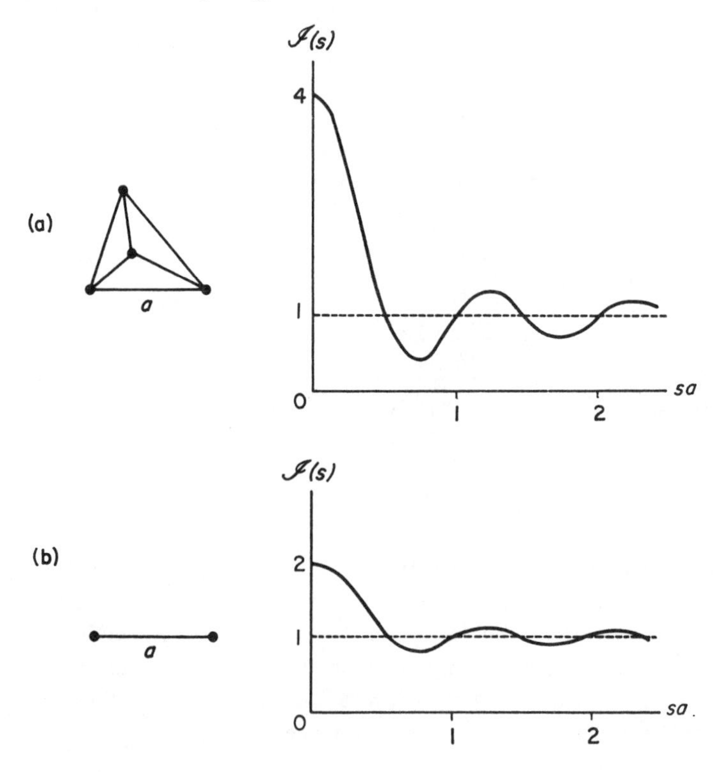

The Debye formula applies, in particular, to a crystalline powder of identical particles. Let us consider grains of a simple cubic crystal having the form of cubes measuring $(n-1)a$, $a$ being the lattice parameter of the unit cell. There are

| | |
|---|---|
| $n^3$ | atoms |
| $3n^2(n-1)$ | [100] pairs of length $a$ |
| $6n(n-1)^2$ | [110] pairs of length $a\sqrt{2}$ |
| $4(n-1)^3$ | [111] pairs of length $a\sqrt{3}$ |
| . . . . . | . . . . . . . . . . . . . . |
| $p(n-h)(n-k)(n-l)$ | $[hkl]$ pairs of length $a\sqrt{h^2+k^2+l^2}$ |

$p$ being the multiplicity factor for the $\{hkl\}$ planes (Section 4.8, Table 4.1).

The scattering power is therefore

$$\mathscr{I}(s) = 1 + \frac{6n^2(n-1)}{n^3}\frac{\sin(2\pi sa)}{2\pi sa} + \frac{12n(n-1)^2}{n^3}\frac{\sin(2\pi sa\sqrt{2})}{2\pi sa\sqrt{2}} + \cdots. \quad (2.56)$$

This equation gives the intensity of the Debye-Scherrer powder diagram, taking into account the shapes of the lines due to the size of the elementary crystal. It is not obvious from this equation that the intensity is zero everywhere, except in the immediate neighborhood of the lines corresponding to the Bragg angles, which are such that

$$s = \frac{2\sin\theta}{\lambda} = \frac{\sqrt{h^2+k^2+l^2}}{a}$$

A complex calculation [9] shows, however, that this expression is equivalent to that derived from the classical expression [Eq. (5.9)]. It is impossible to express the above equation in finite terms but, for very small crystals containing few atoms, it is preferable to calculate the pattern from the Debye formula than to use the procedure described in Section 5.2 [10].

## 2.7. Diffraction by a Group of Particles

We have often assumed until now that all the scattering centers were identical. The case of a random mixture of several different types of atoms or groups of atoms is very complex and the results of diffraction experiments are difficult to interpret because, in general, the statistics of the atomic distribution depend on the nature of the atoms [11].

It is nevertheless possible to analyze a rather general case. We consider a set of particles of various types, for example atoms or molecules, for which we can define statistically, both the positions of the particles and the nature of a particle occupying a given position, the two statistical distributions being *completely independent*. There are real and important cases where these conditions are fulfilled. An example is *disordered solid solutions*, where the atoms are at the nodes of a crystal lattice and where each node is occupied by one of the constituent atoms with a probability which is given by the proportions of the various types of atoms in the mixture. Other examples are gases or liquids composed of polyatomic molecules whose orientations are absolutely random and without influence on the orientations of the neighboring molecules. For given directions of the incident and diffracted rays the different molecules have different structure factors, depending on their orientations, and therefore act as different particles.

Let us therefore consider a set of $N$ particles with scattering factors $f_1, f_2, \cdots, f_n$, which are all functions of $\boldsymbol{s}$. The scattering power for a given configuration is calculated as

$$I_N(\boldsymbol{s}) = \sum_1^N f_n^2 + \sum_{n \neq n'}\sum f_n f_{n'} \cos(2\pi \boldsymbol{s}\cdot\boldsymbol{x}_{nn'}) .$$

The observed intensity is the average of the above expression evaluated over all the possible configurations:

$$\overline{I_N(\boldsymbol{s})} = N\overline{f_n^2} + \sum\sum \overline{f_n f_{n'} \cos(2\pi \boldsymbol{s}\cdot\boldsymbol{x}_{nn'})} .$$

In the second term, $f_n$ and $f_{n'}$ are independent of the value of the cosine, according to the hypotheses stated earlier. The average of the product is therefore equal to the product of the averages and

$$\overline{I_N(\boldsymbol{s})} = N\overline{f_n^2} + \sum\sum \overline{f_n}\,\overline{f_{n'}}\,\overline{\cos(2\pi \boldsymbol{s}\cdot\boldsymbol{x}_{nn'})} ,$$

or

$$\overline{I_N(\boldsymbol{s})} = N[\overline{f_n^2} - (\overline{f_n})^2] + (\overline{f_n})^2[N + \sum\sum \overline{\cos(2\pi \boldsymbol{s}\cdot\boldsymbol{x}_{nn'})}] . \qquad (2.57)$$

The scattering power is thus decomposed into two terms, the second of which is the diffracted intensity for a set of $N$ identical particles with scattering factors $\overline{f_n}$. This reduces to the preceeding case; it is as if we had a single fictitious atomic group with scattering factor $\overline{f_n}$. We now introduce an interference function $\mathscr{I}_1(\boldsymbol{s})$:

$$\mathscr{I}_1(\boldsymbol{s}) = 1 + \frac{1}{N}\sum\sum \overline{\cos(2\pi \boldsymbol{s}\cdot\boldsymbol{x}_{nn'})} ,$$

which corresponds to the diffraction by an object formed of atomic groups with scattering factors $\overline{f_n}$ situated exactly at the positions of the atomic groups of the real object. Equation (*2.57*) shows that one must add to this interference function an additional scattering, since

$$\mathscr{I}(\mathbf{s}) = \frac{I_N(\mathbf{s})}{N(\overline{f_n})^2} = \frac{\overline{f_n^2} - (\overline{f_n})^2}{(\overline{f_n})^2} + \mathscr{I}_1(\mathbf{s}) \,. \tag{2.58}$$

This additional scattering power, $[\overline{f_n^2} - (\overline{f_n})^2]/(\overline{f_n})^2$, results from the differences between the scattering factors of the various scattering centers. Since, in general, $\overline{f_n^2}$ and $\overline{f_n}$ vary slowly with $\mathbf{s}$ if the particles are atoms or small molecules, this "disorder term" gives a general scattering of low intensity with no pronounced maxima but varying slowly with the diffraction angle. This scattering is superimposed over the scattering pattern produced by average identicai particles. We shall see in the following chapters various applications of this general formula which is valid only if the position and the nature of the diffracting centers are independent.

## References


1. P. P. Ewald, *Proc. Phys. Soc. London,* **52** (1940), 167.
2. R. Hosemann, *Z. Phys.,* **128** (1950), 1 and 465.
   R. Hosemann and S. N. Bagchi, *Acta Cryst.,* **5** (1952), 612.
   R. Hosemann, *Ergeb. Exact. Naturwiss.,* **20** (1951), 142.
3. J. Mering and J. Longuet Escard, *C. R. Acad. Sci. Paris,* **236** (1953), 1501 and 1577; *J. Chimie Phys.,* **51** (1954), 416.
4. A. J. C. Wilson, *X-Ray Optics,* Methuen, London, (1949).
5. A. H. Compton and S. K. Allison, *X-Rays in Theory and Experiment* (2nd ed.), Van Nostrand, New York (1951).
6. G. Fournet, *Handbuch der Physik,* Vol. XXXII, Springer Verlag, Berlin (1957), p. 279.
7. T. H. Hill, *Statistical Mechanics,* McGraw-Hill, New York (1956).
8. A. Guinier and G. Fournet, *Small Angle Scattering of X-Rays,* Wiley, New York (1955).
9. R. W. James, *The Optical Principles of X-Ray Diffraction,* Bell, London (1948), p. 515.
10. C. Morozumi and H. L. Ritter, *Acta Cryst.,* **6** (1953), 588.
11. G. Fournet, *Handbuch der Physik,* Vol. XXXII, Springer Verlag, Berlin (1957), p. 296.

# CHAPTER 3

# X-RAY DIFFRACTION FROM AMORPHOUS SUBSTANCES: GASES, LIQUIDS, AND VITREOUS SOLIDS

In the light of the theoretical results established in the preceeding chapter, we shall now examine how X-ray diffraction can be used to investigate the atomic structure of amorphous substances. Some general references are given in [1]. We shall certainly not be able to acquire as deep an understanding of these as is possible in the case of crystalline substances, but the results are nevertheless of great interest and have varied applications. We shall start with perfect gases, and then consider the more complex cases of real gases, liquids, and vitreous solids.

## 3.1. Diffraction by Perfect Gases

### 3.1.1. Monatomic Gases

According to the kinetic theory of gases, a perfect gas is composed of atoms or molecules which exert no influence on each other and which can be represented by mathematical points of zero volume. Consequently, if the origin of the object space is chosen to be at the center of a gas atom, *the probability of finding another atom is constant throughout this space*, despite the presence of the atom at the origin and of all the other atoms, since they do not act on each other, and since their combined volume is negligible. Then the distribution function $P(\boldsymbol{x})$ defined in

Section 2.4.1 is equal to unity for all values of $\boldsymbol{x}$ or, according to Eq. (2.21), the probability for the perfect gas is

$$z(\boldsymbol{x}) = \delta(\boldsymbol{x}) + \frac{1}{v_1},$$

and, from Eq. (2.43), the interference function is

$$\mathscr{I}(s) = 1. \tag{3.1}$$

The scattering power per atom is equal to the square of the atomic scattering factor. The atoms of the gas therefore scatter as if they were alone, and there are no interference effects.

The scattering power per cubic centimeter of a perfect monatomic gas at a temperature $T°K$ and a pressure $p$ is given by

$$I(s) = \frac{f^2}{v_1} = \frac{pf^2}{kT}. \tag{3.2}$$

It was shown in Section 1.2.5 that the scattering factor of an atom has a maximum value equal to the atomic number for $s = (2 \sin \theta)/\lambda = 0$, and that it decreases regularly with $\theta$. The curve of the scattering

**FIGURE 3.1.** *Coherent scattering power for a monatomic gas (neon).*

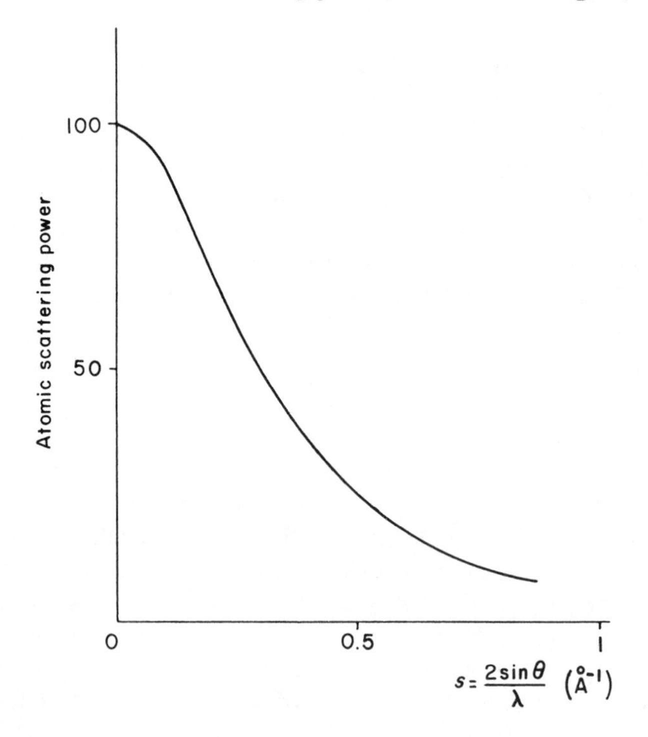

**FIGURE 3.2.** *Total scattering power of helium.*

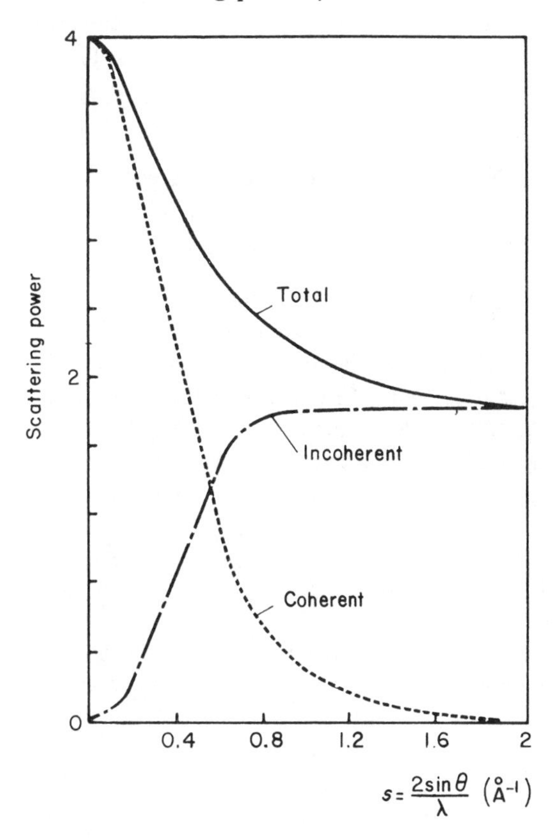


power of a gas given in Figures 3.1 and 3.2 shows the dependence of the square of the atomic scattering factor on $s$.

This result can also be found in a direct and elementary manner by applying Eq. (2.53), which gives the intensity diffracted by a group of $N$ moving atoms:

$$I_N(\boldsymbol{s}) = Nf^2 + f^2 \sum \sum \overline{\cos(2\pi \boldsymbol{s} \cdot \boldsymbol{x}_{nn'})} \,. \tag{3.3}$$

Since the atomic distribution is perfectly random, *the average value of the cosine term is zero,* and we have again Eq. (*3.1*). However, if $|\boldsymbol{s}|$ tends to zero, all the cosine terms tend to unity and, since there are $N(N-1)$ terms in the double sum, the intensity tends to $N^2f^2$ and not $Nf^2$, as from Eq. (*3.1*).

This apparent paradox is found in a more or less hidden form in all X-ray diffraction calculations. It comes from the fact that the diffracting object has a finite size, and this is shown by the rigorous calculations of Chapter 2. In fact, if we set $P(\boldsymbol{x}) = 1$ in Eq. (*2.42*), we have not only the term unity, which alone appears in Eq. (*3.1*), but also a central peak $(1/Vv_1)|\Sigma(\boldsymbol{s})|^2$, whose maximum value at $s = 0$ is $V/v_1 = N$ [Eq. (*2.37*)]. Therefore

$$I_N(0) = Nf^2\mathscr{I}(0) = N^2f^2 .$$

It has been shown in Section 2.4.2 that the width of the peak is of the order of

$$s_0 = V^{-1/3}.$$

When $\boldsymbol{s}$ is smaller than $\boldsymbol{s}_0$, the arguments of the cosine terms in Eq. (*3.3*) are all small, even for the pairs of atoms which are furthest from each other. Their average value is therefore not zero. It is zero, however, for larger angles of diffraction, and this was assumed in the derivation of Eq. (*3.1*). The complete curve of diffracted intensity is thus as shown schematically in Figure 3.3. As has been already mentioned, the central peak is limited to such small angles that it is usually unobservable.

In practice, the facts are thus *completely* represented by Eq. (*3.1*), which supposes, as is verified experimentally, that the intensity diffracted

**FIGURE 3.3.** *Schematic diagram for the scattering power of N atoms of a monoatomic gas.*

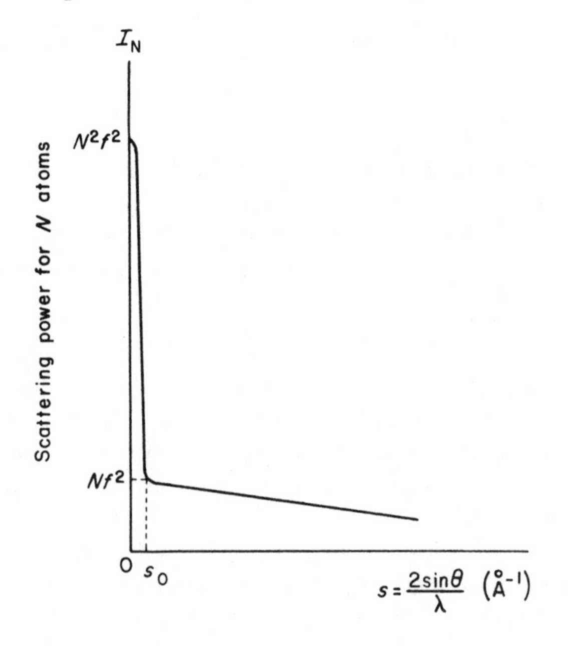

by a homogeneous substance is proportional to the volume of the object and independent of its shape.

### 3.1.2. Mixture of Perfect Monatomic Gases

If the $N$ atoms are not identical, Eq. (*3.3*) becomes

$$I_N(\boldsymbol{s}) = \Sigma_1^N f_n + \sum_{n \neq n'}\sum f_n f_{n'} \overline{\cos(2\pi \boldsymbol{s} \cdot \boldsymbol{x}_{nn'})}\,.$$

If the distribution of all the atoms, regardless of their nature, is the same as for a perfect gas (Dalton's law), the average value of the cosines is still zero and the observed intensity for the $N$ atoms is

$$I_N(s) = \Sigma_1^N f_n^2$$

or

$$I(s) = \overline{f_n^2} \tag{3.4}$$

per atom, where $\overline{f_n^2}$ is the average value of the square of the atomic scattering factors.

When the atoms are dispersed in a perfectly irregular manner in a volume large compared to their own volume, *the observed intensity is simply the sum of the intensities scattered by each of the atoms.*

### 3.1.3. Perfect Polyatomic Gases

In this case the diffracting centers are identical molecules arranged like the atoms of a perfect gas, but the scattering factor of an individual molecule varies from one to another because it depends on the orientation of the molecule with respect to the incident and diffracted rays. This is the case studied in Section 2.7. If $\overline{F_n^2}$ and $\overline{F_n}$ are respectively the average values of the intensity and of the amplitude diffracted by the molecule for a given value of $s = (2\sin\theta)/\lambda$ and for all possible positions of the molecule, the general equation [Eq. (2.58)] applied to the case of a perfect molecular gas gives

$$\mathscr{I}(s) = \frac{\overline{F_n^2} - |\overline{F_n}|^2}{|\overline{F_n}|^2} + 1 + \frac{1}{N}\Sigma\,\Sigma\,\overline{\cos(2\pi \boldsymbol{s} \cdot \boldsymbol{x}_{nn'})}\,.$$

Since the gas is perfect, the average value of the cosine is zero and

$$I(s) = |\overline{F_n}|^2 \mathscr{I}(s) = \overline{F_n^2}\,, \tag{3.5}$$

where $I(s)$ is the intensity diffracted per molecule. This is the same expression as that of Eq. (*3.4*).

The Debye formula (*2.54*) gives $\overline{F_n^2}$ as a function of the structure of the molecule:

$$\overline{F_n^2} = \sum \sum f_m f_{m'} \frac{\sin (2\pi s x_{mm'})}{2\pi s x_{mm'}} \,. \tag{3.6}$$

The scattering factors of the different atoms are designated by $f_m$, and the distance between the atoms $m$ and $m'$ by $x_{mm'}$.

Let us consider a diatomic molecule composed of two atoms of scattering power $f$, at a distance $a$ from each other. Applying Eq. (3.6),

$$F^2 = 2f^2 + 2f^2 \frac{\sin (2\pi sa)}{2\pi sa} = 2f^2 \left(1 + \frac{\sin (2\pi sa)}{2\pi sa}\right). \tag{3.7}$$

This is the average value of the intensity diffracted by a single molecule in rapid rotation or by a group of molecules oriented in an arbitrary manner and forming a perfect gas. The intensity for $s = 0$ is equal to $4f^2$ and after several oscillations tends to $2f^2$, which is the sum of the intensities diffracted by the two atoms of the molecule. At large angles the intensity diffracted by the diatomic gas is the same as if the molecules were completely dissociated, but at small angles it is twice as large.

*If the atoms were quasi-punctual*, $f^2$ would be constant, regardless of the value of $s$, and the diffraction curve of Figure 3.4a would present

**FIGURE 3.4.** *Scattering power per molecule of a diatomic gas.* (a) *Interference function* $1 + (\sin 2\pi sa)/(2\pi sa)$, *where a is the interatomic distance.* (b) *Scattering power for two atoms of nitrogen,* $2f^2$. (c) *Scattering power for the nitrogen molecule.*

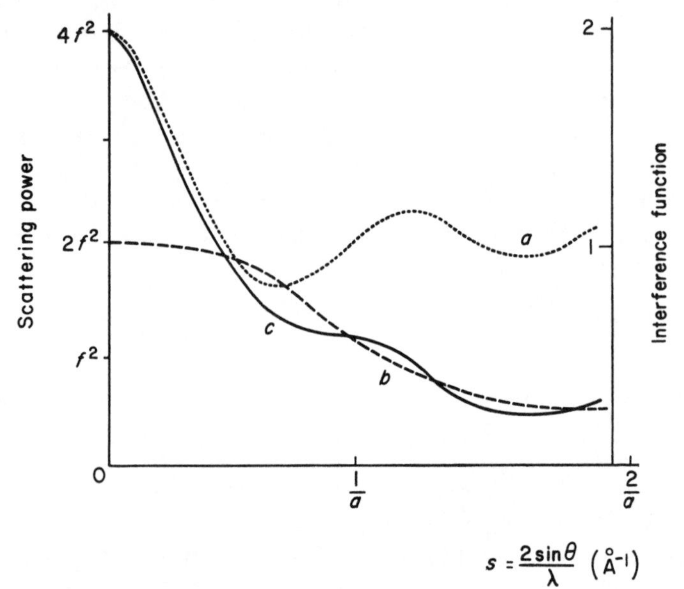

a series of maxima, the first and highest of which would correspond to a value $s_m$; this is found by letting the derivative of Eq. (*3.7*) be equal to zero:

$$as_m = 1.23.$$

That is to say,

$$1.23\lambda = 2a \sin \theta_m. \tag{3.8}$$

The diffraction angle for the first maximum is related to the interatomic distance $a$ by the Bragg relation, which we have multiplied by a correction factor of 1.23. The conditions under which Eq. (3.8) has been established by Ehrenfest [2] are important because the equation has often been applied to cases where its validity is very doubtful.

When the molecule is made up of real atoms, the observed diffraction curve is obtained from the curve of Figure 3.4a by multiplying it by the factor $f^2$ (Fig. 3.4b), which decreases strongly with $s$; this produces both a toning down and a *displacement* of the maxima [3]. Figure 3.4c gives the curve for the molecule $N_2$, which is quite different from Ehrenfest's curve.

In order to deduce the interatomic distance of a diatomic molecule from the diffraction pattern using Ehrenfest's equation, *it is necessary to begin by dividing the measured intensity at each angle by the value of the square of scattering factor of the atom for that angle.* The maximum of this corrected curve is given by Eq. (*3.8*).

The diffraction curve of a polyatomic molecule is given by the sum of functions such as those of Eq. (*3.7*) for each pair of atoms. Unless there is a single interatomic distance, as in the case of the four Cl atoms at the corners of a regular tetrahedron in $CCl_4$, the broad maxima given by Eq. (*3.7*) overlap and give a curve which decreases smoothly and continuously. It is then very difficult to deduce the interatomic distances. X-Ray diffraction therefore gives little information on the structure of gaseous molecules, except for the most simple ones [1].

### 3.1.4. Experimental Measurement of the Diffraction by Gases

The curve of the intensity as a function of angle, which we have just calculated, accounts only for the coherent radiation. In the observed scattering there is, in addition, the incoherent radiation scattered by the Compton effect (Section 1.2.3), which does not give rise to interference phenomena. Its intensity is equal to the sum of the Compton intensity scattered by each atom. In the particular case of perfect gases, the coherent radiation also has an intensity equal to the sum of the intensities scattered by each atom. It follows that the two types of radiation are of comparable intensity, their ratio varying with the scattering angle. At small angles, the coherent radiation is preponderant; at large angles, the inverse occurs. This is shown in Figure 3.2 for helium.

Since the intensity varies slowly with the angle of diffraction, it is permissible to use a rather large sample [4] and a counter which accepts radiation extending over an appreciable solid angle. This compensates for the small number of atoms irradiated due to the low density of gases.

To be more specific, let us consider the orders of magnitude involved. Let us suppose that a beam of X-rays traverses a layer of argon 1 cm thick under normal conditions of temperature and pressure and that the counter is placed at a distance of 10 cm and behind a window measuring $5 \times 5$ mm. The number of photons received by the counter at small angles, divided by the number of photons received in the direct beam, is $2 \times 10^{-6}$. Let us compare the intensity scattered by the same small volume filled with gaseous or solid argon in the form of a crystalline powder. If the counter were placed on the first diffraction line of the powder, the measured intensity would be 5,000 times greater than with the gas. The measurement of diffraction by gases is thus rendered difficult by the fact that the radiation scattered by the walls of the container can be more intense than that scattered by the gas itself.

In practice, one rarely makes diffraction measurements on gases. In the case of monatomic gases, such measurements can only be used to determine the scattering factor, but theoretical calculations are more accurate. In the case of polyatomic gases it is possible to study the structure of the molecule, with the limitation mentioned above, but it is preferable to use electron diffraction [5]. Theoretically electron diffraction gives the same results as X-rays but it is easier to use since the scattering is much more intense than with X-rays, and the scattering curve can be determined to much larger values of $s$.

It is important to remember that air scattering can be quite appreciable. In fact, the volume of the irradiated air can be much larger than that of the sample, and the air may be close to the detector. Because of the $1/r^2$ law, this considerably increases the intensity of the scattering. This difficulty is encountered, in particular, in the study of diffraction at small angles (Figure 3.5). If the sample scatters weakly, it is essen-

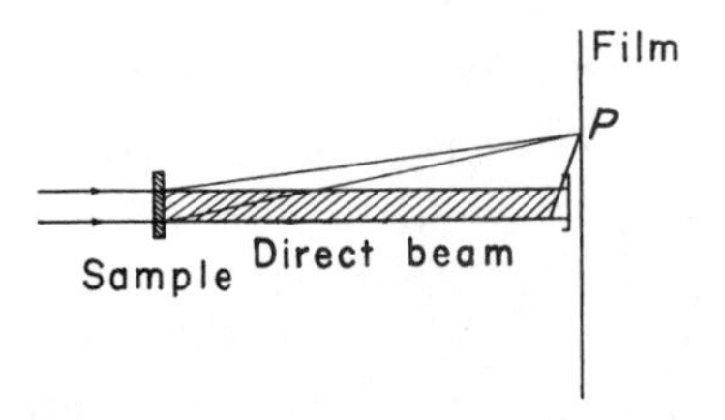


**FIGURE 3.5.** *Air scattering from the hatched region produces background radiation at P.*

tial to eliminate air scattering, either by evacuating the apparatus or by operating in an atmosphere of hydrogen or helium. For these light gases $f^2$ is only 1.6% and 6%, respectively, of the value of $f^2$ for oxygen.

## 3.2. Diffraction by a Real Gas, a Liquid, or a Vitreous Solid

### 3.2.1. The Case of a Group of Identical Atoms

It was shown in the preceding paragraph that when atoms are distributed as in a perfect gas, they behave toward X-rays as if they were isolated. In other words, perfect disorder causes all interatomic interferences to disappear. Whenever the atoms or, more generally speaking, the particles are sufficiently disordered, there is *diffraction of the gaseous type* and the observed intensity is the sum of the intensities diffracted by each individual particle. But it will be recalled that when the atoms are perfectly well ordered, as in crystals, and when the Bragg conditions are exactly satisfied, the amplitudes add up, giving an intensity proportional to the square of the number of particles.

When the disorder is not that of a perfect gas, either because the density of the system is no longer negligible or because the atoms have appreciable effects on their neighbors, we have *interatomic interference.* The distribution function $P(\boldsymbol{x})$ is then no longer constant and equal to unity, and the interference function is given by Eq. (*2.43*),

$$\mathscr{I}(\boldsymbol{s}) = 1 + \frac{1}{v_1}\int [P(\boldsymbol{x}) - 1] \exp(-2\pi i \boldsymbol{s}\cdot\boldsymbol{x})\, dv_x \,,$$

or, for an isotropic body, by Eq. (*2.47*),

$$\mathscr{I}(s) = 1 + \frac{2}{s v_1}\int_0^\infty [P(x) - 1] \sin(2\pi s x) x\, dx \,. \tag{3.9}$$

By introducing the atomic density at a distance $x$, $\rho_a(x) = P(x)/v_1$ and the average atomic density $\rho_0 = 1/v_1$ , Eq. (*3.9*) can be rewritten in the form of Eq. (*2.49*):

$$\mathscr{I}(s) = 1 + \int_0^\infty 4\pi x^2 [\rho_a(x) - \rho_0] \frac{\sin(2\pi s x)}{2\pi s x}\, dx \,. \tag{3.10}$$

Since the interference function for diffraction by perfect gases is equal to unity, the result of interatomic interaction is represented by the difference $\mathscr{I}(s) - 1$, which corresponds to the integral of Eqs. (*3.9*) and (*3.10*).

It is possible to determine $P(x) - 1$ , which is the difference between the real statistics and perfect disorder, by Fourier inversion of Eq. (*3.9*) [Eq. (*2.48*)] :

$$P(x) - 1 = \frac{2v_1}{x}\int[\mathscr{I}(s) - 1]\sin(2\pi sx)s\,ds\,. \tag{3.11}$$

Because of its fundamental importance for the diffraction by amorphous bodies, it will be useful to give a direct and elementary demonstration of this equation. The difficult points have already been discussed in the rigorous calculations of Section 2.4.2.

Let us consider an amorphous object of volume $V$ containing $N$ atoms, each with an atomic scattering factor $f$. The total diffracted power [Eq. (2.8)] is

$$I_N(\mathbf{s}) = Nf^2 + f^2 \sum \sum \cos(2\pi \mathbf{s}\cdot\mathbf{x}_{nn'})\,.$$

However, it is known from experience that the diffraction pattern is the same whether the isotropic amorphous object is fixed or whether it rotates during the experiment. It is thus possible to use the Debye equation [Eq. (2.54)],

$$\overline{I_N(s)} = Nf^2 + 2f^2 \sum \sum \frac{\sin(2\pi sx_m)}{2\pi sx_m}\,.$$

If $4\pi x^2\rho_a(x)\,dx$ is the average number of atoms in the interval $dx$ at the distance $x$ from one atom taken as the origin, the number of pairs of length $x$ is $(N/2)4\pi x^2\rho_a(x)\,dx$ and the double sum in the Debye equation can be replaced by an integral:

$$\overline{I_N(s)} = Nf^2 + Nf^2\int_{(V)} 4\pi x^2\rho_a(x)\frac{\sin(2\pi sx)}{2\pi sx}\,dx\,. \tag{3.12}$$

Introducing the average atomic density $\rho$, Eq. (3.12) can be written as

$$\begin{aligned}\mathscr{I}(s) &= \frac{\overline{I_N(s)}}{Nf^2}\\ &= \left[1 + \int_{(V)} 4\pi x^2[\rho_a(x) - \rho_0]\frac{\sin(2\pi sx)}{2\pi sx}\,dx\right] + \rho_0\int_{(V)} 4\pi x^2\frac{\sin(2\pi sx)}{2\pi sx}\,dx\,.\end{aligned}$$

The last term can be eliminated because this integral represents the diffraction pattern of an object of the same form as the given object but with a rigorously uniform density; it gives rise to the central diffraction peak which cannot be observed, as was shown in Section 2.4.2. In the first integral we can use infinity as the upper limit because $\rho_a(x) - \rho_0$ is zero outside a region which is very small compared to the dimensions of the object. Equation (3.10) is thus re-established.

### 3.2.2. Distribution Function $P(x)$ for Different Amorphous States

For X-rays, all amorphous structures are characterized by a function $P(x)$. There are no clear differences between the functions corresponding to highly compressed gases and liquids, or between liquids and vitreous solids. The function $P(x)$ varies progressively with the degree of condensation and with the degree of organization of the sample.

One case in which the distribution function can be rigorously calculated is that of an arrangement of atoms in one dimension [6]. The atoms, represented by rigid spheres of diameter $a$, are distributed at random along a straight line with $N$ spheres in a length $L$, $N$ varying from 0 to

**FIGURE 3.6.** *The distribution function $P(x)$ for a linear arrangement of objects of length a and of increasing compactness, according to Zernicke and Prins. The ratio between a and the length $l_1$ varies from 0.5 to 0.90.*

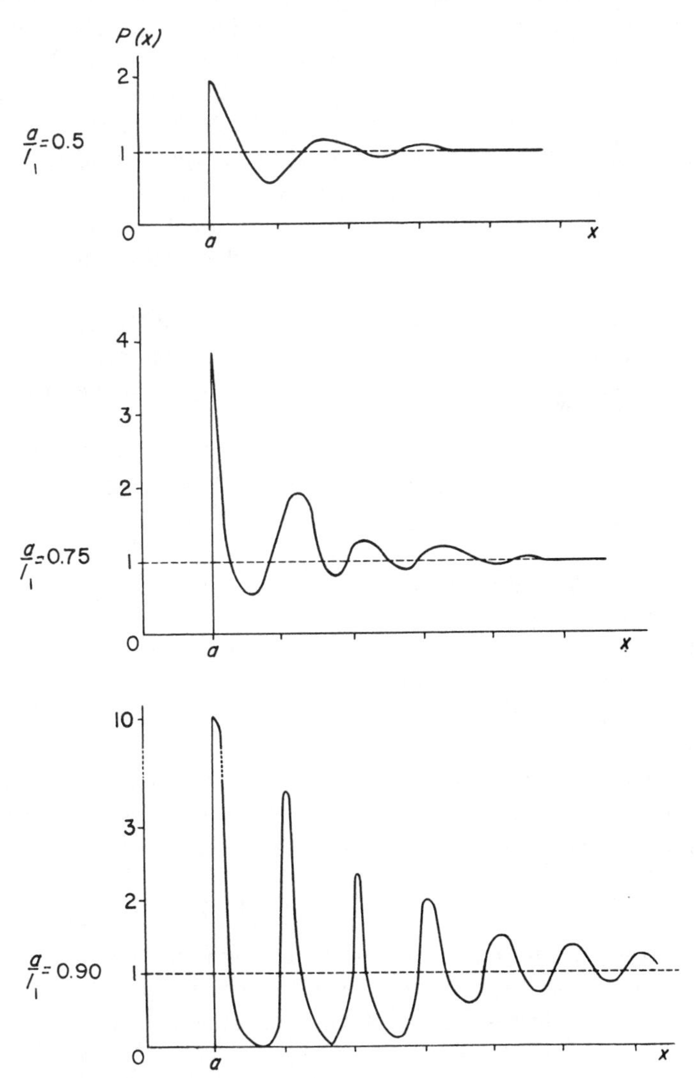

the maximum value $L/a$. The average length alloted to each atom is $l_1 = L/N$. In a one-dimensional space, $[P(x)]\,dx/l_1$ is the probability of finding an atom on the segment $dx$ at a distance $x$ from the atom at the origin. Since $P(x)$ is even, Eq. (*2.43*) for the scattering power then becomes

$$\mathscr{I}(s) = 1 + \frac{2}{l_1}\int_0^\infty [P(x) - 1]\cos(2\pi s x)\,dx\,. \qquad (3.13)$$

Figure 3.6 shows curves of $P(x)$ as a function of the concentration, according to Zernicke. For small values of $N$ (and large values of $l_1/a$), the function $P(x)$ is equal to 0 when $x<a$, since the spheres are impenetrable, and it is close to unity outside this forbidden zone. In Eq. (*3.13*) the term corresponding to interatomic interference is negative for small $s$; it has the effect of reducing the diffracted intensity at the center, especially for large concentrations, since the integral is divided by $l_1$. As the atoms approach each other, $P(x)$ undergoes more and more marked oscillations until, the packing being maximized, the atomic order becomes perfect. We then have a linear lattice of period $a$, and sharp diffraction maxima for $s = 1/a,\ 2/a, \cdots$

With this linear model one passes progressively from the *gaseous state*, where the atoms are dispersed without any order, to the *crystalline state*, where the order is perfect. Between these two states the *liquid state* is characterized by the presence of short-range order and high density; it is thus possible to follow the evolution of the distribution function between the extreme cases of the gaseous and crystalline states. The diffraction pattern corresponding to the intermediate case of the liquid state is characterized by a low intensity at the center and several large maxima at small angles; see Figure 3.9b. At large angles, liquid scattering is similar to gaseous scattering.

It is much more difficult to foresee a priori the atomic arrangements in three-dimensional space. Even if we consider the atoms as rigid spheres without interaction, as Zernicke did for the linear model, the calculation of the distribution function $P(x)$ is extremely complex and it has been impossible to determine it without certain approximations. It is nevertheless possible to give general indications on this function and on the diffraction pattern that results from it, for the various states of matter.

3.2.2.1. REAL GASES. The atoms can no longer be considered as mathematical points, but they still remain widely separated from each other. In a gas at ten atmospheres and at 0°C, there is one atom for every $(15\ \text{A})^3$, while the diameter of the atom is of the order of 1 Å. It is thus possible to use, as Debye [7] did, a function $P(x)$ having the form shown in Figure 3.7a. This function is zero within the forbidden sphere of radius $a$ equal to the diameter of the atom, and equal to unity outside. Substituting this function $P(x)$ in Eq. 3.9, we obtain

$$\mathscr{I}(s) = 1 - \frac{1}{v_1}\int_0^a \frac{\sin(2\pi sa)}{2\pi sa} 4\pi a^2\, da\,. \qquad (3.14)$$

Setting $v_0 = \pi a^3/6$ as the volume of the atom with a diameter $a$, this integral can be put in the form

$$\mathscr{I}(s) = 1 - 8\frac{v_0}{v_1}\left[3\frac{\sin(2\pi as) - 2\pi as\cos(2\pi as)}{(2\pi as)^3}\right] = 1 - 8\frac{v_0}{v_1}\Phi(2\pi as)\,, \qquad (3.15)$$

where $8v_0\Phi(2\pi as)$ is the Fourier transform of the form factor for a sphere of radius $a$ (Appendix A, Table I, 12).

The diffracted intensity at the center is reduced with respect to that of the perfect gas by the factor $1-(8\Omega/V)$, where $\Omega$ is the effective volume of all the atoms in the volume $V$ (Figure 3.7b). This formula is valid only for gases and when $\Omega/V$ is small. For high densities such as those encountered in liquids, a function $P(x)$ of the type described by Figure 3.6a is no longer acceptable. Thus, when $\Omega > (V/8)$, Eq. (3.15) leads to a negative intensity for $s=0$, which is absurd. We must remember that $\Omega/V$ can be as large as 0.74 for a compact arrangement of spheres.

**FIGURE 3.7.** (a) *The distribution function P(x) of Debye.* (b) *Unit scattering power for a set of spheres. Solid curve: low-density gas. The dash curve is calculated from the P(x) function of Debye for a density $\Omega/V = 0.06$.*

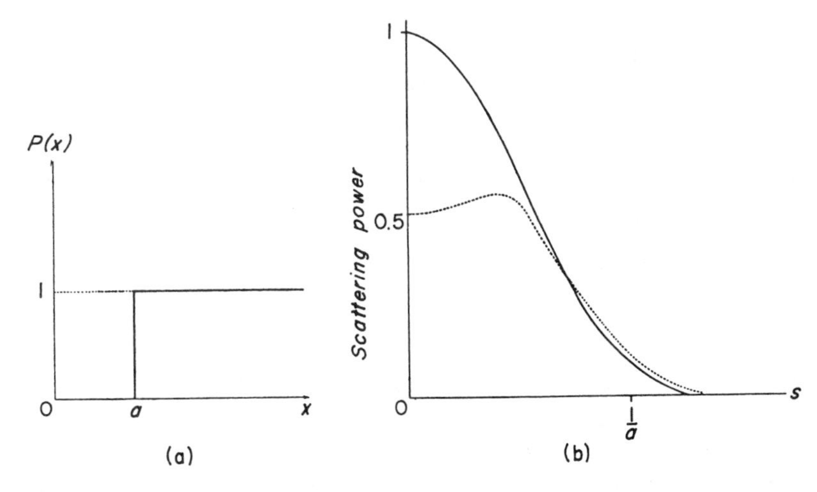

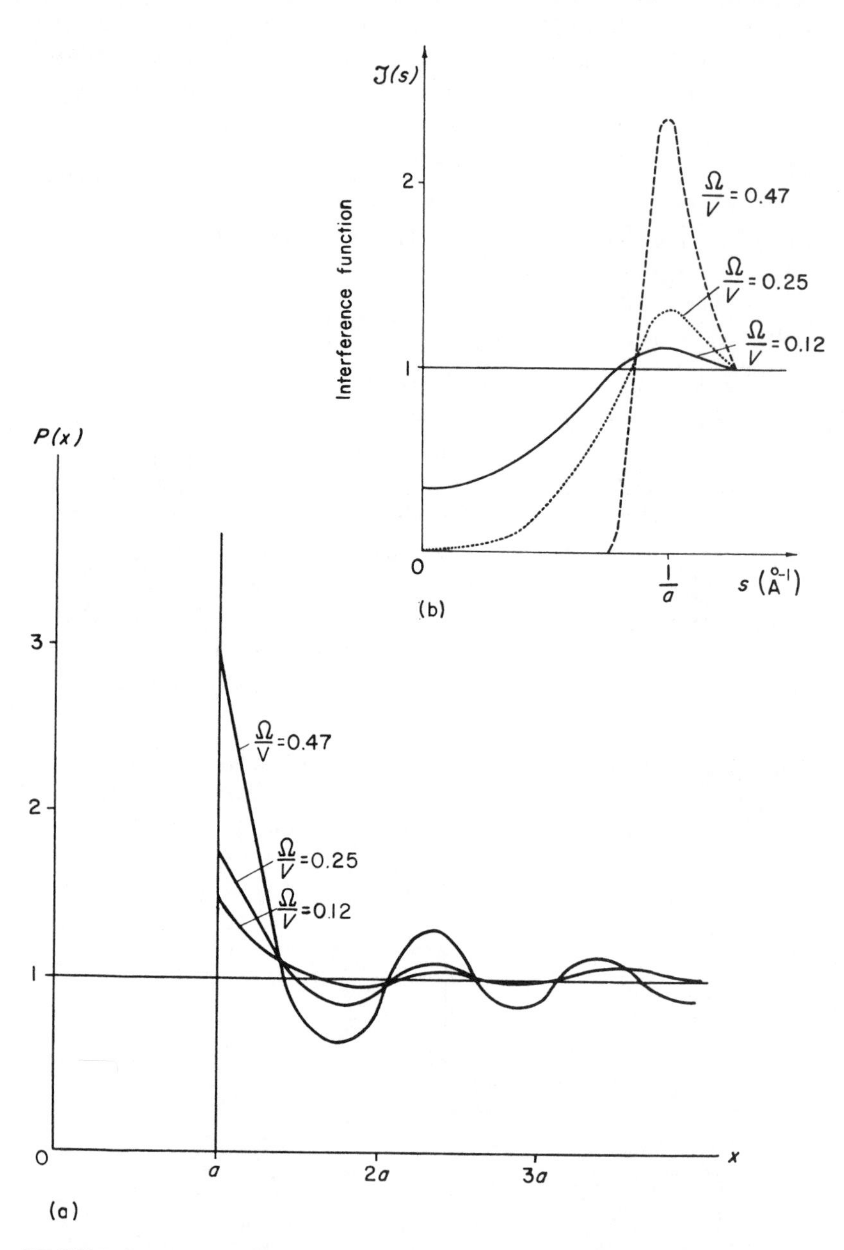


**FIGURE 3.8.** (a) *Distribution function for hard spheres at three concentrations (Kirkwood, Maun, and Alder* [8]). (b) *Interference function for a set of hard spheres calculated with the* $P(x)$ *function of Kirkwood (Fournet* [16]). *For* $\Omega/V = 0.47$, *the calculated values become negative for small values of s because of the approximations used for* $P(x)$.

3.2.2.2. LIQUIDS. As indicated by their density, the atoms of a liquid are pressed against each other so that each is in contact, or nearly so, with a number of its neighbors. Kirkwood et al. [8] have calculated $P(x)$ for atoms having the form of hard spheres (Figure 3.8a). As in the linear model, the oscillations of the function are accentuated as the concentration increases. The first and largest maximum corresponds to a distance $x$ equal to $a$, the diameter of the atom.

The intensity curve (Figure 3.8b) that has been deduced from these functions has the following characteristics.

(a) The intensity at small angles is low because the integral of Eq. (*3.9*) is negative for small values of $s$. This is predicted by Eq. (*2.52*), because of the low compressibility of liquids.

(b) There is a large maximum followed by several less clear maxima; the abscissa $s_m$ for the first maximum is $0.95(1/a)$. As the concentration is increased, its position does not change but it becomes larger.

(c) When $s$ increases, the interference function tends to unity, which is the value for a perfect gas.

Thus the effect of interatomic interference is at first *negative* for small diffraction angles, and the intensity diffracted by a liquid is smaller than that produced by a gas for the same number of atoms. *Positive interference* occurs when $s = (2 \sin \theta)/\lambda$ is of the order of the inverse of the minimum distance for a pair of atoms. Finally, for large angles, there is no longer any appreciable interference even if there is a certain degree of order. This can be seen from Eq. (*2.8*): unless the interatomic distances are fixed according to a rigorous law, the phases of the cosine terms become random for large values of $|\mathbf{s}|$. In the absence of long-range order, slight displacements can completely modify the phase $(2\pi \mathbf{s} \cdot \mathbf{x}_{nn'})$.

These characteristics of the pattern which are predicted theoretically for a collection of hard spheres are in fact found in the patterns for monatomic liquids. Figure 3.9a shows the interference function determined from the pattern for mercury, from which it is possible to deduce the function $P(x)$ particular to liquid mercury, as will be described later. The function $P(x)$ for mercury evidently resembles the curves of Kirkwood, without being entirely identical to them; the differences can be explained by the fact that the atoms of mercury are not merely hard spheres but exert forces on each other. This produces a characteristic short-range order. When mercury crystallizes, all the atoms are found at well-defined distances from each other and the function $P(x)$ then has a series of peaks, shown in Figure 3.9b as dashes whose length is pro-

**FIGURE 3.9.** (a) *Interference function for liquid mercury.* (b) *Distribution $P(x)$ in liquid mercury.*

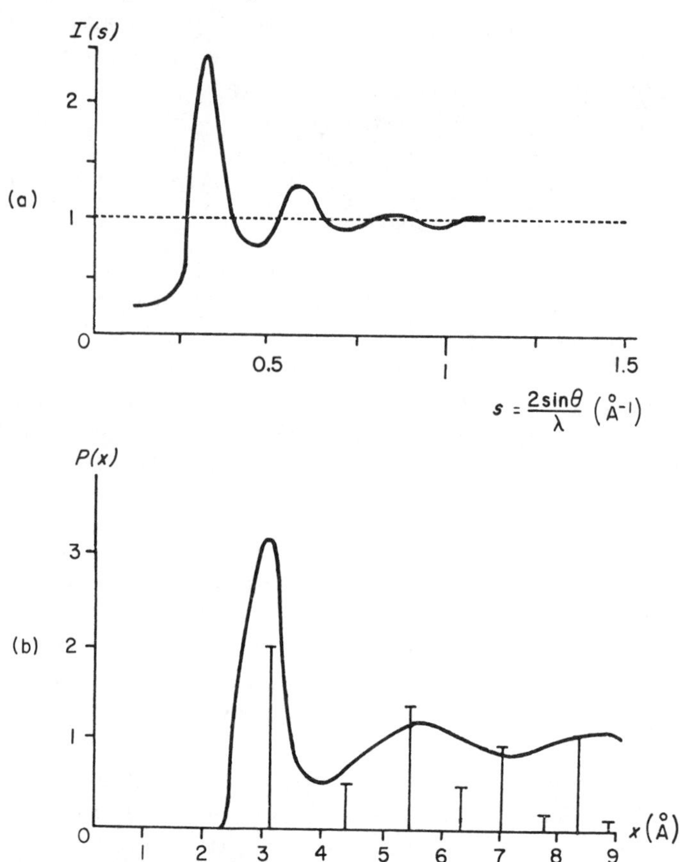


portional to the number of neighbors of each type.

The maxima of $P(x)$ for the liquid thus correspond, in general, to the interatomic distances in the crystal. The sharp peaks of $P(x)$ for the crystal are replaced by blurred maxima which broaden as $x$ increases, so that they merge and their superposition gives a constant sum. This occurs frequently: the atoms in the liquid tend to distribute themselves as in the solid, since this arrangement keeps the energy at a minimum, but this order remains largely imperfect so that there is no long-range order. The conditions are the same in vitreous solids which can be considered as liquids with a very large viscosity.

A fluid is intrinsically defined by the number of atoms per unit volume $1/v_1$ and by the law of interaction between the atoms or, more exactly, the interaction potential $\varphi(x)$ of two atoms at a distance $x$ from each other. The distribution function $P(x)$ depends on these two parameters. For given atoms—that is, for a given law $\varphi(x)$—the form of $P(x)$ depends on the concentration, as shown in Figure 3.8; for a given concentration, $P(x)$ is not the same—for example, for hard spheres and for atoms of mercury.

Until now, calculations of $P(x)$ from the interaction potential $\varphi(x)$ and from the concentration $1/v_1$ have only been approximate. Fournet [9] has used the theory of Born and Green [10]. He avoids the calculation of $P(x)$ by relating the intensity diffracted by a fluid with $\varphi(x)$ and $v_1$ directly. It is possible in this way to foresee the modification of the diffraction pattern (Figure 3.10) for a given collection of atoms as a function of the concentration. Inversely, it would be theoretically possible to find the law of interaction $\varphi(x)$ from the experimental curves of the intensity for collections of atoms at various concentrations [11], but this requires very accurate measurements.

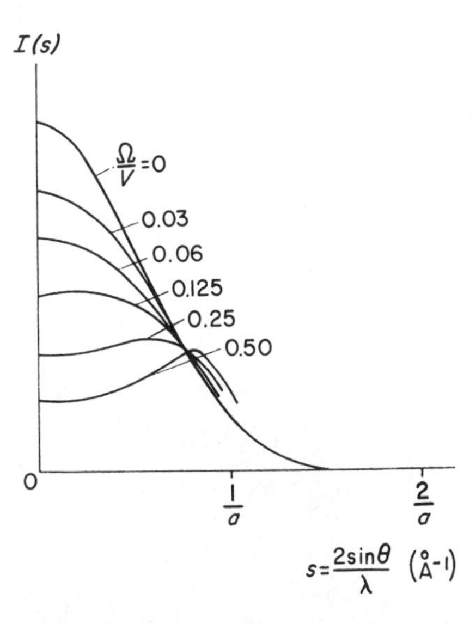


**FIGURE 3.10.** *Scattering power for a set of hard spheres of diameter a for various concentrations Ω/V, where Ω is the volume of the sphere and V the volume which is available to it, according to Fournet* [*1*].

It has in some cases been possible to utilize a distribution function assumed a priori to predict the effects of interatomic interference in a dense collection through Eq. (*3.9*). This function can be very simple (Debye [7]), or it can be similar to those found experimentally [12]; it is sometimes chosen to permit the integration of Equation 3.9. Such a procedure may present difficulties for the function must satisfy certain conditions in order to constitute a function $P(x)$. We do not as yet know how to express these conditions completely, but one case has already been examined in Section 2.5: $P(x)$ is related to the compressibility of a fluid by the relation

$$\frac{1}{v_1}\int_0^\infty [P(x) - 1]4\pi x^2 dx = \frac{kT\beta}{v_1} - 1 \, .$$

More generally, it is possible to show that $P(x)$ is related to the equation of state of the fluid [9]. Thus if we assume the function $P(x)$ chosen by Debye (Figure 3.6a), we effectively admit that the fluid obeys the equation of state

$$p = \frac{kT}{8v_0} \log \frac{1}{1 - (8v_0/v_1)} ,$$

where $v_1$ is the volume available to the molecule of volume $v_0$. At low concentrations, when $v_0/v_1 = \Omega/v$ tends to 0, we have the equation for the perfect gas.

Finally, functions are often used which are independent of the concentration, which is not correct. The accuracy of calculations based on functions $P(x)$ chosen a priori is often difficult to evaluate.

### 3.2.3. The Ring in the Diffraction Patterns of Liquids and Amorphous Solids

For amorphous substances in the condensed state, the essential feature of the diffraction pattern is the presence of a ring or sharp maximum of intensity for a certain angle of diffraction. This is illustrated in Figure 3.11. There are sometimes also other rings which are less intense and less well defined. Attempts have been made to relate the radii of these rings with the structure of the sample. This method of interpretation of amorphous patterns is still often used, mostly by non-specialists, because of its simplicity. It is thus useful to discuss its validity.

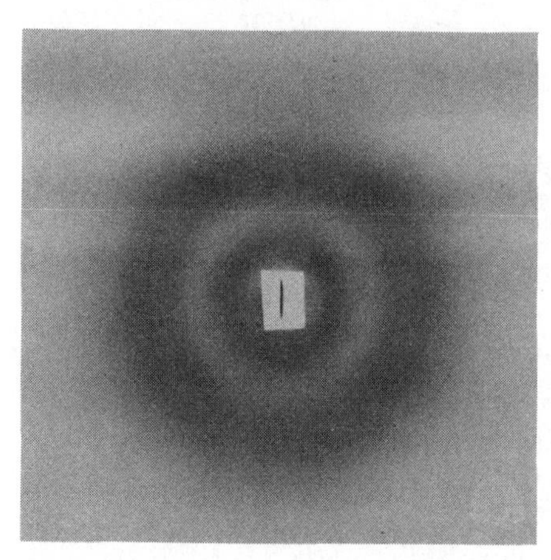

**FIGURE 3.11.** *Diffraction pattern for an amorphous body, cellulose acetate, with monochromatic CuKα radiation. Plane film at 40 mm from the sample (natural size).*

Let us consider again the Debye formula for the diffracted intensity:

$$\mathscr{I}(s) = 1 + \frac{1}{N} \Sigma \begin{Bmatrix} \text{number of pairs} \\ \text{at the distance } x \end{Bmatrix} \times \frac{(\sin 2\pi s x)}{2\pi s x} .$$

According to Eq. (*3.8*), the terms $[\sin(2\pi s x)]/2\pi s x$ have their *first* maximum for

$$s_m = \frac{2 \sin \theta_m}{\lambda} = \frac{1.23}{x_m} . \qquad (3.16)$$

Accordingly, if there exists a particular value of $x$, namely $x_m$, corresponding to the separation of a large number of pairs, and if for values near $x_m$ the number of pairs is relatively small and uniform, a maximum at the value $(2 \sin \theta_m)/\lambda = 1.23/x_m$ will appear in the interference function curve. This is the ratio of the scattering power per atom to the square of the atomic scattering factor. When the atoms are piled against each other, many are in contact and *it is possible to assume that the value $x_m$ deduced from Eq. (3.16) represents the smallest distance of approach of the atoms, i.e. their diameters, if they are in contact.*

This rule which is so frequently used is far from being rigorous. Two examples will show what can be expected from it.

Let us consider atoms arranged in a perfectly close-packed arrangement at the nodes of a hexagonal or face-centered cubic lattice. The powder pattern of these crystals is given by the Debye formula of Section 2.6 and the rule discussed above can be applied here, as in the case of liquids.

For the face-centered cubic lattice, the distance between atoms in contact is $x_m = a/\sqrt{2}$ and the first line with indices 111 corresponds to the angle $\theta_m$ which is such that $\lambda = (2a \sin \theta_m)/\sqrt{3}$, according to the Bragg equation. Thus

$$\frac{2 \sin \theta_m}{\lambda} = \sqrt{\frac{3}{2}} \frac{1}{x_m} = \frac{1.22}{x_m} .$$

This is Eq. (*3.16*), within 1%. By contrast, a rigorous calculation for the hexagonal lattice to which the Debye equation also applies gives $(2 \sin \theta_m)/\lambda = 1.15/x_m$ . The difference is 8% in this case.

One can expect perhaps still larger differences in disordered structures. It is therefore possible to deduce *the order of magnitude of the distance of closest approach of the atoms* from the diffraction angle corresponding to the maximum of the principal ring, using a formula of the type $K\lambda = 2x_m \sin \theta_m$, where $K$ is a little larger than unity, generally 1.1 or 1.2. It is not possible, however, to find an exact general formula, because $K$ depends on the atomic arrangement and thus on the *complete diffraction curve*, the measurement of the position of the maximum giving insufficient information.

We cannot therefore rely on the atomic dimensions deduced from Eq. (*3.16*). When there are several rings, this equation should not be applied to determine, for example, the length and the diameter of an elongated molecule (Stewart [13]). We must remember, finally, that we have only considered isotropic liquids, i.e. liquids whose molecules are approximately spherical.

### 3.2.4. Collection of Polyatomic Molecules

It is possible to arrive at a simple result in this case, only by using the hypothesis in Section 2.7. In particular, the mutual orientation of a pair of molecules is assumed to be absolutely independent of their separation. Equation (2.57) applies:

$$I(s) = \overline{F^2_n} - |\overline{F_n}|^2 + |\overline{F_n}|^2\left[1 + \frac{1}{N}\Sigma\Sigma\cos(2\pi s x_{nn'})\right],$$

where $I(s)$ is the scattering power for each molecule. The quantities $\overline{F^2_n}$ and $|\overline{F_n}|^2$ are, respectively, the average of the square and the square of the average of the molecular structure factor for a given value of $s = (2\sin\theta)/\lambda$, the orientation of the molecule being arbitrary with respect to the incident and diffracted rays. The second term is analogous to the expression for the intensity produced by a monatomic liquid whose fictitious atoms would have a scattering factor $|\overline{F_n}|$. It is thus possible to write it in the form of Eqs. (*3.9*) or (*3.10*). Therefore

$$\begin{aligned} I(s) &= \overline{F_n^2} + \frac{2|\overline{F_n}|^2}{sv_1}\int_0^\infty [P(x)-1]\sin(2\pi sx)x\,dx \\ &= \overline{F_n^2} + |\overline{F_n}|^2\int_0^\infty 4\pi x^2[\rho_a(x)-\rho_0]\frac{\sin(2\pi sx)}{2\pi sx}dx\,. \end{aligned} \tag{3.17}$$

Consider a molecule containing $n$ atoms at positions defined by the vectors $\boldsymbol{x}_j$ measured from an arbitrary origin. The average of the structure factor is the average of

$$\Sigma_1^n f_j \exp(-2\pi i\boldsymbol{s}\cdot\boldsymbol{x}_j)$$

for all orientations of the vector $\boldsymbol{s}$ of a given length. It is then found (Section 2.6) that

$$|\overline{F_n}| = \Sigma_1^n f_j\frac{\sin(2\pi s x_j)}{2\pi s x_j}\,.$$

As to the average $\overline{F_n^2}$, it is the scattering power of the gas constituted by the molecules. It is given by the Debye equation [Eq. (*2.54*)]:

$$\overline{F_n^2} = \Sigma_1^n\Sigma_1^n f_j f_k\frac{\sin(2\pi s x_{jk})}{2\pi s x_{jk}}\,.$$

Let us apply this formula to a diatomic molecule, $a$ being the distance between the two identical atoms of scattering factor $f$. If we consider the center of the molecule as the origin, the two atoms are at the extremities of vectors of length $a/2$. Thus

$$|\overline{F_n}|^2 = 4f^2\left(\frac{\sin(\pi sa)}{\pi sa}\right)^2,$$

$$\overline{F_n^2} = 2f^2\left(1 + \frac{\sin(2\pi sa)}{2\pi sa}\right).$$

From Figure 3.12, which shows $|\overline{F_n}|^2$ and $\overline{F_n^2}$ as functions of $s$, it is seen that the coefficient $|\overline{F_n}|^2$ becomes zero, as well as the second term of Eqs. (*3.17*), as $\overline{F_n^2}$ approaches $2f^2$. At large angles the intermolecular

**FIGURE 3.12.** *Case of a diatomic molecule formed of two atoms, with scattering factors f and at a distance a.* (a) *The square of the average structure factor of the molecule as a function of s.* (b) *The average of the square of the structure factor as a function of s.*

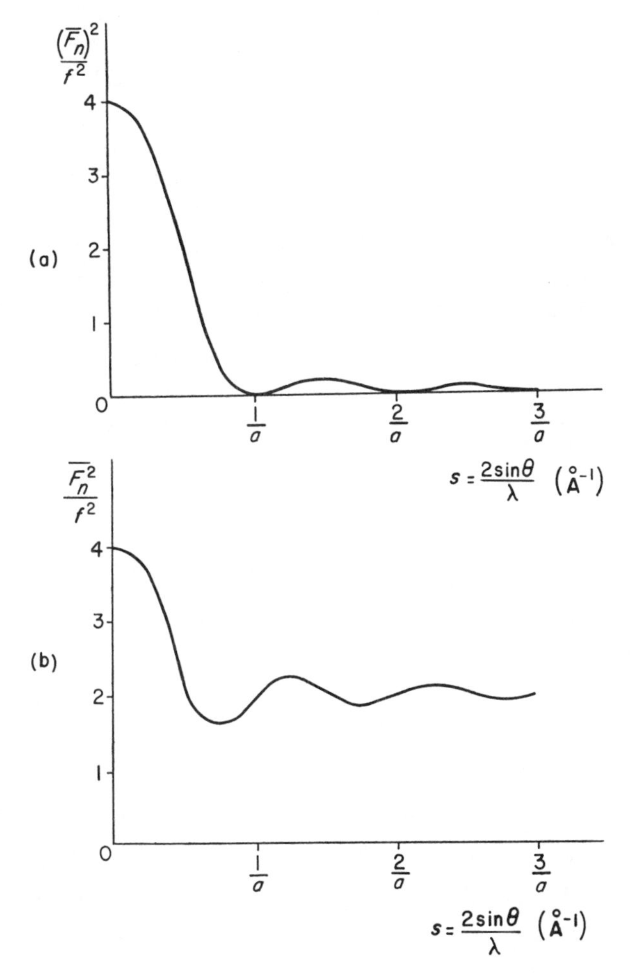

interference decreases rapidly and the pattern tends toward that of a molecular gas. In general, the interference between unlike particles is less important than that between identical particles except at small angles where, as in all liquids, the intensity is diminished by destructive interference.

In the case where the molecule contains a single type of atom, one can use the general Eq. (*3.9*), but in this case $P(x)$ represents the atomic distribution of atoms *in one molecule and in its neighbors at the same time.*

Figure 3.13b shows the function $4\pi r^2\rho_a(x)$ $(\rho_a(x) = [P(x)]/v_1)$ deduced by Eq. (*3.10*) from the diffraction pattern of liquid nitrogen at 89°K (Gingrich [14]). The shaded area under the first peak, which has an abscissa of 1.3 Å and is well detached from the rest of the curve, is 1.03. According to the definition of $\rho_a(x)$, this means that *each atom has one neighbor at an average distance of 1.3* Å, the other neighbors being considerable further away. The presence of the $N_2$ molecule in the liquid and its structure are both clearly shown. It is well to note the excellent agreement between the experimental value 1.03 and the theoretical value of unity. This illustrates the precision of this method of interpretation of amorphous patterns. The rest of the curve is more difficult to interpret, since it depends both on the distribution of the neighboring molecules and on that of the atoms in each molecule.

**FIGURE 3.13.** *Liquid nitrogen at 89°K.* (a) *Scattering power per atom. The curve c shows $f^2$, the square of the atomic scattering factor, as a function of s.* (b) *Average number of atoms at a distance x from a given atom.*

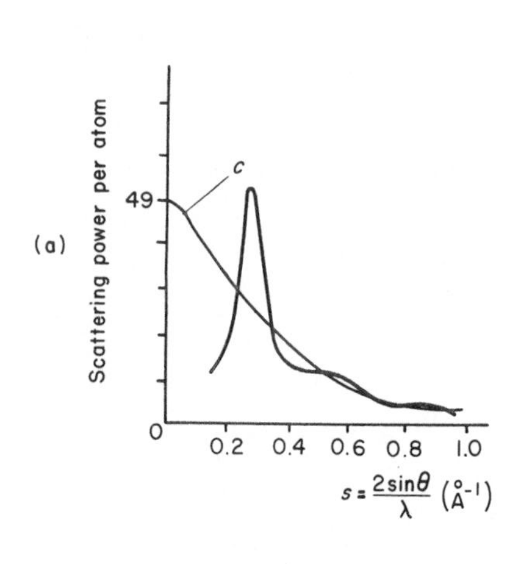


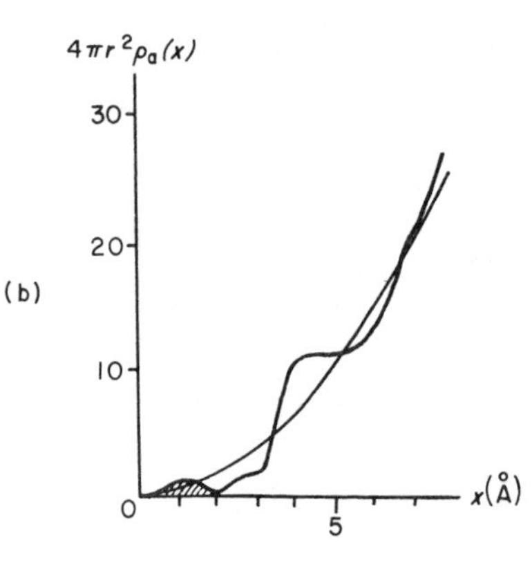

### 3.2.5. Experimental Study of the Diffraction by Liquids and Amorphous Bodies

It was just after the discovery of the diffraction pattern of crystalline powders that Debye and several others noted that, when a crystallized sample in the diffraction camera was replaced by an amorphous solid or by a liquid, the diffraction pattern did not show peaks. They found a continuous blackening with a ring at a diffraction angle corresponding to a value of $s$ of the order of 2-5 $\text{Å}^{-1}$. As we have seen, that gives the order of magnitude of the smallest distance of approach of the atoms. However, since atomic diameters differ little from each other, except for the lighter ones, the rings for all amorphous bodies have nearly the same radius; yet these rings have no characteristic structure. It is therefore not possible to use X-ray diffraction as a method of identification of amorphous bodies, as is done for crystals.

In certain cases, as when comparing samples, it is sufficient to measure the position of the ring, but for a complete study of the experimental data it is necessary to determine the distribution function $P(x)$. This requires a lengthy calculation of a Fourier transform. *The final result is significant only if the pattern has been determined with great care.* A Geiger counter is preferable to photography in this case, since accurate measurements of intensity are required. It is possible to use the same apparatus as for powder patterns, but with a wider slit at the counter. This increases the intensity without spoiling the angular measurement, since the intensity curve has no sharp peaks. The double-filter technique is recommended for isolating the chosen radiation, a monochromator being essential with photographic methods. It is necessary to know the curve of $I(s)$, where $s = (2 \sin \theta)/\lambda$, in as broad a region as possible. Geometric focusing is unimportant because of the breadth of the rings.

When the substance under consideration is a solid, the sample preparation is the same as for crystalline powders, but liquids must be placed in a tight thin-walled capsule which does not absorb X-rays and which does not produce an annoying pattern. A sheet of mica or of beryllium produces spots or lines which stand out from the amorphous pattern without ambiguity; on the contrary, glass and certain plastic materials such as collodion may cause very annoying diffraction rings. Polystyrene in very thin sheets is acceptable. Finally, there are ways of eliminating the container entirely: the rays can be reflected from the free surface of a liquid, or a drop of a very viscous liquid may be deposited in a hole drilled in a plate 1 mm thick, or the liquid can be made to flow as a column of convenient cross-section.

The problem is even more complex when the liquid has to be maintained at a high or at a low temperature. The description of several instruments will be found in a monograph by Gingrich [14] and in the general paper of Frost [15]. Diffraction experiments with liquids are never simple and sometimes require highly elaborate experimental methods.

Once the intensity curve has been obtained, either directly with a counter or after measurement of a film with a microphotometer, the first step is to correct for the absorption corresponding to the shape of the sample and the geometry of the experiment. In order to determine the scattering power involved in the theoretical formulas, it is necessary to introduce the intensity scattered by an isolated electron, taking into account the state of polarization of the primary beam. The experimental curve for the scattering power, $I(s)$, is plotted as a function of $s = (2 \sin \theta)/\lambda$.

We must remember that the measured intensity is the sum of the intensities of the coherent scattering, which has been calculated in this chapter, and of the Compton scattering. But we have at our disposal only relative measurements of the variation of intensity with angle, since absolute measurements require the measurement of the primary intensity and are very complicated. Hence the following method of treating the experimental curves.

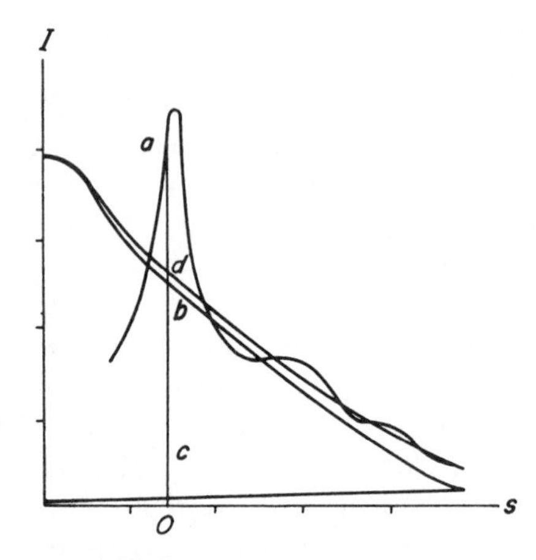


**FIGURE 3.14.** *Calculation of i(s) of Eq. (3.18) from experimental data.* (a) *Experimental curve.* (b) *Coherent scattering power per molecule.* (c) *Compton, or incoherent, scattering power per molecule.* (d) *Total (coherent plus incoherent) scattering power per molecule.*

The experimental curve is drawn using a arbitrary scale for the intensities. We calculate the curve for coherent "gaseous" scattering, $I_{coh}$, i.e. $f^2$ if the liquid is monatomic, and $\overline{F}^2$ if the liquid is molecular and if the molecular structure is known. The Compton scattering, $I_{incoh}$, is then calculated from the data of the *International Tables of Crystallography*; for a molecule this is simply the sum of the incoherent intensities produced by the constituent atoms.

Advantage is taken of the fact that toward large angles the interatomic or intermolecular interference becomes negligible (Section 3 2.2.2). Thus, for large

$s$, the end of the experimental curve gives the sum of $I_{\text{coh}}$ and $I_{\text{incoh}}$. A scale is established such that the extremities of the theoretical and experimental curves overlap. The curves $I_{\text{coh}}$ and $I_{\text{coh}} + I_{\text{incoh}}$ are drawn on this scale (Figure 3.14). For an arbitrary value of $s$,

$$i(s) = \mathscr{I}(s) - 1 = \frac{I(s) - I_{\text{incoh}}}{I_{\text{coh}}} - 1 = \frac{Oa - Od}{Ob}.$$

**FIGURE 3.15.** *The distribution function $4\pi x^2 \rho_a(x)$ for liquid metals.* (a) *Sodium at 400°C.* (b) *Lithium at 200°C. The curves corresponding to a uniform distribution and the positions of the neighbors in the crystalline state are also shown.*

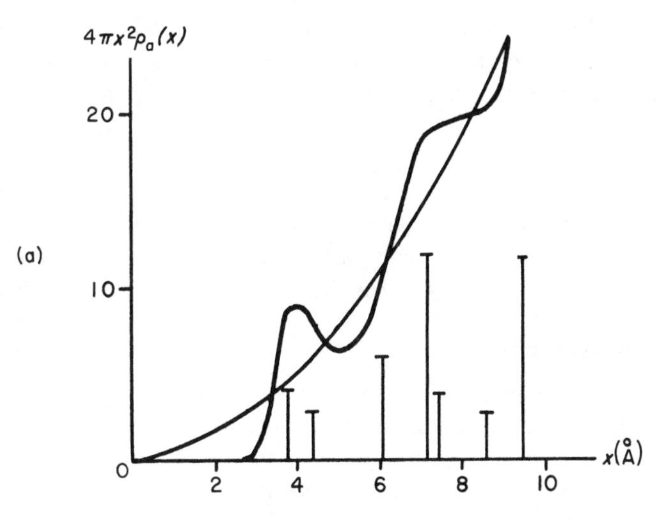


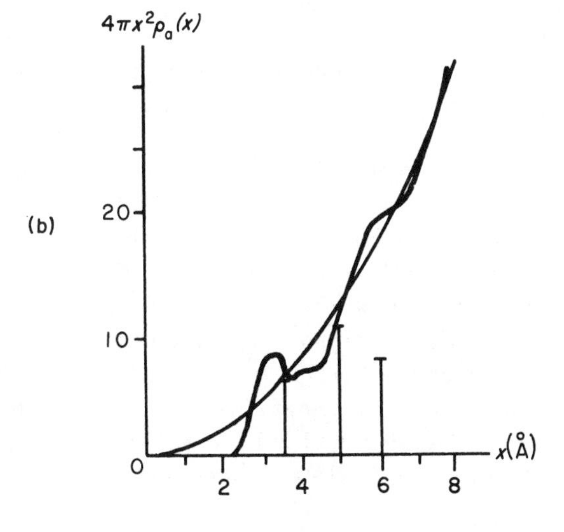

Without making any absolute measurements it is thus possible to find the function required to determine the distribution function $P(x)$ or $\rho_a(x)$ according to Eqs. (*3.9*) or (*3.10*). In fact, from Eq. (*3.11*),

$$P(x) = 1 + \frac{2v_1}{x}\int_0^\infty si(s)\sin(2\pi sx)\,dx \tag{3.18}$$

or, from Eq. (*2.50*),

$$4\pi x^2\rho_a(x) = 4\pi x^2\rho_0 + 8\pi x\int_0^\infty si(s)\sin(2\pi sx)\,dx\,. \tag{3.18a}$$

Practical methods for obtaining a numerical solution of Eq. (*3.18*) will be found in Appendix A on the Fourier transformation (Section A.6.1). The function to be transformed is the product $si(s)$, which can be appreciable for large values of $s$, even if $i(s)$ is small. This explains why it is necessary to determine the intensity curve with accuracy at large angles.

Figure 3.15 shows the results for liquid sodium and lithium obtained by this method. At high temperatures the atomic arrangement approaches a uniform distribution, short-range order being diminished by thermal agitation. The first maximum of the distribution function corresponds to the first-neighbor distance in crystallized sodium.

### 3.2.6. Conclusion

X-Ray diffraction permits an accurate description of the statistical distribution of the atoms in a substance showing disordered structure, if (a) the substance is homogeneous, (b) the forces of interaction are isotropic, and (c) the substance is monatomic or made up of molecules of known structure whose orientation is completely random. If the internal structure of the molecule is unknown and complex, the interpretation of the pattern becomes difficult because intramolecular interference adds to the intermolecular interference and it is impossible to separate the two effects.

There are thus many cases of real liquids for which the interpretation of X-ray diffraction experiments still remains difficult and doubtful—for example liquids whose molecules are elongated or have other very anisotropic shapes. If the molecules are elongated, it is reasonable to expect that neighboring molecules will have a tendency to align themselves parallel to each other. Calculations of intermolecular interference then become very complex and there is little hope, in the general case, of obtaining the distribution function of the molecules from the diffraction curve alone [16]. But when a definite model may be expected to be valid, as for a single liquid crystal (nematic state), X-ray data may be used to determine the unknown parameters [17].

Liquid mixtures may also be studied in some particular cases: an investigation of liquid mixtures of sodium and potassium [18] illustrates the methods of calculation.

## REFERENCES

1. J. T. ANDALL, *The Diffraction of X-Rays and Electrons by Amorphous Solids, Liquids and Gases*, Chapman and Hall, New York (1934).
   M. H. PIRENNE, *The Diffraction of X-Rays and Electrons by Free Molecules*, Cambridge Univ. Press, Cambridge (1946).
   G. FOURNET, *Handbuch der Physik*, Vol. XXXII, Springer Verlag, Berlin (1957), pp. 238–320.
2. P. EHRENFEST, *Proc. Amst. Acad.*, **17** (1919), 1132.
3. G. FOURNET, *Trans. Faraday Soc.*, **11** (1951), 121.
4. E. O. WOLLAN, *Phys. Rev.*, **38** (1931), 15.
   G. HERZOG, *Z. Phys.*, **70** (1931), 583 and 590.
5. Z. G. PINSKER, *Electron Diffraction*, Butterworths, London (1953).
6. F. ZERNICKE and J. A. PRINS, *Z. Phys.*, **41** (1927), 184.
7. P. DEBYE, *Phys. Z.*, 28 (1927), 135; **31** (1930), 348.
   P. DEBYE and H. MENKE, *Ergeb. Techn. Röntgenkde.*, **2** (1931), 1.
8. J. G. KIRKWOOD, E. K. MAUN, and B. J. ALDER, *J. Chem. Phys.*, **85** (1952), 777.
9. G. FOURNET, *Handbuch der Physik*, Vol. XXXII, Springer Verlag, Berlin (1957), p. 273.
10. M. BORN and H. S. GREEN, *Proc. Roy. Soc. London, A*, **188** (1946), 10.
11. R. CHIN-YU LING, unpublished Ph. D. thesis, University of Missouri (1952).
12. R. HOSEMANN, *Ergeb. Exact. Naturwiss.*, **24** (1951), 142.
    R. M. FRANK and K. L. YUDOWITCH, *Phys. Rev.*, **88** (1952), 759.
    G. OSTER and D. P. RILEY, *Acta Cryst.*, **5** (1952), 1.
13. G. STEWART, *Rev. Mod. Phys.*, **2** (1930), 116.
14. N. S. GINGRICH, *Rev. Mod. Phys.*, **15** (1943), 90.
15. B. R. T. FROST, *Progress in Metal Physics*, Vol. 5, *Pergamon*, London (1954), p. 143.
16. G. FOURNET, *J. Phys. Rad.*, **12** (1951), 592.
17. J. FALGUEIRETTES, *Bull. Soc. Franc. Minér. Cryst.*, **82** (1959), 171.
18. B. R. ORTON, B. A. Shaw, G. I. Williams, *Acta Met.*, **8** (1960), 177.

CHAPTER

# 4

# DIFFRACTION OF X-RAYS BY CRYSTALS

The present importance of X-ray diffraction arose from the striking phenomena observed in crystalline matter which, as is well known since the work of Laue and Bragg, often permit a complete analysis of crystalline structure on the atomic scale.

The fundamental laws of X-ray diffraction by crystals can be found by elementary methods, and they are described in most of the classical works on X-rays.

We shall adopt here a different point of view. We shall deduce the laws which are valid for crystals, starting from the general formulas established in Chapter 2. We shall of course arrive at well-known results, but the calculations will be interesting in themselves, for they will illustrate the general formulas and show how they can be used in cases where no elementary method is available.

We shall assume that the elementary laws of geometrical crystallography are known[1].

## 4.1. The Electron Density in a Crystal. The Crystal Lattice

The object which we shall consider is a small crystal, small enough that absorption can be neglected, each atom receiving the same incident intensity. We shall nevertheless assume that the object contains a large number of unit cells. These conditions will be satisfied, for instance, if

[1] See, for example, the elementary books listed at the end of Chapter 1.

the crystal has dimensions of the order of one micron.

To define the local electron density $\rho_c(\boldsymbol{x})$ according to the general method of Section 2.4.2, we require two quantities.

(a) We first require the *form factor* $\sigma(\boldsymbol{x})$, which defines the exterior shape of the object.

(b) We also require the distribution of matter within the crystal, assumed to be *homogeneous* and *infinite*. From the very definition of the crystalline state, the electron density of an infinite crystal is a triply periodic function, the three periodicities being defined by the three fundamental vectors $\boldsymbol{a}, \boldsymbol{b}, \boldsymbol{c}$ on which a *unit cell* is built. This is a parallelepiped containing the smallest atomic group of the crystal which repeats itself identically throughout space in the case of the infinite crystal by translations equal to lattice vectors. We shall call $V_c$ the volume of the unit cell.

A vector of the crystal lattice is of the form

$$\boldsymbol{x}_{pqr} = p\boldsymbol{a} + q\boldsymbol{b} + r\boldsymbol{c}\,, \tag{4.1}$$

where $p$, $q$, and $r$ are positive or negative whole numbers or zero. Mathematically, it is possible to describe the lattice by a series of *Dirac functions* centered on the lattice points:

$$z(\boldsymbol{x}) = \textstyle\sum_p \sum_q \sum_r \delta(\boldsymbol{x} - \boldsymbol{x}_{pqr})\,. \tag{4.2}$$

As to the atomic group in the unit cell, we shall describe it by the function $\rho(\boldsymbol{x})$, representing the electron density inside a unit cell. This quantity exists only within the unit cell and is zero outside.

The crystal will be represented by a unit cell centered on each lattice point. For the infinite crystal this is expressed by a faltung of the form

$$\rho_\infty(\boldsymbol{x}) = \rho(\boldsymbol{x}) * z(\boldsymbol{x})\,. \tag{4.3}$$

In a finite crystal the density is written as

$$\rho_c(\boldsymbol{x}) = \rho(\boldsymbol{x}) * [z(\boldsymbol{x})\sigma(\boldsymbol{x})]\,, \tag{4.4}$$

$\sigma(\boldsymbol{x})$ being the form factor defined in Section 2.4.1.

According to the general Eq. (*2.3*), the amplitude diffracted by the object is represented by the Fourier transform of the function $\rho_c(\boldsymbol{x})$, and to perform this calculation we require the transforms of all three terms which enter in the right-hand side of the above equation.

(a) We have already calculated the transform $\Sigma(\boldsymbol{s})$ of $\sigma(\boldsymbol{x})$ in Eq. (*2.27*).

(b) The transform of $\rho(\boldsymbol{x})$ is simply the structure factor for the unit cell, as defined in Section 1.2.6.

(c) We now have to calculate the transform of $z(\boldsymbol{x})$. This will permit us to introduce one of the fundamental concepts of crystallography, namely the *reciprocal lattice*.

## 4.2. The Reciprocal Lattice

The transform of a set of functions $\delta(\boldsymbol{x})$ at the lattice points is another set of functions $\delta(\boldsymbol{s})$ at the nodes of another lattice, which is called the *reciprocal lattice*.

Because of the importance of this lattice in crystallography, we shall now give a complete demonstration of this result. We have to calculate

$$Z(\boldsymbol{s}) = \int z(\boldsymbol{x}) \exp(-2\pi i \boldsymbol{s} \cdot \boldsymbol{x}) dv_x ,$$

where $z(\boldsymbol{x})$, according to Eq. (*4.2*), is equal to the triple sum

$$\Sigma \Sigma \Sigma \delta(\boldsymbol{x} - \boldsymbol{x}_{pqr}) .$$

According to the definition of the Dirac function, $Z(\boldsymbol{s})$ can be written as

$$\begin{aligned} Z(\boldsymbol{s}) &= \int \Sigma \Sigma \Sigma \delta(\boldsymbol{x} - \boldsymbol{x}_{pqr}) \exp(-2\pi i \boldsymbol{s} \cdot \boldsymbol{x}) dv_x \\ &= \Sigma_p \Sigma_q \Sigma_r \exp(-2\pi i \boldsymbol{s} \cdot \boldsymbol{x}_{pqr}) . \end{aligned} \tag{4.5}$$

We write the vector of the crystal lattice as a function of the fundamental vectors (4.1), which permits us to transform the above triple sum into a product of three sums:

$$Z(\boldsymbol{s}) = \{\Sigma_p \exp[-2\pi i p(\boldsymbol{s} \cdot \boldsymbol{a})]\} \{\Sigma_q \exp[-2\pi i q(\boldsymbol{s} \cdot \boldsymbol{b})]\} \{\Sigma_r \exp[-2\pi i r(\boldsymbol{s} \cdot \boldsymbol{c})]\} . \tag{4.6}$$

Let us first consider a finite volume of the crystal lattice defined by a parallelepiped centered on the origin and extending along the three axes from $-N\boldsymbol{a}/2$ to $N\boldsymbol{a}/2$, $-N\boldsymbol{b}/2$ to $N\boldsymbol{b}/2$, $-N\boldsymbol{c}/2$ to $N\boldsymbol{c}/2$.

As is shown in Appendix *B* (Eq. *B.2*), each one of the above summations can be written in the following manner:

$$\begin{aligned} \Sigma_{-N/2}^{N/2} \exp[-2\pi i p(\boldsymbol{s} \cdot \boldsymbol{a})] &= 1 + 2 \Sigma_1^{N/2} \cos[2\pi p(\boldsymbol{s} \cdot \boldsymbol{a})] \\ &= \frac{\sin[(N+1)\pi \boldsymbol{s} \cdot \boldsymbol{a}]}{\sin[\pi(\boldsymbol{s} \cdot \boldsymbol{a})]} . \end{aligned} \tag{4.7}$$

The function (*4.7*) has sharp peaks of height $N+1$ and width $1/(N+1)$ for integral values of $(\boldsymbol{s} \cdot \boldsymbol{a})$.

We can now go from a finite to an infinite lattice by letting $N$ tend to infinity. When $(\boldsymbol{s} \cdot \boldsymbol{a})$ is exactly equal to a whole number $h$, the right-hand side of Eq. (*4.7*) tends to infinity, and when $(\boldsymbol{s} \cdot \boldsymbol{a})$ is different from a whole number, this expression is smaller, in absolute value, than $1/\sin[\pi(\boldsymbol{s} \cdot \boldsymbol{a})]$. As $N$ increases, it becomes so rapidly oscillating that, for any value of $\boldsymbol{s}$, the only physically observable value is its average, which is zero. The quantity $Z(\boldsymbol{s})$ is therefore zero for all values of $\boldsymbol{s}$, except if $h$, $k$, and $l$ are simultaneously whole numbers and if $\boldsymbol{s}$ satisfies the three conditions

$$\begin{aligned} \mathbf{s} \cdot \mathbf{a} &= h \,, \\ \mathbf{s} \cdot \mathbf{b} &= k \,, \\ \mathbf{s} \cdot \mathbf{c} &= l \,. \end{aligned} \tag{4.8}$$

When this triple condition is satisfied $Z(\mathbf{s})$ is infinite and it can be represented by a set of functions proportional to Dirac functions $\delta(\mathbf{s})$ at the points of the $\mathbf{s}$-space or reciprocal space which satisfy the above conditions.

We shall now show that these points form a triply periodic lattice. We shall use three vectors $\mathbf{a}^*$, $\mathbf{b}^*$, $\mathbf{c}^*$ such that

$$\begin{array}{lll} \mathbf{a}^* \cdot \mathbf{a} = 1 & \mathbf{b}^* \cdot \mathbf{a} = 0 & \mathbf{c}^* \cdot \mathbf{a} = 0 \\ \mathbf{a}^* \cdot \mathbf{b} = 0 & \mathbf{b}^* \cdot \mathbf{b} = 1 & \mathbf{c}^* \cdot \mathbf{b} = 0 \\ \mathbf{a}^* \cdot \mathbf{c} = 0 & \mathbf{b}^* \cdot \mathbf{c} = 0 & \mathbf{c}^* \cdot \mathbf{c} = 1 \end{array} \tag{4.9}$$

Multiplying the first column by $h$, the second by $k$, and the third by $l$ and adding, we find that

$$\begin{aligned} (h\mathbf{a}^* + k\mathbf{b}^* + l\mathbf{c}^*) \cdot \mathbf{a} &= h \,, \\ (h\mathbf{a}^* + k\mathbf{b}^* + l\mathbf{c}^*) \cdot \mathbf{b} &= k \,, \\ (h\mathbf{a}^* + k\mathbf{b}^* = l\mathbf{c}^*) \cdot \mathbf{c} &= l \,, \end{aligned} \tag{4.10}$$

which are identical to Eqs. (4.8), if we identify $\mathbf{s}$ with the vector

$$\mathbf{r}_{hkl} = h\mathbf{a}^* + k\mathbf{b}^* + l\mathbf{c}^* \,. \tag{4.11}$$

The vectors $\mathbf{a}^*$, $\mathbf{b}^*$, $\mathbf{c}^*$ define the *reciprocal lattice*. Setting the vector $\mathbf{r}_{hkl}^*$ as any vector in this lattice, Eq. (4.6) can be written as

$$Z(\mathbf{s}) = K \sum \sum \sum \delta(\mathbf{s} - \mathbf{r}_{hkl}^*) \,. \tag{4.12}$$

We now have to calculate the value of the coefficient $K$. To do this we must calculate the integral $\int Z(\mathbf{s}) dv_s$ in a small region around a node of the reciprocal lattice. We shall write $\mathbf{s}$ as a function of its coordinates with respect to the axes of the reciprocal lattice:

$$\mathbf{s} = s_1\mathbf{a}^* + s_2\mathbf{b}^* + s_3\mathbf{c}^* \,.$$

As to the vector of the crystal lattice $\mathbf{x}_{pqr}$, we shall write it as a function of the fundamental vectors $\mathbf{a}$, $\mathbf{b}$, $\mathbf{c}$ of this lattice. The scalar product $\mathbf{s} \cdot \mathbf{x}_{pqr}$ which appears in Eq. (4.5) can then be rewritten as

$$\mathbf{s} \cdot \mathbf{x}_{pqr} = ps_1 + qs_2 + rs_3 \,, \tag{4.13}$$

because of Eqs. (4.9). If we now substitute Eq. (4.13) in Eq. (4.6), we find that

$$Z(\boldsymbol{s}) = \{\sum_{p=-N/2}^{p=+N/2} \exp(-2\pi i p s_1)\}\{\sum_{q=-N/2}^{q=+N/2} \exp(-2\pi i q s_2)\} \times \{\sum_{r=-N/2}^{r=+N/2} \exp(-2i\pi r s_3)\} \quad (4.14)$$

or, from Eq. (4.7),

$$Z(\boldsymbol{s}) = \frac{\sin[(N+1)\pi s_1]}{\sin \pi s_1} \frac{\sin[(N+1)\pi s_2]}{\sin \pi s_2} \frac{\sin[(N+1)\pi s_3]}{\sin \pi s_3} . \quad (4.15)$$

We can write the element $dv_s$ which appears in the integral $\int Z(\boldsymbol{s})dv_s$ as a function of the coordinates which we have used for $\boldsymbol{s}$: then $dv_s$ is the volume of the parallelepiped with sides $\boldsymbol{a}^*ds_1$, $\boldsymbol{b}^*ds_2$, $\boldsymbol{c}^*ds_3$, and since $V_c^*$ is the volume of the unit cell in reciprocal space of sides $\boldsymbol{a}^*, \boldsymbol{b}^*, \boldsymbol{c}^*$,

$$dv_s = V_c^* ds_1 ds_2 ds_3 . \quad (4.16)$$

We must calculate the integral over the region around a particular node $hkl$. The parameters $s_1, s_2, s_3$ thus vary from $h-\varepsilon$ to $h+\varepsilon$, from $k-\varepsilon$ to $k+\varepsilon$, and from $l-\varepsilon$ to $l+\varepsilon$. Now

$$\int_{h-\varepsilon}^{h+\varepsilon} \frac{\sin[N+1)\pi s_1]}{\sin \pi s_1} ds_1 = \int_{-\varepsilon}^{+\varepsilon} \frac{\sin[(N+1)\pi s]}{\sin \pi s} ds . \quad (4.17)$$

*Whatever be the value of* $\varepsilon$, *when* $N \to \infty$ *the limiting value of the definite integral is unity.* Thus, from Eqs. (4.15) and (4.16),

$$\int Z(\boldsymbol{s})dv_s = V_c^* = \frac{1}{V_c} , \quad (4.18)$$

where we have written $V_c^*$ in terms of the volume of the unit cell of the crystal lattice $V_c$, according to a relation which we shall demonstrate later on in this section. Finally,

$$Z(\boldsymbol{s}) = \frac{1}{V_c} \sum \sum \sum \delta(\boldsymbol{s} - \boldsymbol{r}_{hkl}^*) . \quad (4.19)$$

We have considered the reciprocal lattice to be the Fourier transform of the crystal lattice, but crystallographers always define the reciprocal lattice directly from Eqs. (4.9). Let us recall certain properties of the reciprocal lattice which follow from these equations.

(a) Each one of the vectors of the reciprocal lattice, for example $\boldsymbol{a}^*$, is perpendicular to two of the vectors of the crystal lattice $\boldsymbol{b}$ and $\boldsymbol{c}$. The three vectors $\boldsymbol{a}^*$, $\boldsymbol{b}^*$, $\boldsymbol{c}^*$ are therefore respectively normal to the planes of the faces

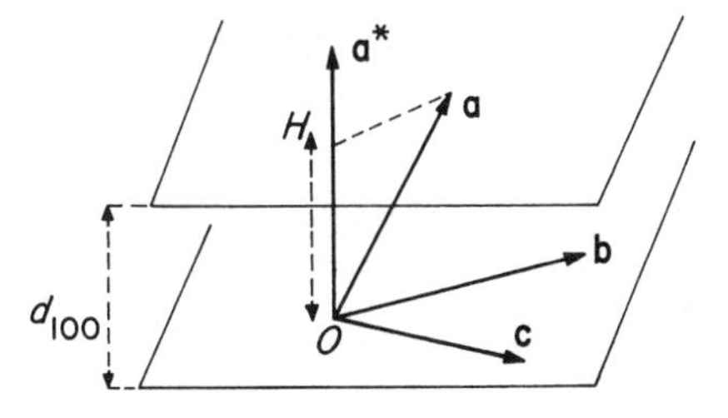


**FIGURE 4.1.** *Construction of one of the fundamental vectors of the reciprocal lattice, the vector* $\boldsymbol{a}^*$.

of the unit cell, as in Figure 4.1. The modulus of $\boldsymbol{a}^*$, on the other hand, is the inverse of the projection of $\boldsymbol{a}$ on this normal, i.e. the inverse of the spacing of the lattice planes parallel to the face. We can write this spacing under the form $V_c/|\boldsymbol{b} \times \boldsymbol{c}|$, whence

$$\boldsymbol{a}^* = \frac{\boldsymbol{b} \times \boldsymbol{c}}{V_c} . \tag{4.20}$$

The angles between the axes of the reciprocal lattice are equal to the angles between the faces of the crystal lattice, and they can therefore be calculated by the usual formulas of spherical trigonometry.

In the important case of the orthorhombic lattice, the axes of the reciprocal lattice are parallel to those of the crystal lattice and

$$a^* = \frac{1}{a}, \quad b^* = \frac{1}{b}, \quad c^* = \frac{1}{c},$$

so that

$$a^*b^*c^* = V_c^* = \frac{1}{abc}\,\frac{1}{V_c} .$$

The volume of the unit cell in reciprocal space is therefore the inverse of the volume of the cell in the crystal lattice. This result is general and applies to any lattice. This can be shown by projecting the vectors $\boldsymbol{a}$, $\boldsymbol{b}$, $\boldsymbol{c}$ and the vectors of reciprocal space $\boldsymbol{a}^*$, $\boldsymbol{b}^*$, $\boldsymbol{c}^*$ on any three rectangular axes. The volumes of the two unit cells are then

$$V_c = \begin{vmatrix} a_1 & a_2 & a_3 \\ b_1 & b_2 & b_3 \\ c_1 & c_2 & c_3 \end{vmatrix} \text{ and } V_c^* = \begin{vmatrix} a_1^* & a_2^* & a_3^* \\ b_1^* & b_2^* & b_3^* \\ c_1^* & c_2^* & c_3^* \end{vmatrix} = \begin{vmatrix} a_1^* & b_1^* & c_1^* \\ a_2^* & b_2^* & c_2^* \\ a_3^* & b_3^* & c_3^* \end{vmatrix} .$$

If we multiply the determinant for $V_c$ by the second one for $V_c^*$, taking into account Eqs. (*4.9*), we find that

$$V_c V_c^* = 1 . \tag{4.21}$$

(b) One characteristic of the reciprocal lattice is that any vector $\boldsymbol{r}^*_{hkl}$ is normal to the planes of the crystal lattice with indices $(hkl)$, and that its modulus is the inverse of the lattice spacing $d_{hkl}$. To demonstrate this fact, we draw in the crystal lattice the plane $(hkl)$, which is next to that passing through the origin. Then, according to the definition of the Miller indices,

$$OH = \frac{a}{h}, \quad OK = \frac{b}{k}, \quad OC = \frac{c}{l},$$

as in Figure 4.2. Let us calculate the scalar products $\boldsymbol{r}^*_{hkl} \cdot \boldsymbol{HK}$ and $\boldsymbol{r}^*_{hkl} \cdot \boldsymbol{HL}$, or

$$\left(h\boldsymbol{a}^* + k\boldsymbol{b}^* + l\boldsymbol{c}^*\right) \cdot \left(\frac{\boldsymbol{b}}{k} - \frac{\boldsymbol{a}}{h}\right),$$

$$\left(h\boldsymbol{a}^* + k\boldsymbol{b}^* + l\boldsymbol{c}^*\right) \cdot \left(\frac{\boldsymbol{c}}{l} - \frac{\boldsymbol{a}}{h}\right).$$

Using Eqs. (*4.9*), we find that the scalar products are zero, or that the vector $\boldsymbol{r}^*_{hkl}$ is normal to two straight lines drawn in the lattice plane, and therefore normal to that plane.

On the other hand, if we calculate the scalar product $\boldsymbol{r}^*_{hkl} \cdot \boldsymbol{OH}$ we find that

$$\boldsymbol{r}^*_{hkl} \cdot \boldsymbol{OH} = (h\boldsymbol{a}^* + k\boldsymbol{b}^* + l\boldsymbol{c}^*) \cdot \frac{\boldsymbol{a}}{h} = 1$$

$$= r^*_{hkl} d_{hkl} .$$

Thus

$$r^*_{khl} = \frac{1}{d_{hkl}} . \qquad (4.22)$$

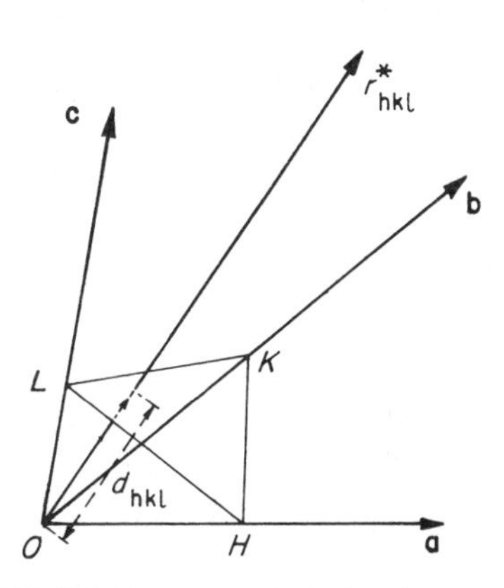


**FIGURE 4.2.** *Relation between the vector $\boldsymbol{r}^*_{kkl}$ of the reciprocal lattice and the (hkl) plane of the crystal lattice.*

*Each of the nodes of reciprocal space therefore corresponds to a family of lattice planes of the crystal, and gives their direction and spacing.*

(c) In any rectangular coordinate system, but only in rectangular systems, the scalar product of two vectors $\boldsymbol{X} \cdot \boldsymbol{Y}$ can be written under the simple form $x_1y_1 + x_2y_2 + x_3y_3$ as a function of the coordinates of the two vectors.

We can retain this simple form for any system of axes if, for the coordinates of the vectors, we express one with respect to the fundamental vectors of the crystal lattice, and the other with respect to the axes of the reciprocal lattice. If

$$\boldsymbol{X} = x_1\boldsymbol{a} + x_2\boldsymbol{b} + x_3\boldsymbol{c}$$

and

$$\boldsymbol{Y} = y_2\boldsymbol{a}^* + y_2\boldsymbol{b}^* + y_3\boldsymbol{c}^* ,$$

then, because of Eqs. (*4.9*),

$$\boldsymbol{X} \cdot \boldsymbol{Y} = x_1y_1 + x_2y_2 + x_3y_3 . \qquad (4.23)$$

(d) It is possible to invert the relation between the crystal lattice and the reciprocal lattice and, from Eqs. (*4.9*), *the reciprocal lattice of the reciprocal lattice is the crystal lattice.*

## 4.3. Reciprocal Space Corresponding to a Finite Lattice

Let us recall that our aim is to calculate the Fourier transform of $\rho_c(\boldsymbol{x})$ as defined in Eq. (*4.4*). We start by evaluating the transform of the product $[z(\boldsymbol{x}) \cdot \sigma(\boldsymbol{x})]$. Using the faltung theorem, we can write this transform as a function of the transforms of the two terms of the product which we have already calculated, namely $Z(\boldsymbol{s})$ and $\Sigma(\boldsymbol{s})$:

$$R(\boldsymbol{s}) = \text{transf}\,[z(\boldsymbol{x}) \cdot \sigma(\boldsymbol{x})] = Z(\boldsymbol{s}) * \Sigma(\boldsymbol{s})$$
$$= \frac{1}{V_c} \sum \sum \sum \delta(\boldsymbol{s} - \boldsymbol{r}^*_{hkl}) * \Sigma(\boldsymbol{s}) \,.$$

The case of a faltung involving a Dirac function is discussed in Appendix A.5, and the result is

$$R(\boldsymbol{s}) = \frac{1}{V_c} \sum \sum \sum \Sigma(\boldsymbol{s} - \boldsymbol{r}^*_{hkl}) \,. \tag{4.24}$$

Physically, this is the amplitude diffracted by an object which has a single electron at each one of the nodes of the crystal lattice, and whose external shape is defined by the form factor $\sigma(\boldsymbol{x})$.

The reciprocal space corresponding to $R(\boldsymbol{s})$ is given by the set of functions $\Sigma(\boldsymbol{s})$ repeated around each of the nodes of the reciprocal lattice and extending out to infinity. We have already discussed the behavior of the function $\Sigma(\boldsymbol{s})$: it decreases from its maximum value at $\boldsymbol{s} = 0$ and is different from zero only in a small region around the origin, the volume of which is of the order of $V_c^*/N$, $N$ being the number of cells in the crystal.

The reciprocal space of the crystal is obtained by the superposition of regions which are similar to each other and which are centered on each of the nodes of the reciprocal lattice. For each point of reciprocal space there is only one term of the above summation which is important—that corresponding to the node which is nearest to that point; all the other terms in the Eq. (4.24) are negligible. As the size of the crystal increases, these reflection regions where $R(\boldsymbol{s})$ is different from zero become smaller and smaller and shrink around the nodes of the reciprocal lattice.

It is easy to deduce the value of the intensity from the expression obtained for the amplitude diffracted by a hypothetical crystal containing one electron per unit cell. The function $I_N(\boldsymbol{s})$ is the square of the modulus of $R(\boldsymbol{s})$ [Eq. (*4.24*)]. Since the function $\Sigma(\boldsymbol{s})$ is limited to the immediate neighborhood of the nodes, there is absolutely no overlapping between the functions centered on the various nodes of the lattice. The double products in the square of the amplitude are therefore all zero since there is no region in space where two of these functions belonging to different nodes are simultaneously different from zero. Then

$$I_N(\boldsymbol{s}) = \frac{1}{V_c^2} \sum_h \sum_k \sum_l |\Sigma(\boldsymbol{s} - \boldsymbol{r}^*_{hkl})|^2 .$$

If the object has a volume $V$, it contains $V/V_c$ unit cells and the diffracted intensity per unit cell, or the interference function, is given by

$$I(\boldsymbol{s}) = \frac{1}{VV_c} \sum \sum \sum |\Sigma(\boldsymbol{s} - \boldsymbol{r}^*_{hkl})|^2 . \quad (4.25)$$

This is exactly what we would have obtained from the general Eq. (*2.35*) for $I(\boldsymbol{s})$.

The expression (4.25) describes the distribution of the diffracted intensity around a node of the reciprocal lattice. However, if the peak is very narrow, the resolution of the instruments is inadequate to register its shape and the only measurable quantity is the integral value $\int I(\boldsymbol{s})dv_s$. From Eq. (*2.38*), we find that

$$\int I(\boldsymbol{s})dv_s = \frac{1}{V_c} . \quad (4.25a)$$

This result may be interpreted as follows: the integral is equal to the product of the average value of the interference function, $I(\boldsymbol{s})$—unity according to Section 2.3—by the volume of the unit cell of the reciprocal lattice, $1/V_c$ [Equation (*4.21*)].

## 4.4. The Reciprocal Space of a Crystal. Structure Factor for a Lattice Plane

Let us now consider the crystal itself. Its electron density is defined by Eq. (*4.4*) as

$$\rho_c(\boldsymbol{x}) = \rho(\boldsymbol{x}) * [z(\boldsymbol{x})\sigma(\boldsymbol{x})] ,$$

and the diffracted amplitude is the Fourier transform of $\rho_c(\boldsymbol{x})$. According to the faltung theorem stated in Section 2.4.2, this is the *algebraic* product of the transforms of the two factors appearing in Eq. (*4.4*). The second of these is $R(\boldsymbol{s})$, which we have already calculated in the preceding section. The first is the amplitude diffracted by the atomic group contained within the unit cell and which we have called the *structure factor* of the unit cell (Section 1.2.6):

$$F(\boldsymbol{s}) = \int \rho(\boldsymbol{x}) \exp(-2\pi i \boldsymbol{s} \cdot \boldsymbol{x}) dv_x .$$

The amplitude diffracted by the crystal is therefore

$$A(\boldsymbol{s}) = F(\boldsymbol{s})R(\boldsymbol{s}) . \quad (4.26)$$

But $R(\boldsymbol{s})$ and $A(\boldsymbol{s})$ are different from zero only in the immediate neighborhood of the nodes of the reciprocal lattice and, in each of the reflection regions, we can replace the slowly varying function $F(\boldsymbol{s})$ by its value at the center of the region, the lattice point *hkl*:

$$F_{hkl} = F(\boldsymbol{r}^*_{hkl}) = \int \rho(\boldsymbol{x}) \exp(-2\pi i \boldsymbol{r}^*_{hkl} \cdot \boldsymbol{x}) dv_x \,. \tag{4.27}$$

*The function $F_{hkl}$ is called the structure factor of the crystal for the node hkl. The effect on reciprocal space of the atomic group within the unit cell is then to multiply the diffracted amplitude around each node of the reciprocal lattice by the factor $F_{hkl}$, and the intensity diffracted by the crystal is multiplied by the factor $F^2_{hkl}$.* From Eq. (*4.25*), the scattering power per unit cell is

$$I(\boldsymbol{s}) = \frac{F^2_{hkl}}{V V_c} \sum_h \sum_k \sum_l |\Sigma(\boldsymbol{s} - \boldsymbol{r}^*_{hkl})|^2 \,. \tag{4.28}$$

From Eq. (*4.25a*), the integral value of the scattering power in one node is

$$\int I(\boldsymbol{s}) d V_s = \frac{F^2_{hkl}}{V_c} \,. \tag{4.28a}$$

This is the expression which is involved in the theoretical calculation of experimentally measurable quantities such as the reflecting power of a small crystal (Section 4.6.2) or the intensity of a powder diffraction line (Section 4.8).

In the case of a simple crystal the structure factor is equal to the atomic scattering factor $f$ for $s = r^*_{hkl}$. If the unit cell contains $n$ atoms whose coordinates are $x_i, y_i, z_i$ ($\boldsymbol{x}_i = x_i\boldsymbol{a} + y_i\boldsymbol{b} + z_i\boldsymbol{c}$) and with scattering factor $f_i$, taking into account Eq. (*1.28*), Eq. (*4.27*) becomes

$$F_{hkl} = \textstyle\sum_1^n f_i \exp[-2\pi i(hx_i + ky_i + lz_i)] \,. \tag{4.29}$$

At the center of the reciprocal lattice, which we can identify as the node (000), the structure factor is $F_{000} = \sum_1^n f_i$. The diffracted ray then coincides with the incident ray and it is experimentally impossible to observe. The atomic scattering factor at zero scattering angle is equal to the number of electrons in the atom, and $F_{000}$ is therefore equal to the total number of electrons contained in the unit cell of the crystal lattice. Since the scattering factor for any atom always decreases with the scattering angle, the structure factors necessarily decrease, on the average, for increasing indices *hkl*. *It is only the nodes of the reciprocal lattice which are near the center which have an appreciable structure factor.*

The reciprocal space of a crystal is therefore formed of a series fo points situated at the nodes of the reciprocal lattice and with coefficients whose moduli tend to zero with increasing distance from the center. The structure factors are equal to unity for all the nodes only in the fictitious case, which we have already examined [Eq. (*4.25*)], of a lattice containing a single electron per unit cell. It is only in this case that the reciprocal space is the reciprocal lattice extending to infinity in all directions.

It is interesting to notice that, for a given atomic group defined by the function $\rho(\boldsymbol{x})$ whose transform is $F(\boldsymbol{s})$, Eq. (*4.28*) involves only the values of this function at the nodes of the reciprocal lattice. Thus different functions $F(\boldsymbol{s})$ correspond to the same reciprocal space if they have the same values at the nodes of the reciprocal lattice, despite the fact that their transforms are different or, in other words, that the elementary groups of atoms are different. This corresponds to the fact that it is possible to have different atomic groups which, when repeated in the crystal lattice, give the same crystal. As an example, we can consider a one-dimensional lattice of period $a$ and a unit cell containing two atoms. As shown in Figure 4.3, it is possible to imagine several different atomic groups which, when repeated with the same period $a$, give the same crystal on the whole. It is easy to verify that these atomic groups, having very different Fourier transforms, nevertheless give the same structure factors for each lattice point.

**FIGURE 4.3.** (a) *Schematic diagram of a one-dimensional crystal with two atoms per unit cell represented by a circle and by a dot.* (b) *and* (c) *Two possible elementary atomic groups.*

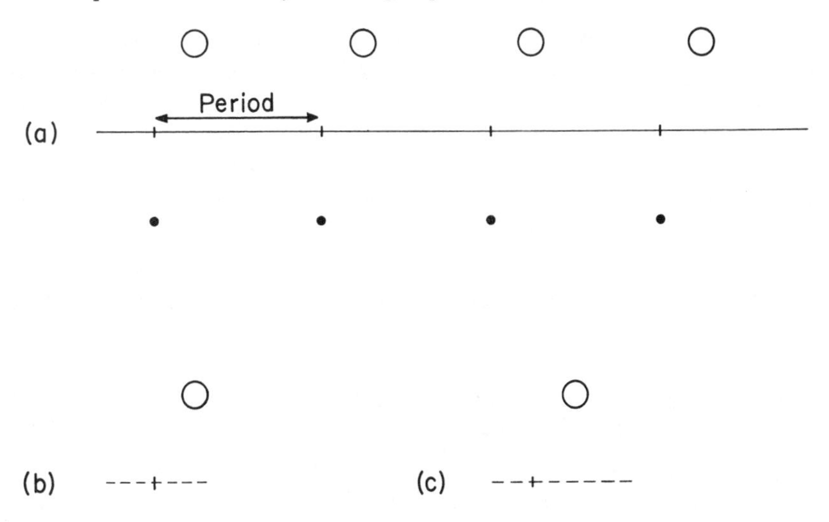

## 4.5. Reflection of X-Rays on Crystals

Now that we have discussed the reciprocal space of a crystal, we can deduce the elementary results for the diffraction of X-rays by crystals.

Let us first consider the case where the crystal is large enough that the diffraction regions around each lattice point of reciprocal space are much smaller than the minimum volume which can be explored in reciprocal space under the given experimental conditions. It is obvious that, in actual fact, incident or diffracted rays are never rigorously parallel nor monochromatic, so it is impossible to measure accurately the value of $I(\boldsymbol{s})$ at a single point of reciprocal space. It is possible to define a resolving power for a given experiment in terms of the volume in reciprocal space where the extremities of the vectors $\boldsymbol{s} = (\boldsymbol{S} - \boldsymbol{S}_0)/\lambda$ are located for the various $\boldsymbol{S}$, $\boldsymbol{S}_0$ and $\lambda$ effectively used in the measurement.

When the crystal is very small, this region of uncertainty is smaller than the domain of reflection surrounding each node of the reciprocal lattice, and these domains may thus be investigated in detail. This will be done in the next chapter.

It will be sufficient, for the moment, to assume that the reflection region is a geometrical point at the node of the reciprocal lattice.

### 4.5.1. The Bragg Law. The Ewald Construction

Since the diffracted intensity in reciprocal space is practically zero outside the nodes of the reciprocal lattice, a diffracted ray is formed only if *one of the nodes of the reciprocal lattice is situated on the surface of the Ewald sphere* discussed in Section 1.2.1.1. We have shown that a given diffraction experiment corresponds to a section of reciprocal space by this sphere: if there is no node on the sphere, the diffracted intensity in any direction is zero.

The *Ewald construction* illustrated in Figure 4.4 permits us to find the direction of the rays diffracted by a crystal. Let us consider the reciprocal lattice for a given crystal. We draw the Ewald sphere corresponding to a given experiment, which is the sphere of radius $1/\lambda$ centered at O, such that $\boldsymbol{OS}_0 = \boldsymbol{S}_0/\lambda$, where $\boldsymbol{S}_0$ is the unit vector parallel to the incident ray. If the node *hkl* of the reciprocal lattice is on this sphere, there is a diffracted ray in the direction of $\boldsymbol{S}$, such that $\boldsymbol{S}/\lambda = \boldsymbol{OR}_{hkl}$. This is equivalent to the Bragg reflection law. The incident and diffracted rays which are parallel to $\boldsymbol{S}_0$ and $\boldsymbol{S}$ are symmetrical with respect to the vector $\boldsymbol{s}$, which is normal to the plane (*hkl*). Thus when the diffraction conditions are

**FIGURE 4.4.** *The Ewald Construction.*

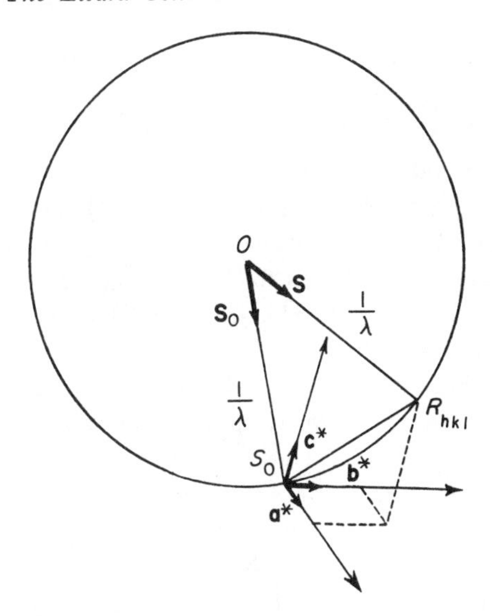


satisfied, the direction of the diffracted ray is the same as if reflection occurred on the (*hkl*) plane. The diffraction angle is given by

$$s = \frac{2\sin\theta}{\lambda} = r^*_{hkl} = \frac{1}{d_{hkl}}, \tag{4.30}$$

which is equivalent to the Bragg law,

$$\lambda = 2d_{hkl}\sin\theta\,. \tag{4.31}$$

### 4.5.2. Definition of the Reflecting Power of a Small Crystal

The directions of reflection being known, we must now deduce from the reciprocal space of a crystal the intensity of diffraction for the various lattice planes. This requires an accurate definition for the *reflecting power* of a crystal such that this quantity can be measured experimentally.

As we have already seen, in the case of an extended crystal the regions where $I(\mathbf{s})$ is different from zero are so small that, experimentally, it is not possible to determine the rapid variations of $I(\mathbf{s})$ within the domain around a lattice point; Eq. (*4.28*) cannot therefore be verified experimentally. It is only possible to measure the values of the intensity integrated over a region containing a lattice point.

Another difficulty comes from the fact that real crystals are never perfect but rather are made up of many small blocks more or less disoriented one with respect to the others—the so-called *mosaic structure*. We therefore require an experiment giving a value of the reflecting power which is independent both of the real structure of a given sample and of the more or less perfect geometry of the incident beam.

We shall use the method first given by W. H. Bragg, which is valid in the case of a very small crystal of negligible absorption.

The crystal is immersed in a beam of intensity $I_0$ and rotated about an axis which is normal to the plane of incidence of the rays on the lattice plane considered. The angular velocity is $\omega$ radians/second, so that the angle $\theta$ between the lattice plane considered and the average ray of the incident beam varies from $\theta_0 - \varepsilon$ to $\theta_0 + \varepsilon$, $\theta_0$ being the Bragg angle and $\varepsilon$ of the order of one degree. This value of $\varepsilon$ is much larger than the region where selective reflection can occur and larger also than the divergence of the incident beam. In this way, for every ray of the incident beam, the sample goes through the position of selective reflection.

If we now draw the curve of the measured diffracted power $R(\theta)$ as a function of the angle $\theta$, we find that the shape of this curve depends both on the divergence of the incident beam and on the degree of perfection of the crystal. However, the ratio between the area under this curve and the intensity of the incident beam is a constant for a sample of a given crystalline type. This ratio is called the *reflecting power* ($P$) *of the crystal*:

$$P = \int_{\theta_0-\varepsilon}^{\theta_0+\varepsilon} \frac{R(\theta)}{I_0} d\theta \,. \tag{4.32}$$

Since $d\theta = \omega dt$, where $t$ is the time,

$$\int R(\theta) d\theta = \omega \int R(\theta) dt \,. \tag{4.33}$$

As $\int (R)\theta\, dt$ is the total energy $E$ diffracted by the crystal while it goes through the reflection position,

$$P = \frac{E\omega}{I_0} \,. \tag{4.34}$$

If a Geiger counter is used as detector, it is first placed in the direct beam and the number of pulses per second is a measure of the incident energy $E_0$. Then, if $S$ is the cross-section of the incident beam at the position of the sample, $I_0 = E_0/S$. The counter is then placed to receive the whole beam diffracted by the small crystal and the energy $E$ is measured from the total number of pulses registered when the crystal goes through the reflection position. The ratio $E/I_0$, and therefore $P$, can thus be obtained experimentally. It is also possible to show that the same determination can be made photographically.

### 4.5.3. Calculation of the Reflecting Power of a Small Crystal

We shall calculate $P$ for a small crystal by examining the method of measurement described above.

We can assume that the incident beam has a definite direction $\boldsymbol{S}_0$ because, during the rotation, each ray of the incident beam goes through the reflection position under the same geometrical conditions and diffraction occurs as if all the incident rays were parallel. We may also assume that the incident beam is perfectly monochromatic.

Let us represent the reciprocal space of the crystal during an experiment designed to measure the reflecting power. Figure 4.5 shows the meridian plane of the Ewald sphere containing the node *hkl* corresponding to the reflection investigated. The crystal rotates about the axis normal to the plane of the figure with the angular velocity $\omega$, and the lattice point *hkl* traverses the Ewald sphere. For a given position of the crystal, the scattering power per unit cell in the direction defined by the vector $\boldsymbol{S}$ is given by Eq. (*4.28*), where $\boldsymbol{s} = (\boldsymbol{S} - \boldsymbol{S}_0)/\lambda$. If $I_e$ is the scattered intensity as given by the Thomson formula (*1.21*) for the scattering by a single electron, the intensity per unit cell diffracted in a small solid angle $d\Omega$ in the direction of the vector $\boldsymbol{S}$ is given by

$$I_e I(\boldsymbol{s})\, d\Omega\ ,$$

$I(\boldsymbol{s})$ being given by the Eq. (*4.28*). The total energy collected by the

**FIGURE 4.5.** *Calculation of the intensity reflected by a small crystal.*

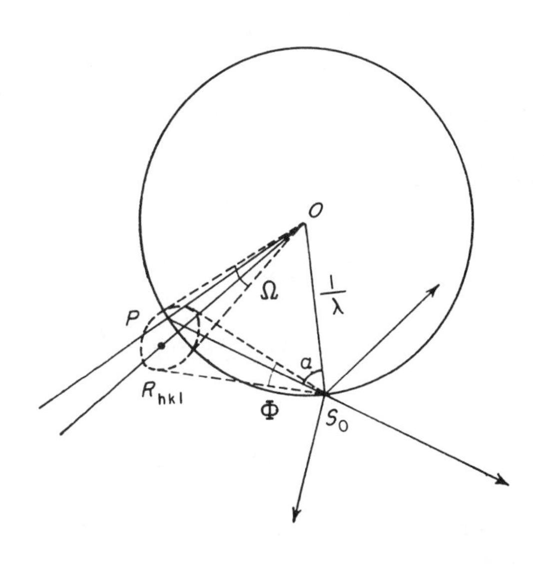

detector per unit time and per unit cell for this position of the crystal is then

$$I_1 = \int_\Omega I_e I(\boldsymbol{s})\, d\Omega\,, \tag{4.35}$$

where the integral is evaluated over the solid angle $\Omega$ subtended at $O$ by the section through the Ewald sphere of the reflection domain surrounding the node under consideration (Figure (4.5).

As the crystal rotates, the surface of the sphere sweeps through this region. If $\alpha$ is the angle between $\boldsymbol{S_0P}$ and $\boldsymbol{S_0O}$, the rotation $d\alpha$ of the crystal takes place during the time $d\alpha/\omega$. Then, while the crystal turns through a very small angle $d\alpha$ about the position $\alpha$, the energy diffracted is $I_1 d\alpha/\omega$ and the total energy diffracted during the rotation $\Phi$ of the crystal while the node crosses the sphere is

$$E_1 = I_e \int_\Phi \frac{d\alpha}{\omega} \int_\Omega I(\boldsymbol{s})\, d\Omega = \frac{I_e}{\omega} \iint I(\boldsymbol{s})\, d\alpha\, d\Omega\,. \tag{4.36}$$

We can express the differential appearing in the above integral as a function of the element of volume in reciprocal space, since the solid angle $d\Omega$ corresponds to an area $d\sigma = d\Omega/\lambda^2$ on the reflection sphere. During the rotation $d\alpha$ this surface sweeps through the volume

$$dv_s = d\sigma\, \frac{2 \sin\theta}{\lambda}\, d\alpha \cos\theta = d\alpha\, d\Omega\, \frac{\sin 2\theta}{\lambda^3} \tag{4.37}$$

and

$$E_1 = \frac{I_e}{\omega} \frac{\lambda^3}{\sin 2\theta} \int I(\boldsymbol{s})\, dv_s\,. \tag{4.38}$$

Near a node of the reciprocal lattice, $I(\boldsymbol{s})$ is proportional to the function $|\Sigma(\boldsymbol{s})|^2$ [(Eq. (4.28)]. From Eq. (2.38), the integral of this quantity over a small volume around the origin is

$$\int |\Sigma(\boldsymbol{s})|^2\, dv_s = V\,.$$

Then, from Eq. (4.28),

$$\int I(\boldsymbol{s})\, dv_s = \frac{F_{hkl}^2}{V V_c} \int |\Sigma(\boldsymbol{s})|^2\, dv_s = \frac{F_{hkl}^2}{V_c} \tag{4.39}$$

and

$$E_1 = \frac{I_e}{\omega} \frac{\lambda^3}{\sin 2\theta} \frac{F_{hkl}^2}{V_c}\,.$$

If the volume of the small crystal is $dV$, it contains $dV/V_c$ unit cells and the reflected energy is therefore $E = E_1 dV/V_c$. For an unpolarized incident beam, we can use the Thomson formula for $I_e$ as given in Eq. (*1.23*):

$$I_e = I_0 r_e^2 \frac{1 + \cos^2 2\theta}{2} .$$

Finally, the reflecting power of a small crystal is given by

$$P = \frac{E\omega}{I_0} = r_e^2 \left( \frac{1 + \cos^2 2\theta}{2 \sin 2\theta} \right) \frac{\lambda^3}{V_c^2} F_{hkl}^2 \, dV = Q_{hkl} \, dV . \qquad (4.40)$$

One important consequence of this equation is that the reflecting power of a small crystal is proportional to its volume, if absorption is negligible. If we compare two similar crystals, one of which has double the volume of the other, the maximum intensity diffracted by the first will be four times larger since it contains twice as many atoms. But, just as the line width in the optical spectra produced by a grating varies inversely as the width of the grating, the angular width of the region of selective reflection varies inversely as its volume. The width of the maximum of the diffracted intensity as a function of the reflection angle for the large crystal is one half as great, so that the area under the curve—which is proportional to the reflecting power—is only twice as large. It is proportional to the volume of the sample.

We now wish to make a few *remarks concerning the expression for the reflecting power*. The reflecting power per unit volume for the *hkl* reflection from a small crystal is written $Q_{hkl}$:

$$Q_{hkl} = r_e^2 \frac{1 + \cos^2 2\theta}{2} \frac{\lambda^3}{\sin 2\theta} \frac{1}{V_c^2} F_{hkl}^2 , \qquad (4.41)$$

where the quantity $Q_{hkl}$ has the dimension of (length)$^{-1}$ and $P$ in Eq. (*4.40*) has the dimensions of an area. Using the centimeter as the unit of length, the coefficient $r_e^2$ of the above equation is $7.90 \times 10^{-26}$.

The factor $\lambda^3/\sin 2\theta$ comes from the integration of the intensity over the sharp peak in the region of a node for given experimental conditions. It is called the *Lorentz factor*.

The term $(1 + \cos^2 2\theta)/2$ is the *polarization factor*. In the Thomson formula (*1.23*), its value corresponds to an *unpolarized* incident beam. This is evidently the general case, but is must be remembered that there are important practical cases where this condition is not fulfilled. This occurs when a crystal monochromator is used for the incident beam. If $\alpha$ is the reflection angle at the monochromator crystal, the two components of the reflected ray which are polarized in the directions parallel and perpendicular to the plane of incidence vary as 1 and $\cos^2 2\alpha$, and their relative intensities are

$$k_{\parallel} = \frac{1}{1 + \cos^2 2\alpha}, \qquad k_{\perp} = \frac{\cos^2 2\alpha}{1 + \cos^2 2\alpha}.$$

This modifies Eq. (*4.41*) as follows, according to Eq. (1.21),

$$Q_{hkl} = r_e^2 F_{hkl}^2 \frac{\lambda^3}{V_c^2 \sin 2\theta} \frac{1 + \cos^2 2\alpha \cos^2 2\theta}{1 + \cos^2 2\alpha}. \tag{4.42}$$

It is found that, for $\alpha = 15°$ and $2\theta = 90°$, the error introduced in the general Eq. (*4.41*) by neglecting the polarization effect is 14%. This correction can be important and it must never be omitted for absolute intensity measurements.

If the primary ray, the ray reflected by the monochromator, and the diffracted ray are not in the same plane, the polarization factor is more complex [1].

*The temperature effect.* It will be recalled that Eq. (*4.41*) was deduced under the hypothesis that thermal agitation was negligible. The atoms of course vibrate about their theoretical positions with an amplitude which increases with temperature. This phenomenon will be investigated in Section 7.1.1. One of the effects of thermal agitation is to reduce the intensity of the selective reflections. We must therefore multiply the reflecting power as given by Eq. (*4.41*) by a coefficient $D$ which is less than unity; this is called the *Debye* or *temperature factor*. The value of $D$ will be given in Eq. (*7.7*); it decreases with temperature and with $s = (2 \sin \theta)/\lambda$. The effect can be important for large values of $\theta$, even at room temperature.

It is possible to calculate the observed diffracted intensity for various usual X-ray experiments, starting from the expression for the reflecting power of a very small crystal. The following sections will be devoted to a few important particular cases.

## 4.6 Reflecting Power of a Crystal Face (Kinematic Theory)

When the object is a large crystal, we must obviously take the absorption into account. In the *kinematic theory of reflection* we assume that the diffracted intensities for each element of volume simply add together. This is equivalent to assuming that the waves emitted by the various regions are incoherent.

**FIGURE 4.6.** *Schematic diagram showing the mosaic crystal structure.*

This hypothesis can be used in the case of the *ideally imperfect* crystal illustrated by the mosaic crystal of

Figure 4.6. This crystal is formed of an aggregate of small blocks, about one micron in size, separated by walls where the crystalline faults such as dislocation lines conglomerate. The blocks can thus be rotated slightly, one with respect to the other, or displaced by a distance which is not exactly a lattice vector, so that the waves diffracted by two neighboring blocks are incoherent. This explains the angular width of the reflections actually observed on most crystals. This width can be as small as a few minutes or as large as one degree, for a crystal rotating in a parallel monochromatic beam. In some cases the misaligned blocks can actually be seen on crystal or cleavage faces which are not exactly plane.

In a large number of cases the ideally imperfect crystal approximation is satisfactory. We shall discuss the theory of *dynamic* diffraction in Section 4.8. This theory is valid only for perfect crystals.

**FIGURE 4.7.** *Intensity reflected by an infinitely thick crystal in the symmetrical position.*

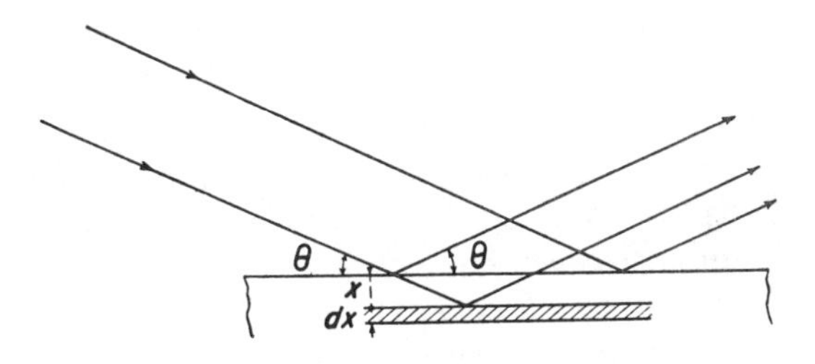


Let us consider the case of symmetrical reflection by lattice planes parallel to the face of a crystal of infinite thickness, as in Figure 4.7. The parallel incident beam has a cross-section $S$ and an intensity $I_0$. At a depth $x$ in the crystal, this intensity is reduced to $I_0 \exp(-\mu\rho x/\sin\theta)$ where $\mu$ is the mass coefficient of absorption, as in Section 1.1.5, and $\rho$ is the density. The volume irradiated in the layer of thickness $dx$ is

$$dv = S\frac{dx}{\sin\theta},$$

and the energy diffracted by $dv$ is $dE'$ such that, according to Eq. (*4.40*),

$$\frac{dE'\omega}{I_0} = QS\frac{dx}{\sin\theta}\exp\left(-\frac{\mu\rho x}{\sin\theta}\right). \tag{4.43}$$

However, because of the absorption, the energy detected outside the crystal is reduced to $dE$ such that

$$\frac{dE\omega}{I_0} = QS\frac{dx}{\sin\theta}\exp\left(-\frac{2\mu\rho x}{\sin\theta}\right). \tag{4.44}$$

For the whole crystal the energy collected while the crystal passes through the reflection position is

$$E = \int_{x=0}^{x=\infty} dE = \frac{I_0}{\omega} \frac{QS}{2\mu\rho}. \tag{4.45}$$

We now set $E_0 = I_0 S$ to represent the total incident energy per unit time. Then, according to Eq. (*4.41*),

$$P_k = \frac{E\omega}{E_0} = \frac{Q}{2\mu\rho} = r_e^2 \frac{1 + \cos^2 2\theta}{2 \sin 2\theta} \frac{\lambda^3}{V_c^2} F_{hkl}^2 \frac{1}{2\mu\rho}. \tag{4.46}$$

The dimensionless number $P_k$ is the reflecting power for the face of a semi-infinite crystal parallel to the planes $(hkl)$.

## 4.7. The Line Intensities in a Powder Pattern

Let us consider a diffracting object of volume $dV$ small enough that absorption can be neglected, composed of a very large number of small crystals whose orientations are perfectly isotropic.

A Debye-Scherrer line of indices $hkl$ is produced by reflection on all the crystal planes having the same spacing, or by all the planes $\{hkl\}$. The number of the lattice planes which are equivalent for reasons of symmetry is called the multiplicity factor $n$ and depends on the nature of the lattice and on the indices $hkl$.

In a triclinic system which has no symmetry elements, there corresponds to a given value of $d$ only the lattice planes $(hkl)$ and $(\bar{h}\bar{k}\bar{l})$. Then $n$ is equal to 2 for all the lines. However, as the symmetry of the system increases, several groups of indices correspond to planes having the same spacing $d$. The number of lines therefore decreases, while their intensity increases. Table 4.1 shows the value of $n$ for the different crystal systems.

TABLE 4.1. Multiplicity Factors for Various Lattice Planes

| Cubic System | | Hexagonal System | | Tetragonal System | | Orthorhombic System | | Monoclinic System | | Triclinic System | |
|---|---|---|---|---|---|---|---|---|---|---|---|
| Indices | $n$ | Indices | $n$ | Indices | $n$ | Indices | $n$ | Indices | $n$ | Indices | $n$ |
| $h00$ | 6 | $00l$ | 2 | $00l$ | 2 | $h00$ | 2 | $0k0$ | 2 | $hkl$ | 2 |
| $hhh$ | 8 | $h00$ | 6 | $h00$ | 4 | $hk0$ | 4 | $h0l$ | 2 | | |
| $hh0$ | 12 | $hh0$ | 6 | $hh0$ | 4 | $hkl$ | 8 | $hkl$ | 4 | | |
| $hhk$ | 24 | $h0l$ | 12 | $h0l$ | 8 | | | | | | |
| $hk0$ | 24 | $hhl$ | 12 | $hhl$ | 8 | | | | | | |
| $hkl$ | 48 | $hk0$ | 12 | $hk0$ | 8 | | | | | | |
| | | $hkl$ | 24 | $hkl$ | 16 | | | | | | |

Moreover, lines corresponding to different indices can in some cases overlap, either systematically like the lines (333) and (511) of the cubic lattice, or accidentally according to the various crystal parameters.

If a crystal gives rise to a Debye-Scherrer line with indices *hkl*, the normal to the lattice plane (*hkl*) must make an angle $(\pi/2) - \alpha$ with the direction $\boldsymbol{S}_0$ of the incident rays, the angle $\alpha$ being within the region of reflection, or very close to the Bragg angle $\theta$. The probability that the normal is within $(\pi/2) - \alpha$ and $(\pi/2) - \alpha + d\alpha$ of $\boldsymbol{S}_0$ is $(\cos\alpha\, d\alpha)/2$. Since there are $n$ equivalent normal directions, the total volume of the crystals having a normal to a plane of the form $\{hkl\}$ in this direction $(n\,dV/2)\cos\alpha\, d\alpha$ or, since $\alpha$ is very close to $\theta$, $(n\,dV/2)\cos\theta\, d\alpha$.

Setting $I(\alpha)$ as the energy reflected per second by the plane (*hkl*) for the angle of incidence $\alpha$, the total energy diffracted per second by the volume $dV$ in the ring *hkl* is

$$\Phi = \frac{n\,dV}{2}\cos\theta \int I(\alpha)\, d\alpha\ .$$

This is the energy which is reflected by the single crystal of volume $(n\,dV/2)\cos\theta$ crossing the reflection position at the angular velocity $\omega = 1$ (Section 4.5.3).

According to Eq. (4.40),

$$\Phi = \frac{n\,dV}{2}\cos\theta\, Q\, I_0\ .$$

The energy $\Phi$ is distributed uniformly over all of the Debye-Scherrer ring. At a distance $r$ from the sample, the length of this ring is $2\pi r \sin 2\theta$ and the diffracted energy per second and per unit length of the ring is therefore

$$I = \frac{\Phi}{2\pi r \sin 2\theta} = \frac{I_0 Q n\,dV}{8\pi r \sin\theta}\ .$$

Finally, replacing $Q$ from Eq. (*4.41*) and introducing the temperature factor $D$ of Section 7.1.1,

$$I = I_0 r_e^2 \frac{1+\cos^2 2\theta}{2}\,\frac{1}{16\pi r \sin^2\theta\cos\theta}\,\lambda^3 F_{hkl}^2 n \frac{1}{V_c^2}\, D\,dV\ . \qquad (4.47)$$

Let us define again the terms of this equation:

$I$ is the power diffracted per unit length of a Debye-Scherrer line at a distance $r$ from the sample;

$I_0$ is the intensity per unit area of the incident beam;

$r_e^2 = 7.9 \times 10^{-26}\ \text{cm}^2$;

$\theta$ is the Bragg angle ($\lambda = 2d_{hkl}\sin\theta$);

the term $(1+\cos^2 2\theta)/2$ is the polarization factor which must be replaced by the expression of Eq. (*4.42*) when the incident beam is partly polarized;

the factor $1/(\sin^2\theta\cos\theta)$ related to the geometry of the experiment is called the Lorentz factor for the Debye-Scherrer pattern;

$F_{hkl} = \sum_1^n f_i \exp[-2\pi i(hx_i + ky_i + lz_i)]$ is the structure factor for the lattice plane $(hkl)$ given by Eq. (*4.29*);

$n$ is the multiplicity factor for the planes $(hkl)$ as in Table 4.1;

$V_c$ is the volume of a unit cell of the crystal;

$D$ is the temperature or Debye factor of Eq. (*7.7*);

$dV$ is the volume of the diffracting powder.

Finally, we note again that Eq. (*4.47*) is valid only if the elementary crystals are so small that there are no extinction phenomena as in Section 4.10 and if their orientations are perfectly isotropic.

When the sample is large enough that absorption becomes important, we must integrate over the volume, taking the absorption into account. The intensity is then given by Eq. (*4.47*), where $dV$ is replaced by the *effective volume.*

The following are useful results.

(a) In a Debye-Scherrer camera with a cylindrical sample, the effective volume is $AV$, where $V$ is the irradiated volume and $A$ is a function of the radius of the cylinder and of the absorption coefficient. The coefficient $A$ is tabulated in the *International Crystallographic Tables*[1].

(b) In the diffractometer the sample is a small plate oriented so that the incident and diffracted rays are symmetrically oriented with respect to its surface. The effective volume is then $S/2\mu\rho$, where $S$ is the cross-section of the incident beam, and the calculation is analogous to that of Section 4.6. As usual, $\mu$ and $\rho$ are, respectively, the absorption coefficient and the density.

In a transmission camera with a thin sample of thickness $a$, the effective volume is

$$\frac{S}{\mu\rho\left[1-\dfrac{\cos\alpha}{\cos(2\theta-\alpha)}\right]}\left[\exp\left\{-\frac{\mu\rho a}{\cos(2\theta-\alpha)}\right\}-\exp\left\{-\frac{\mu\rho a}{\cos\alpha}\right\}\right], \qquad (4.48)$$

the angle of incidence being $\alpha$. This applies to the case where a monochromator is used [2]. When $\alpha=\theta$, this simplifies to

$$\frac{Sa}{\cos\theta}\exp\left(-\frac{\mu\rho a}{\cos\theta}\right). \qquad (4.49)$$

Then the optimum thickness is $a_m = (\cos\theta)/\mu\rho$, or that which reduces the incident intensity by a factor of 1/e, or about 1/3.

[1] Vol. II, p. 291.

## 4.8. The Intensity of the Reflections in a Rotating Crystal Pattern

The experimental setup is similar to that used for the measurement of the reflecting power of a crystal. If we consider an equatorial spot on the pattern, the energy reflected when the crystal goes through the reflection position is simply given by Eq. (*4.41*). If the spot is not equatorial, the reflecting plane does not go through the axis of rotation and the rate of change of the angle of incidence $d\theta/dt$ is then not equal to $\omega$ and

$$\frac{d\theta}{dt} = \omega\left(1 - \frac{\sin^2\mu}{\sin^2 2\theta}\right)^{1/2},$$

where $\mu$ is the angle between the diffracted ray and the equatorial plane.

It follows that, for these extraequatorial reflections, we must introduce into Eq. (*4.41*) the correction factor

$$\left(1 - \frac{\sin^2\mu}{\sin^2 2\theta}\right)^{-1/2}. \tag{4.50}$$

## 4.9. Relation between the Structure of a Crystal and its Reciprocal Space

As in the general case, the diffracted amplitude represented as a function of $\boldsymbol{s}$ in reciprocal space is the Fourier transform of the electron density in the crystal. Since this density is a periodic function, the Fourier integrals can be expressed as sums. We shall discuss them further because they are fundamental for the determination of crystal structure.

The reciprocal space of an infinite crystal is formed of the nodes of the reciprocal lattice, each of which has a coefficient called the structure factor,

$$F_{hkl} = \sum f_i \exp[-2\pi i(hx_i + ky_i + lz_i)],$$

the sum being extended to all the atoms of the unit cell. Instead of considering a discontinuous distribution of atoms repeating from one unit cell to another, we shall define the distribution of matter in the crystal by the electron density $\rho$, which is the number of electrons per unit volume at each point. We shall select as axes the sides $\boldsymbol{a}$, $\boldsymbol{b}$, $\boldsymbol{c}$ of the unit cell, and we shall choose as unit of length in each of these directions the lengths $a$, $b$, $c$ of these three vectors. The density $\rho(X, Y, Z)$ is then a periodic function of each of the three variables, with a period of unity. The structure factor of the crystal can be written as

$$F_{hkl} = \int_{V_c} \rho(X, Y, Z) \exp[-2\pi i(hX + kY + lZ)]\, dv\,, \tag{4.51}$$

where the integral is extended to the volume of the elementary unit cell $V_c$.

We now express the density $\rho(X, Y, Z)$ as a Fourier series:

$$\rho(X, Y, Z) = \sum_p \sum_q \sum_r A_{pqr} \exp[2\pi i(pX + qY + rZ)]\,, \tag{4.52}$$

where $p, q, r$ are whole numbers, positive or negative, and $A_{pqr}$ is in general complex and depends only on $p, q,$ and $r$. Since $\rho(X, Y, Z)$ is real, we must have

$$A_{pqr} = A^*_{\bar{p}\bar{q}\bar{r}} \tag{4.53}$$

Let us calculate the structure factor $F_{hkl}$ by replacing $\rho(X, Y, Z)$ in Eq. (4.51) by its Fourier series as given above:

$$F_{hkl} = \sum_p \sum_q \sum_r \int_{V_c} A_{pqr} \exp[2\pi i(pX + qY + rZ)] \times \exp[-2\pi i(hX + kY + lZ)]\, dv\,. \tag{4.54}$$

Let us write the element of volume $dv$ as a function of the coordinates $X, Y, Z$:

$$dv = V_c\, dX\, dY\, dZ\,.$$

Then $F_{hkl}$ is the sum of terms of the form

$$V_c A_{pqr} \int_0^1 \exp[2\pi i(p-h)X]dX \int_0^1 \exp[2\pi i(q-k)Y)]dY \times \int_0^1 \exp[2\pi i(r-l)Z)]\, dZ\,.$$

*Each one of the integrals is zero unless the exponent is zero; the integral is then equal to unity.* In the above triple sum we therefore have only the term for which $p = h$, $q = k$, $r = l$, and

$$F_{hkl} = V_c A_{hkl}\,. \tag{4.55}$$

*The structure factor for the node is equal to the coefficient of the hkl term of the Fourier series representing the electron density in the crystal, within a multiplying factor, $V_c$.*

This relation is fundamental for the most profound and general interpretation of X-ray diffraction by crystals. To discuss this interpretation we shall rewrite the expression for $\rho$, using the indices $h, k, l$ for numbering the terms, and grouping the $hkl$ and $\bar{h}\bar{k}\bar{l}$ terms. From Eq. (4.53), $A_{hkl} = A^*_{\bar{h}\bar{k}\bar{l}}$ and we can write these complex numbers as functions of the modulus $|A_{hkl}|$ and of the argument $\alpha_{hkl}$. Finally, we note that

$$F_{000} = \int_{V_c} \rho(X, Y, Z) dv$$

represents the total number of electrons in the unit cell; and

$$A_{000} = \frac{F_{000}}{V_c} = \rho_0 ,$$

which is the average electron density in the unit cell.

Equation (4.52) can be written as

$$\rho(X, Y, Z) = \rho_0 + \sum_h \sum_k \sum_l |A_{hkl}| \cos[2\pi(hX + kY + lZ) + \alpha_{hkl}] . \quad (4.56)$$

The *hkl* term represents a periodic fluctuation in the electron density in the form of a plane sinusoidal wave whose wave fronts are parallel to the lattice planes (*hkl*) and whose wavelength is equal to the lattice spacing $d_{hkl}$ because, if $\boldsymbol{x}$ is the vector from the origin to the point $X$, $Y$, $Z$ of the crystal lattice and if $\boldsymbol{r}^*_{hkl}$ is the vector of the reciprocal lattice corresponding to the node *hkl*,

$$hX + kY + lZ = \boldsymbol{x} \cdot \boldsymbol{r}^*_{hkl} ,$$

from Section 4.2 [Eq. (4.23)]. The points $X$, $Y$, $Z$ where the wave has a given phase are therefore situated in a plane perpendicular to $\boldsymbol{r}^*_{hkl}$ and therefore parallel to the lattice plane (*hkl*). Also, two points which are in phase along the direction $\boldsymbol{r}^*_{hkl}$ are separated by a distance [Eq. (4.20)] given by

$$d_{hkl} = \frac{1}{r^*_{hkl}} .$$

In optics it is well known that a grating with a sinusoidal diffracting power gives only two diffracted beams, those of the first order, and that their amplitude is proportional to the amplitude of the variation in the diffracting power of the grating.

Similarly, a sinusoidal wave of electron density reflects X-rays only when the Bragg condition is satisfied for the first order—that is, when $\boldsymbol{s} = (\boldsymbol{S} - \boldsymbol{S}_0)/\lambda$ coincides with the vector $\boldsymbol{r}^*_{hkl}$ or $\boldsymbol{r}^*_{\bar{h}\bar{k}\bar{l}}$. The amplitude of these reflections is then proportional to $|A_{hkl}|$. For this particular reflection position, all the other density waves whose sum gives $\rho(X, Y, Z)$ have no effect. This is the physical interpretation of Eq. (4.55).

One first consequence of this equation is that $\rho(X, Y, Z)$ can be written as

$$\rho(X, Y, Z) = \frac{1}{V_c} \sum \sum \sum F_{hkl} \exp[2\pi i(hX + kY + lZ)] . \quad (4.57)$$

The problem of determining the crystal structure, or of determining $\rho(X, Y, Z)$ in the unit cell, would be resolved in a completely general and rigorous manner if we could determine $F_{hkl}$ for all the nodes of the reciprocal lattice. However, it is only the amplitude $|F_{hkl}|$ which can be determined from the intensity of the reflection from the planes (*hkl*).

**FIGURE 4.8.** *Summation of three sinusoidal waves of photographic blackening obtained by the "photosummator" of Von Eller [3]. The phase angle $\alpha$ is either 0 or $\pi$. The superposition of the three waves in the unit cell gives a total blackening depending on the phase of wave III.*

*Wave I*: $A_{10}\left(1+\cos 2\pi \frac{x}{a}\right)$. *Wave II*: $A_{01}\left(1+\cos 2\pi \frac{y}{b}\right)$.

*Wave III*: $A_{11}\left\{1+\cos\left[2\pi\left(\frac{x}{a}+\frac{y}{b}\right)+\alpha\right]\right\}$.

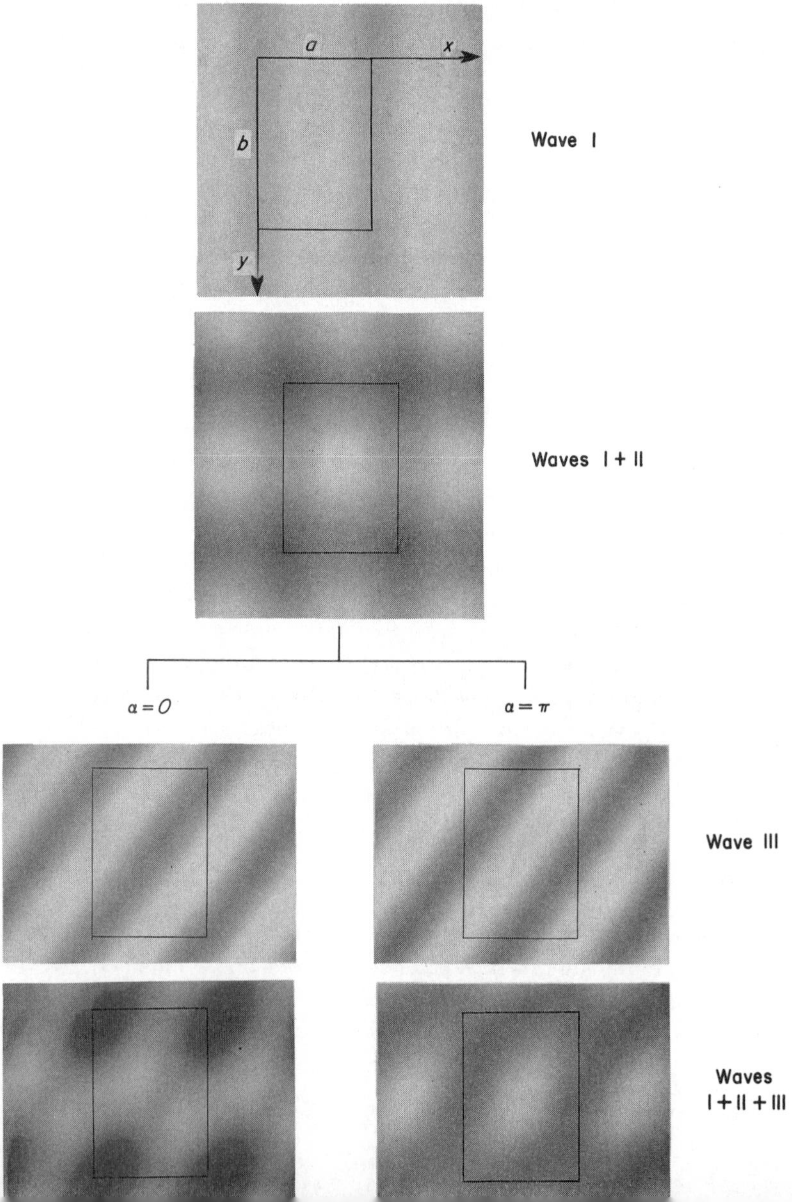

But the number of reflections which can be recorded is limited since, from the Ewald construction discussed in Section 4.5.1,

$$r^*_{hkl} = \frac{1}{d_{hkl}} < \frac{2}{\lambda} .$$

Since the intensity of each reflection is known, it is possible to find the amplitude of each one of the density waves into which $\rho(X, Y, Z)$ can be decomposed. However, if one of these waves is displaced parallel to itself, that is, if we modify $\alpha_{hkl}$, the intensity of the reflection is unchanged. Therefore, two different structures obtained by displacing the waves of different indices give the same diffraction pattern since it is not possible to measure the phase differences between the diffracted waves. Displacing these waves can cause considerable variations in the sum $\rho(X, Y, Z)$. Figure 4.8 shows this fact schematically in two dimensions. This is the fundamental uncertainty which the present techniques for structure determinations attempt to overcome. Real progress has been made toward obtaining more general solutions in recent years [4].

### 4.9.1. The Patterson Series

One of the methods which is much used by crystallographers is that of the Patterson series which, contrary to the Fourier series of Eq. (*4.57*), utilizes only the experimentally determined quantity $|F_{hkl}|^2$.

Let us consider the expression

$$p = \rho(X, Y, Z)\rho(X + U, Y + V, Z + W) .$$

This is the product of the electron densities at $X, Y, Z$ and at the point separated from it by the vector $(U, V, W)$. This product is large when the density is appreciable at both points simultaneously. If we consider atoms which are nearly punctual, the product is different from zero only when the points $X, Y, Z$ and $X + U, Y + V, Z + W$ coincide with atoms, and then $U, V, W$ is an *interatomic vector.* Let us now calculate the integral of the product $p$ in the unit cell. We find that

$$\begin{aligned} P(U, V, W) &= \int_{V_c} \rho(X, Y, Z)\rho(X + V, Y + V, Z + W)dv \\ &= V_c \int \rho(X, Y, Z)\rho(X + U, Y + V, Z + W)dXdYdZ \qquad (4.58) \end{aligned}$$

In the case of point atoms, $P(U, V, W)$ is different from zero only when $U, V, W$ is a vector relating the positions of any two atoms. In particular, it is maximum for $U = V = W = 0$.

Let us replace the densities $\rho$ in Eq. (*4.58*) by their Fourier series as given in Eq. (*4.57*):

$$P(U, V, W)$$
$$= \frac{1}{V_c} \sum_h \sum_k \sum_l \sum_{h'} \sum_{k'} \sum_{l'} \int_0^1 \int_0^1 \int_0^1 F_{hkl} F_{h'k'l'} \exp[2\pi i(hX + kY + lZ)]$$
$$\times \exp[2\pi i(h'(X + U) + k'(Y + V) + l'(Z + W)] dX dY dZ .$$

All the terms of this sum are zero, except those where $h = -h$, $k' = -k$, $l' = -l$. Thus, in the remaining terms,

$$F_{h'k'l'} = F_{\bar{h}\bar{k}\bar{l}} = F^*_{hkl}$$

and

$$P(U, V, W) = \frac{1}{V_c} \sum \sum \sum |F_{hkl}|^2 \exp[-2\pi i(hU + kV + lW)] . \qquad (4.59)$$

We have thus found directly a result which is also demonstrated in Appendix A.5.1 by means of the faltung theorem. The function $P(U, V, W)$

**FIGURE 4.9.** *Two-dimensional crystal containing three identical atoms in the unit cell.* (a) *Crystal structure.* (b) *Patterson function for the crystal.*

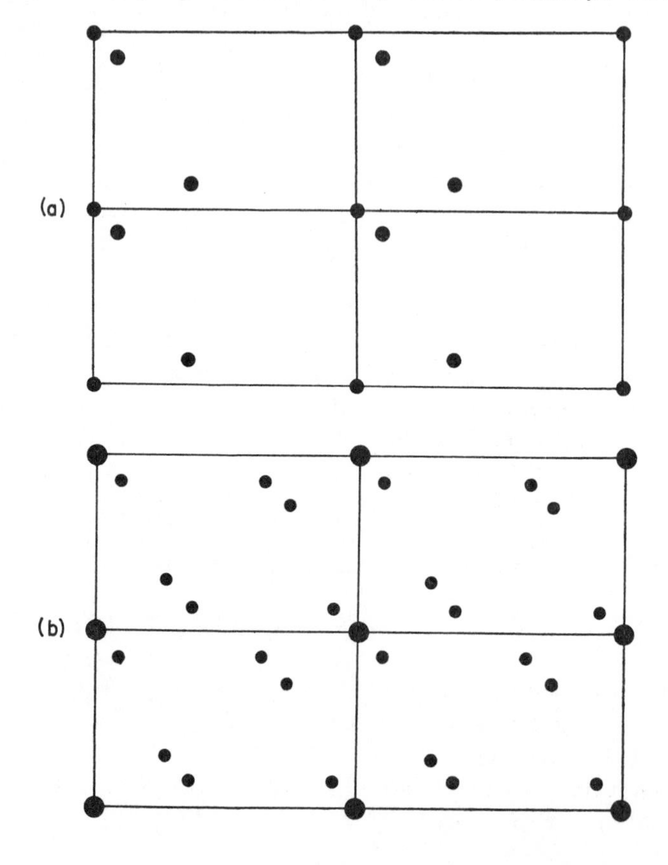

is called the *autocorrelation function for the electron density*. This function is also called the Patterson function and, from Eq. (*4.59*), it can be calculated directly from the experimental results. Figure 4.9 shows a Patterson function in a very simple two-dimensional case. It shows characteristics of the real structure and, in particular, the interatomic distances. However, in complex cases—in particular when the electrons are not part of well-separated atoms—the Patterson function does not give an explicit picture of the real structure.

The Patterson method in its present form provides one of the basic tools for determining crystal structures, but it is well known that it does provide an easy solution in all cases.

The diffraction pattern can give directly only the average value of the product of the electron densities, or the *simultaneous density* at any two points separated by a given vector. It is simply the application of a general result to the case of crystals.

## 4.10. Dynamic Theory of Diffraction by a Perfect Crystal

The *kinematic theory* which we have discussed above is based on interference calculations for the waves scattered by the individual atoms. It is assumed that the diffracted amplitude is always small so the interaction between the incident and the scattered waves can be neglected. This is not the case when the object is a perfect crystal whose lattice is coherent over a large region. In such a case a large number of waves are scattered in phase and the total scattered wave is then not negligible compared to the incident wave.

Darwin, Ewald, and von Laue [5] have proposed the *dynamic theory*, to take into account the conditions of propagation of X-rays in a perfect crystal. The results of this theory are quite different from those of the kinematic theory. It has been verified, in particular, for silicon and germanium, which can now be obtained in the form of rigorously perfect crystals. However, these are very special types of crystals and the dynamic theory is rather restricted in its applications. Since, on the other hand, it is rather complex, we shall merely state the basic assumptions and the results of Darwin which are based on the simplest form of this theory. We shall then give a few interesting applications to the physics of solids.

Let us consider the face of a crystal which is parallel to an infinite set of identical lattice planes rigorously parallel and equidistant from each other. We consider a parallel and monochromatic beam of X-rays incident at an angle $\theta$ on this face and we calculate the intensity of the beam reflected at the angle $\theta$, as a function of $\theta$.

Inside the crystal we have two waves, one propagating in the direction of the incident ray and another propagating in the direction of the reflected ray. At each lattice plane, each wave produces both transmitted and reflected waves. The wave $S_p$ of Figure 4.10 between the planes $p$ and $p+1$ therefore results from the superposition of the transmitted wave corresponding to $S_{p-1}$ and of the wave reflected by the plane $p$ corresponding to the incident wave $T_p$. Since the reflected ray is $\pi/2$ radians out of phase with the incident wave, a wave which is reflected twice is $\pi$ radians out of phase with the incident wave. The incident wave which had first lost energy through reflection is thus further reduced in amplitude by the doubly reflected wave. The result is that the incident wave is attenuated much more rapidly than one would expect from the usual photoelectric absorption: there is *extinction*. Thus, in the case of the reflection of the Mo K$\alpha$ radiation on the planes (200) of NaCl, the intensity is reduced by a factor of 2 after crossing 2,000 lattice planes, or in a distance of 0.56 $\mu$, while the ordinary type of absorption for this distance is only 7/1,000. The active layer of the crystal is therefore very thin and the reflecting power calculated according to this theory is consequently much smaller than with the kinematic theory. It is even permissible to neglect, as a first approximation, the usual absorption. The results of Darwin are the following.

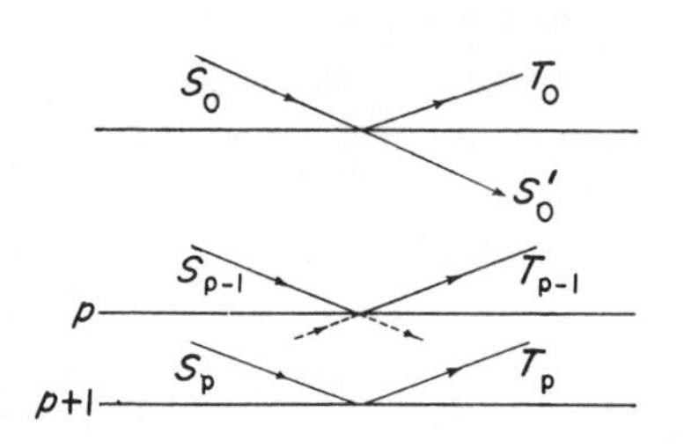


**FIGURE 4.10.** *Wave propagation in a perfect crystal.*

(a) As $\theta$ is varied, the reflected intensity is zero except in the immediate neighborhood of angles $\theta$ defined by the Bragg relation $n\lambda = 2d \sin \theta$.

(b) In these regions we have total reflection[1] and all the incident energy appears in the reflected beam. The width of the region of total reflection is proportional to the reflecting power of a single lattice plane. It is always very small—of the order of a few seconds of arc.

(c) The average angle of reflection is given by the Bragg formula corrected for the refraction effect, but this correction is always very small.

As we have seen previously (Section 1.1.2), the index of refraction for

[1] When absorption is taken into account, the reflection is not total but is still quite large.

X-rays is slightly less than unity: $n = 1 - \delta$ [Eq. (*1.1*)]. A ray incident at an angle $\theta$ therefore penetrates inside at an angle $\theta - d\theta$ such that, from Snell's law,

$$\cos\theta = n\cos(\theta - d\theta) = (1 - \delta)(\cos\theta + \sin\theta\, d\theta)\,. \qquad (4.60)$$

The waves diffracted by the successive planes are in phase if

$$p\lambda = 2nd\sin(\theta - d\theta)\,, \qquad (4.61)$$

where $p$ is a whole number. From Eqs. (*4.60*) and (*4.61*), we have

$$p\lambda = 2d\sin\theta\left(1 - \frac{\delta}{\sin^2\theta}\right). \qquad (4.62)$$

The correction term is of the order of $10^{-5}$.

The reflecting power of a crystal, as defined by Eq. (*4.34*), is proportional to the area under the curve of the reflected intensity as a function of $\theta$. Since the region of total reflection is very narrow, the reflecting power is small, despite the fact that the crystal is a perfect mirror for one particular angle.

We shall use the following formulas.

(a) Consider a small crystal of volume $dV$ cut in the form of a thin plate parallel to the $(hkl)$ reflecting planes and of thickness $t$ [6]; its reflecting power, according to the kinematic theory, is $Q\,dV$, where $Q$ is given by Eq. (*4.40*). According to the dynamic theory, $Q$ must be replaced by $Q'$:

$$Q' = Q\,\frac{\tanh\left\{2t\left[\dfrac{Q\cot\theta}{\lambda(1+\cos^2 2\theta)}\right]^{1/2}\right\}}{2t\left[\dfrac{Q\cot\theta}{\lambda(1+\cos^2 2\theta)}\right]^{1/2}}\,. \qquad (4.63)$$

**FIGURE 4.11.** *Correction factor for the primary extinction effect. Ratio of the reflecting powers of a plate of NaCl according to the dynamic and kinematic theories as a function of the plate thickness for the 200, 400, and 600 reflections.*

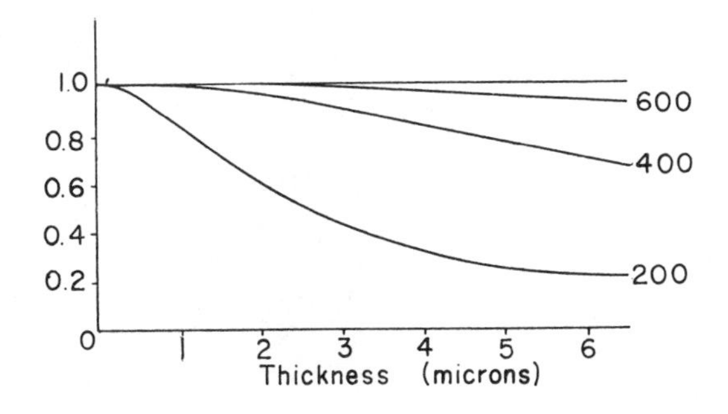

The correction factor, which is equal to unity for small $t$, decreases with increasing $t$, especially when $Q$ is large (Figure 4.11). This phenomenon is called *primary extinction*, as opposed to another extinction phenomenon which we shall meet later on.

(b) For a semi-infinite crystal, the reflecting power of the face parallel to the $(hkl)$ planes is

$$P_d = \frac{4}{3\pi}\frac{\lambda^2}{V_c}\left|F_{hkl}\right| r_e^2 \frac{1 + |\cos(2\theta)|}{\sin 2\theta}, \tag{4.64}$$

where $V_c$ is the volume of the unit cell;

$|F_{hkl}|$ is the modulus of the structure factor for the lattice plane $(hkl)$;

$\theta$ is the angle of reflection;

$r_e^2$ comes from the Thomson formula (*1.23*) for the scattering by a free electron;

$1 + |\cos 2\theta|$ is the polarization factor, the primary beam being assumed to be unpolarized.

The reflecting powers for a crystal face, as predicted by the kinematic [Eq. (*4.46*)] and dynamic [Eq. (*4.64*)] theories are fundamentally different; they involve different quantities under different forms. Figure 4.12 shows

**FIGURE 4.12.** *Reflecting power $P_k$ and $P_d$ for a cleavage plane of calcite according to the kinematic and dynamic theories, as a function of wavelength. The points show experimental values.*

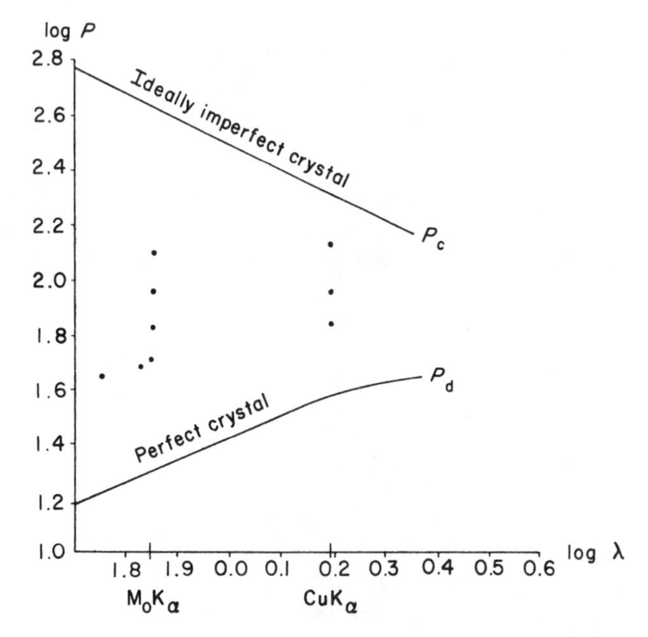

these two reflecting powers calculated as a function of $\lambda$ for the cleavage face of calcite. *The dynamic theory gives a much smaller reflecting power than the kinematic theory*—about 20 times less for Mo K$\alpha$ radiation.

How do these results compare with experiment? With a perfectly pure crystal of germanium containing a very low density of dislocations, absolute measurements of the reflecting power agree with the dynamic theory within a few percent [7].

In the case of calcite we find very different results, depending on the sample and especially on the condition of its surface, as shown in Figure 4.12. It is found that the simple fact of rubbing the surface lightly produces a considerable increase in the reflecting power. However, the experimental results are always situated somewhere between the values given by the dynamic and the kinematic theories for most types of crystals. In the case of distorted or broken crystals, the results agree with those of the kinematic theory.

In conclusion, the two theories apply to two limiting cases which are the only ones accessible to calculations; experimental results range from one to the other. There are crystals which satisfy one or the other theory, but the perfect crystal is very rare, while the "ideally imperfect" crystal is a fair approximation for most samples. But there are often cases where the reflecting power is lower than predicted by the kinematic theory, especially for the most intense reflections corresponding to small Bragg angles. This was shown, for example, by the measurements of James et al. [8] on an aluminum crystal, as illustrated in Table 4.2.

TABLE 4.2. Reflecting Power for the Lattice Planes of an Aluminum Crystal (Mo K$\alpha$ Radiation)

| Reflection Planes ($hkl$) | Bragg Angle $\theta$ | Reflecting Power | | |
|---|---|---|---|---|
| | | Dynamic Theory, $P_d$ Eq. (4.64) | Kinematic Theory, $P_k$ Eq. (4.46) | Experimental Value |
| 111 | 8°43′ | 19.6 | 818 | 580 |
| 200 | 10° 6′ | 16.2 | 619 | 436 |
| 222 | 17°40′ | 6.3 | 158 | 144 |
| 400 | 20°32′ | 4.47 | 91 | 86.4 |
| 333 | 27° 6′ | 2.19 | 28.3 | 26.2 |
| 600 | 31°45′ | 1.31 | 12.0 | 12.2 |
| 444 | 37°23′ | 0.76 | 5.14 | 4.9 |
| 800 | 44°33′ | 0.40 | 2.09 | 2.1 |
| 555 | 49°23′ | 0.37 | 1.39 | 1.43 |

**4.10.1. The Corrections for Primary and for Secondary Extinction.** It is possible to explain the fact that the reflecting power of a real crystal is smaller than that of an "ideally imperfect" crystal if we use the model of the mosaic structure,

(a) If the individual blocks are too large, their reflecting power is reduced by extinction inside each block. As we have said, this is primary extinction. The formula (4.63) gives the order of magnitude of the reduction of the reflected intensity as a function of the thickness $t$ of the mosaic blocks. It can be seen from Figure 4.11 that the correction is large for planes which have a large reflecting power (low indices), when the block size exceeds one micron.

(b) Let us assume that the mosaic blocks are very small so there is no primary extinction. We assume that they are all parallel, although irregularly spaced as in Figure 4.13. Under these conditions all the blocks are simultaneously in the reflection position for a given orientation of the sample. The incident intensity is reduced in the thickness $dx$, not only because of ordinary absorption, but also because part of the energy has been reflected by the blocks situated in the layer $dx$. For a beam of unit cross-section of intensity $I_0$ incident at an angle $\theta$, the energy absorbed is $I_0\mu\rho(dx/\sin\theta)$. Since the irradiated volume is $dx/\sin\theta$, the reflected energy is then $I_0Q(dx/\sin\theta)$. As a first approximation, it is as if the effective linear absorption coefficient were $\mu\rho + Q$ instead of $\mu\rho$. From Eq. (*4.46*) the reflecting power of the crystalline face is then

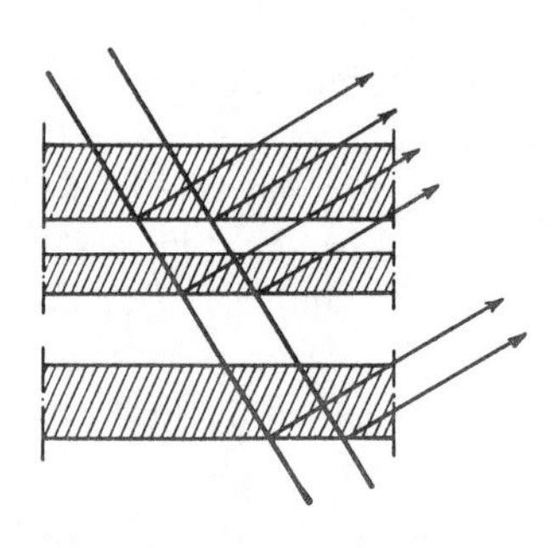

**FIGURE 4.13.** *Crystal with parallel but irregularly spaced layers.*

$$P = \frac{Q}{2(\mu\rho + Q)}\,. \tag{4.65}$$

It is therefore smaller than for the ideally imperfect crystal.

This effect, which is called *secondary extinction*, is more important when the reflecting power of the planes is large, as for primary extinction.

In a mosaic structure such as that illustrated in Figure 4.6, the blocks are not parallel, and when one block is in the reflection position only a small fraction of the others are oriented likewise and act as a shield as indicated above. In this case the effect of secondary extinction is written as

$$P = \frac{Q}{2(\mu\rho + gQ)} . \tag{4.66}$$

where $g$ is a coefficient which is equal to unity if all the blocks are parallel and equal to zero when their orientation is random.

In the case of primary extinction, the lattice planes close to the surface act as a shield for the others because of the energy which they reflect back. In the case of secondary extinction, the same effect is produced by whole blocks. Their action is not as large since they are not coherent, while there is perfect coherence between the lattice planes of a perfect crystal. It is the intensities and not the amplitudes which subtract. When both primary and secondary extinction exist, $Q$ must be replaced by $Q'$ [Eq. (*4.63*)] in the above formula (*4.66*).

If $P_0$ is the reflecting power of the ideally imperfect crystal, we have

$$P_0 = \frac{Q}{2\mu\rho} .$$

If only secondary extinction occurs, the measured reflecting power is

$$P = \frac{Q}{2(\mu\rho + gQ)} ,$$

and thus

$$P_0 = \frac{P}{1 - 2gP} . \tag{4.67}$$

When $g$ is known, Eq. (*4.67*) permits us to deduce the corrected value from the experimental value. The different methods for determining the secondary extinction coefficients $g$ will be found in the treatise by James [6].

The results of the measurements of Bragg, James, and Bosanquet [9] for rock salt will show the importance of these corrections. In the case of the Rh $K\alpha$ radiation, $\mu\rho$ is equal to 10.70 and the effective coefficient of absorption, $(\mu\rho + gQ)$, was found experimentally to be equal to 16.30 for the 200 reflection, 12.66 for the 400 reflection, and 10.72 for the 600 reflection. It is only in this last case that the extinction is negligible.

### 4.10.2. Applications of the Dynamic Theory

One first consequence of the dynamic theory is that there can be large errors in the determination of structure factors when there are no particular reasons for expecting the extinction effects to be negligible. Now, in the determination of crystal structures it is often necessary to use fairly large crystals ($\sim$1 mm). The intensities of the most intense reflections can thus be considerably reduced and it is very difficult to make proper corrections; this is one factor which limits the accuracy of structure determinations [10].

On the other hand, dynamic effects have an interesting application in the field of solid-state physics. Since the reflecting power of the crystal varies widely with its perfection, one can imagine that such a measurement could permit the detection of crystal faults. In fact, one reaches nearly the ideally imperfect state when the size of the blocks of coherent lattice is less than a few microns; from then on, a larger distortion in the crystal has no effect. This method will therefore be most sensitive to the first changes in a crystal which was originally very perfect. However, Batterman [7] has observed that the reflecting power of germanium remains practically equal to that of a perfect crystal so long as the number of dislocation lines remains less than $10^3/cm^2$.

The most important use has been that of localizing regions of imperfections in a "good" crystal. There are several methods for obtaining images of a crystal by diffraction, one of the most important being that of Berg-Barrett, illustrated in Figure 4.14 [11]. Let us consider, for example, a crystal face illuminated by a linear source situated at a large distance—say, 30 cm—the crystal being oriented so that the characteristic radiation $K\alpha$ is reflected. If we place a photographic film parallel to the surface and as close as possible to the crystal, the rays reflected by a point in the crystal traverse the film over a very small area so that the film gives a true image of the surface of the crystal, with a resolving power which can be of the order of a few microns. It is also possible to obtain the image of a thin crystal plate by transmission [12].

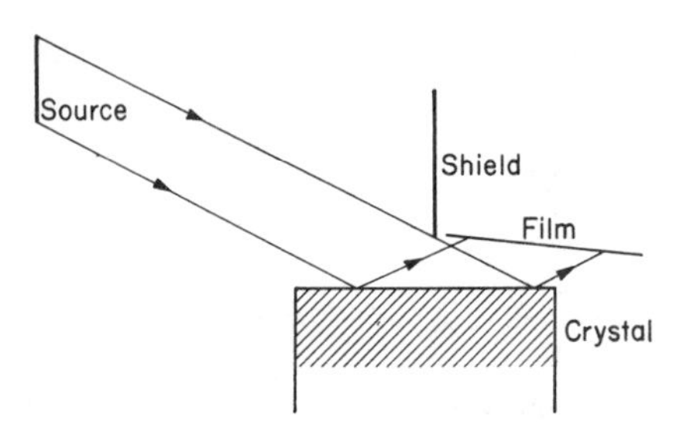


**FIGURE 4.14.** *The Berg-Barrett method for obtaining an image of the face of a crystal.*

The regions which contain imperfections appear black, because the energy reflected at these points is larger than in the good part of the crystal. The contrast can be very great and the slightest scratch shows up clearly. Lang has even been able to observe individual dislocation lines in a silicon crystal [13]. The subboundaries formed by a dislocation wall and the slip planes are also visible [14]. However, in the present state of the theory, these experiments can only be qualitative, since it it is impossible to deduce the nature of the crystal imperfection from the change in reflecting power.

REFERENCES

1. E. J. W. WITTAKER, *Acta. Cryst.*, **6** (1953), 222. L. V. AZAROFF. *Acta Cryst.*, **8** (1955), 701.
2. A. GUINIER, *X-Ray Crystallographic Technology*, Hilger and Watts, London (1952), p. 166.
3. *Handbuch der Physik*, Vol. XXXII, Springer Verlag, Berlin (1957), p. 58.
4. M. J. BUERGER, *Crystal Structure Analysis*, WILEY, New York (1960).
5. M. VON LAUE, *Die Interferenzen von Röntgen und Elektronenstrahlen*, J. Springer, Berlin (1935). R. W. JAMES, *The Optical Principles of the Diffraction of X-Rays, Bell*, London (1948). W. H. ZACHARIASEN, *Theory of X-Ray Diffraction by Crystals*, Wiley, New York (1945).
6. R. W. JAMES, *The Optical Principles of the Diffraction of X-Rays, Bell*, London (1948), p. 272.
7. B. W. BATTERMAN, *J. Appl. Phys.*, **30** (1959), 508. J. R. PATEL, R. S. WAGNER, and S. MOSS. *Acta Met.*, **10** (1962), 759.
8. R. W. JAMES, G. W. BRINDLEY and R. C. WOOD, *Proc. Roy. Soc. London, A*, **125** (1929), 401.
9. W. L. BRAGG, R. W. JAMES and C. H. BOSANQUET, *Phil. Mag.*, **41** (1921), 309; **42** (1921), 1; **44** (1922), 433.
10. K. LONSDALE, *Miner. Mag.*, **28** (1947), 14. L. GATINEAU and J. MERING, *C.R. Acad. Sci., Paris*, **242** (1956), 2018. V. VAND. *J. Appl. Phys.*, **26** (1955), 1191. S. CHANDRASEKHAR, *Acta Cryst..*, **9** (1956), 954.
11. C. S. BARRETT, *Structure of Metals*, McGraw-Hill, New York (1952), p. 96.
12. A. R. LANG, *Acta Met.*, **5** (1957), 358; *J. Appl. Phys.*, **29** (1958), 597.
13. J. B. NEWKIRK, *J. Appl. Phys.*, **29** (1958), 995.
14. Z. NISHIYAMA and S. HAYAMI, *Trans. Jap. Inst. Met.*, **1** (1960), 9.

CHAPTER

# 5

# DIFFRACTION BY VERY SMALL CRYSTALS

As in the preceding chapter, our discussion will again be based on the results of calculations on the reciprocal space corresponding to the object [Section 4.4, Eq. (*4.28*)]. We shall now consider the case where the diffraction domain surrounding each lattice point is large enough to be investigated, despite the limitations inherent to any experiment. This will be the case if the crystal is at most of the order of $0.1\mu$ in size.

The diffraction spots obtained with single crystals, as well as the lines given by crystalline powders, become *broader* as the crystal size decreases. This phenomenon provides an experimental method for determining the size of submicroscopic crystals. Let us first establish in a very elementary way the effect of crystal size on the diffraction pattern.

## 5.1. Elementary Theory. The Scherrer Equation

The effect of crystal size on the diffraction pattern is analogous to a well-known phenomenon connected with optical gratings: the position of the diffraction maximum depends only on the groove spacing, but the width of the line decreases as the number of grooves increases [1]. In the case of X-rays it is the number of reflecting lattice planes which is the important factor. The following elementary calculation is identical to that for the resolving power of optical gratings [2].

Let us consider a crystal having the form of a thin plate of large area consisting of $N$ lattice planes of spacing $d$, as in Figure 5.1. We con-

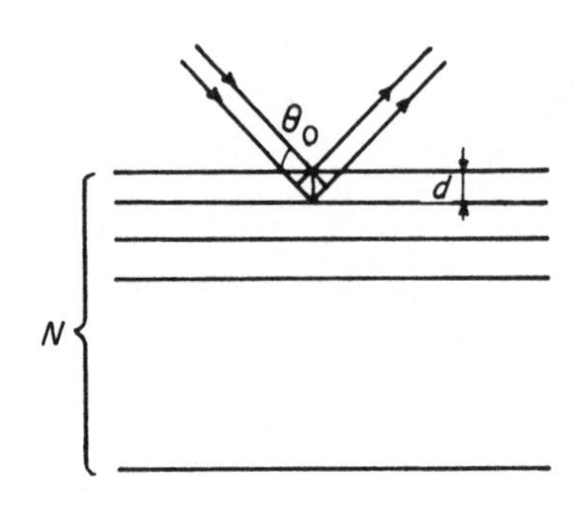


**FIGURE 5.1.** *Diffraction by N lattice planes.*

sider an incident ray of wavelength $\lambda$ at the Bragg angle of incidence $\theta_0$, such that $\lambda = 2d \sin \theta_0$. If $A$ is the amplitude diffracted by a single lattice plane, then the reflected ray has an amplitude $NA$ since the waves originating at the $N$ planes are exactly in phase. Now let the angle of incidence change to $\theta_0 + \varepsilon$. The path difference from one plane to the following is then

$$2d \sin(\theta_0 + \varepsilon) = \lambda + 2d\varepsilon \cos \theta_0 .$$

We must therefore add together $N$ waves of the same amplitude but with phase differences of

$$2\Phi = \frac{4\pi d}{\lambda} \varepsilon \cos \theta_0 . \tag{5.1}$$

The Fresnel construction of Figure 5.2 shows that the vectors representing the wavelets determine a circle of radius $A/(2 \sin \Phi)$, each vector subtending an angle $2\Phi$ at the center. The sum of all these vectors subtends an angle of $2N\Phi$, and its modulus, which is equal to the required amplitude, is

$$\mathscr{A} = A \frac{\sin N\Phi}{\sin \Phi} = A \frac{\sin(2\pi N d\varepsilon \cos \theta_0/\lambda)}{\sin(2\pi d\varepsilon \cos \theta_0/\lambda)} . \tag{5.2}$$

**FIGURE 5.2.** *Fresnel construction for the amplitude resulting from the diffraction by N lattice planes.*

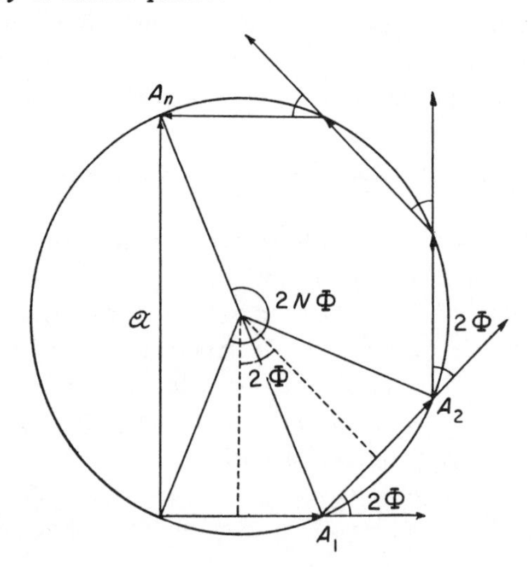

**FIGURE 5.3.** *The function* $\frac{\sin^2 N\Phi}{(N\Phi)^2} \simeq \frac{1}{N^2}\frac{\sin^2 N\Phi}{\sin^2 \Phi}$ *for small values of* $\Phi$.

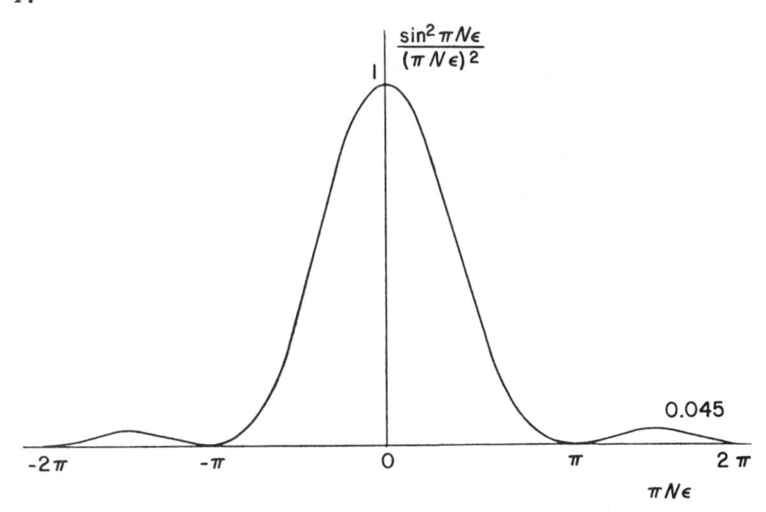


The intensity reflected in the direction $\theta_0 + \varepsilon$ is

$$I(\varepsilon) = A^2 \frac{\sin^2 N\Phi}{\sin^2 \Phi} .$$

This function, which is found in many diffraction calculations, is shown in Figure 5.3. It becomes nearly zero for $N\Phi > \pi$ or for

$$\varepsilon > \frac{\lambda}{2Nd \cos \theta_0} .$$

The width of the angular domain of reflection is thus inversely proportional to $Nd$, or to the thickness of the crystal.

The line width is defined in two different ways. First, we can consider the value of $\varepsilon$, or of $\Phi$ as given in Eq. (*5.1*), such that the intensity is one half of its maximum value. This $\Phi$ is such that

$$\frac{\sin^2 N\Phi}{\sin^2 \Phi} \cong N^2 \frac{\sin^2 (N\Phi)}{(N\Phi)^2} = \frac{N^2}{2} .$$

From tables for this function we find that

$$N\Phi = 0.444\pi ,$$

and therefore

$$\varepsilon_{(I=1/2)} = \frac{0.222\lambda}{Nd \cos \theta_0} .$$

If the angle of incidence increases by $\varepsilon$, the angle of diffraction between the incident and reflected rays increases by $2\varepsilon$. The angular width at half maximum is then $\Delta'(2\theta) = 4\varepsilon$ and, if $L = Nd$ is the thickness of the crystal,

$$\Delta'(2\theta) = \frac{0.9\lambda}{L \cos \theta_0} . \tag{5.3}$$

The second definition of line width is more convenient for theoretical calculations: the *integral width of the line* is that of a line of rectangular profile which would have the maximum and integral values of the observed line. Thus

$$\Delta(2\theta) = \frac{2\int I(\varepsilon)d\varepsilon}{I_{\max}} = \frac{2\lambda}{2\pi d \cos \theta_0} \cdot \frac{\int I(\Phi)d\Phi}{I(0)}$$

Now

$$I(0) = N^2 A^2$$

and

$$\int I(\Phi)d\Phi = A^2 \int_{-\infty}^{+\infty} \frac{\sin^2 N\Phi}{\Phi^2} d\Phi = N\pi A^2 ,$$

so that

$$\Delta(2\theta) = \frac{\lambda}{L \cos \theta_0} . \tag{5.4}$$

It will be seen from Eqs. (*5.3*) and (*5.4*) that the two definitions are quite similar, because of the shape of the function $I(\varepsilon)$.

The second formula, which we shall normally use, is called the *Scherrer formula*. We have demonstrated it in the special case of a thin plate of crystal cut parallel to the lattice planes. In the case of a line in a powder pattern whose observed integral width is $\Delta(2\theta)$, the quantity $\Delta(2\theta)$ corresponds to a length $L$, according to Eq. (*5.4*). The calculations of the following paragraphs will show that this is a good approximation for the size of the powder grains, and we shall call it the *apparent size* of the grains.

For $\theta = 20°$, $\lambda = 1.54$Å (Cu K$\alpha$), and with a camera 60 mm in radius, the line width is 0.2 mm for $L = 500$Å. Because of the line width arising from geometrical factors, this is about the minimum width which can be observed without special precautions. It is therefore not easy to use the widths of the Debye-Scherrer lines to determine the sizes of small crystals larger than a few hundred Ångströms. For very small crystals, however, the lines become very broad and weak, and therefore difficult to observe. For example, with the above experimental conditions and $L = 10$Å, the line has a width of 8 mm: it is a broad diffuse ring which can hardly be distinguished from the background. The method can therefore be applied to crystals of approximately 10 to 1,000 Å in size (Figure 5.8).

## 5.2. Case of an Isolated Small Crystal

The exact calculation of the effect of the small size of a crystal on its diffraction pattern is to be found in the general theory discussed in Section 4.3. It will be recalled that the reciprocal space is obtained by the superposition of similar domains centered on each of the nodes of the reciprocal lattice, as in Figure 5.4. Equation (*4.28*) gives the scattering power per unit cell:

$$I(\mathbf{s}) = \frac{F_{hkl}^2}{VV_c} \sum_h \sum_k \sum_l |\Sigma(\mathbf{s} - \mathbf{r}_{hkl}^*)|^2, \tag{5.5}$$

where $V$ is the volume of the diffracting crystal containing $N$ unit cells of volume $V_c$, $\Sigma(\mathbf{s})$ is the Fourier transform of $\sigma(\mathbf{x})$, the form function which is equal to unity inside the crystal and to zero outside, $\mathbf{r}_{hkl}^*$ is the vector defining the node of indices $hkl$ of the reciprocal lattice, and $F_{hkl}$ is the structure factor related to this node.

From the general properties of the function $\Sigma(\mathbf{s})$, described in Section 2.4.2, we can draw several important conclusions which are independent of the shape of the small crystal. Since the form function $\sigma(\mathbf{x})$ is real, its transform $\Sigma(\mathbf{s})$ is such that

$$|\Sigma(\mathbf{s})| = |\Sigma(-\mathbf{s})|$$

and *the reflection domain in reciprocal space is always centrosymmetric.*

**FIGURE 5.4.** *Schematic diagram of the reflection domain at the nodes of the reciprocal lattice in the case of small crystals.*

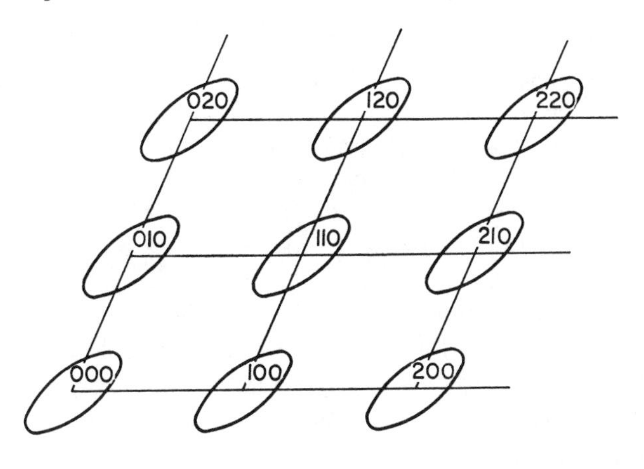

We have shown [Eq. (*2.37*)] that the maximum value of $\Sigma(\boldsymbol{s})$ just at the node is equal to $V$, so the maximum scattering power at the node is

$$I_{\max} = \frac{VF^2_{hkl}}{V_c} = NF^2_{hkl} \,. \tag{5.6}$$

The integral value of $\Sigma(\boldsymbol{s})$ in the domain surrounding the node, according to Eq. (*2.38*), is equal to $V$. Thus the integral of the scattering power is

$$\int I(\boldsymbol{s})\, dv_s = \frac{F^2_{hkl}}{V_c} \,. \tag{5.7}$$

If we define the width of the reflection domain by generalizing the definition of the integral width of a line, we can write

$$\varepsilon = \frac{\int I(\boldsymbol{s}) dv_s}{I_{\max}}$$

or, from Eqs. (*5.6*) and (*5.7*),

$$\varepsilon = \frac{1}{V} \,. \tag{5.8}$$

The reflection domain therefore becomes narrower as the crystal increases in size. Let us compare $\varepsilon$, which has the dimensions of a volume in reciprocal space, to the volume $V^*$ of the unit cell in the reciprocal lattice. From Eq. (*4.21*), $V^* = 1/V_c$ and therefore

$$\frac{\varepsilon}{V^*} = \frac{\varepsilon}{1/V_c} = \frac{1}{N} \,.$$

Thus, if the crystal contains at least about ten unit cells, the reflection domains become small enough to justify the assumption that they are well separated.

The Scherrer formula given in Eq. (*5.3*) shows that the broadening of the diffraction lines increases with the Bragg angle $\theta_0$, and we find that the shape of the reflection domain is independent of the lattice point indices. There is no contradiction between these results. Let us consider, as a very schematic example, that the reflection domain has the form of a small sphere of diameter $a$. In reciprocal space we then have, corresponding to the crystalline powder, the superposition of a large number of lattices obtained by rotating one of them about the center. It is its intersection with the Ewald sphere during this motion which determines the observed line. Figure 5.5 shows how identical domains can produce increasingly broad lines as the angle $\theta_0$ approaches $\pi/2$. The angular width of the line is equal to

$$\frac{MM'}{OM} = \frac{a}{\cos\theta_0(1/\lambda)} = \frac{\lambda a}{\cos\theta_0}$$

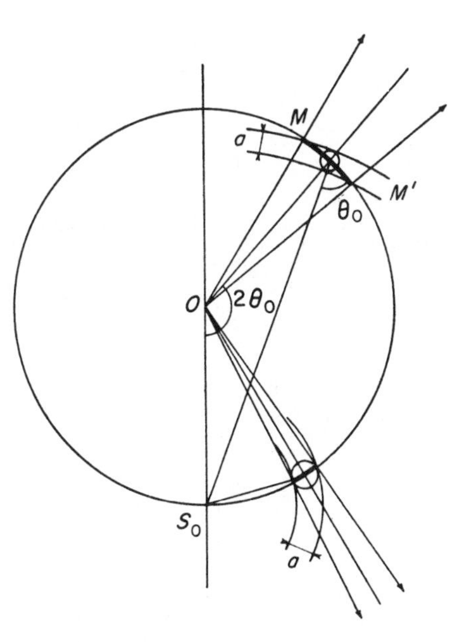


**FIGURE 5.5.** *The width of the Debye-Scherrer line corresponding to a given reflection domain as a function of the Bragg angle.*

and, since $a$ is inversely proportional to the size of the crystal, we again have the Scherrer formula.

It is interesting to note that the center of reciprocal space, which is the lattice point (000), is surrounded by the same domain as the other lattice points (Figure 5.4). One would expect the small crystal to produce scattered radiation in the direction close to that of the incident beam. However, this small-angle scattering is not characteristic of the crystalline structure of the object but rather comes from the existence of a sharp central peak in the diffraction pattern (Section 2.4.2). This peak depends on the shape of the object, as shown in the general Eq. (*2.42*), which is also valid for noncrystalline bodies. We shall investigate this small-angle scattering in Chapter 10.

Let us calculate the reciprocal lattice for a cubic crystal having the shape of a parallelepiped with sides $N_a\boldsymbol{a}$, $N_b\boldsymbol{b}$, $N_c\boldsymbol{c}$ $(a = b = c)$. Let us call $x_a$, $x_b$, $x_c$ the projections of the vector $\boldsymbol{x}$ on the crystal axes. The form function $\sigma(\boldsymbol{x})$ is equal to unity in the volume where, simultaneously,

$$|x_a| < \frac{N_a a}{2},$$

$$|x_b| < \frac{N_b a}{2},$$

$$|x_c| < \frac{N_c a}{2}.$$

To calculate the Fourier transform,

$$\Sigma(\boldsymbol{s}) = \int \sigma(\boldsymbol{x}) \exp(-2\pi i \boldsymbol{s}\cdot\boldsymbol{x})\, dv_x ,$$

we define $\boldsymbol{s}$ by its projections $s_a$, $s_b$, $s_c$ on the axes of the reciprocal lattice:

$$\Sigma(\mathbf{s}) = \int_{\text{crystal}} \exp\left[-2\pi i(s_a x_a + s_b x_b + s_c x_c)\right] dv_x$$

$$= \int_{-N_a a/2}^{+N_a a/2} \exp(-2i\pi s_a x_a)\, dx_a \int_{-N_b a/2}^{+N_b a/2} \exp(-2\pi i s_b x_b)\, dx_b \times \int_{-N_c a/2}^{+N_c a/2} \exp(-2\pi i s_c x_c)\, dx_c\,.$$

This gives

$$|\Sigma(\mathbf{s})|^2 = \frac{\sin^2(\pi N_a s_a a)}{\pi^2 s_a^2} \frac{\sin^2(\pi N_b s_b a)}{\pi^2 s_b^2} \frac{\sin^2(\pi N_c s_c a)}{\pi^2 s_c^2}$$

and

$$\mathscr{I}(\mathbf{s}) = \frac{1}{VV_c} |\Sigma(\mathbf{s})|^2$$

$$= \frac{1}{N_a N_b N_c} \frac{\sin^2(\pi N_a s_a a)}{(\pi s_a a)^2} \frac{\sin^2(\pi N_b s_b a)}{(\pi s_b a)^2} \frac{\sin^2(\pi N_c s_c a)}{(\pi s_c a)^2}\,. \tag{5.9}$$

Equation (*5.9*) is obtained by neglecting all the terms of the rigorous Eq. (*5.5*) except that which is related to the nearest node. The exact calculation [3] leads to a formula similar to (*5.9*) but containing in the denominator $[\sin(\pi s_a a)]^2$ instead of $(\pi s_a a)^2$, and so on. For not too small crystals the results are practically equivalent, because each one of the

**FIGURE 5.6.** *The reflection domain for a small crystal having the form of a needle, at* (a), *and of a small plate, at* (b).

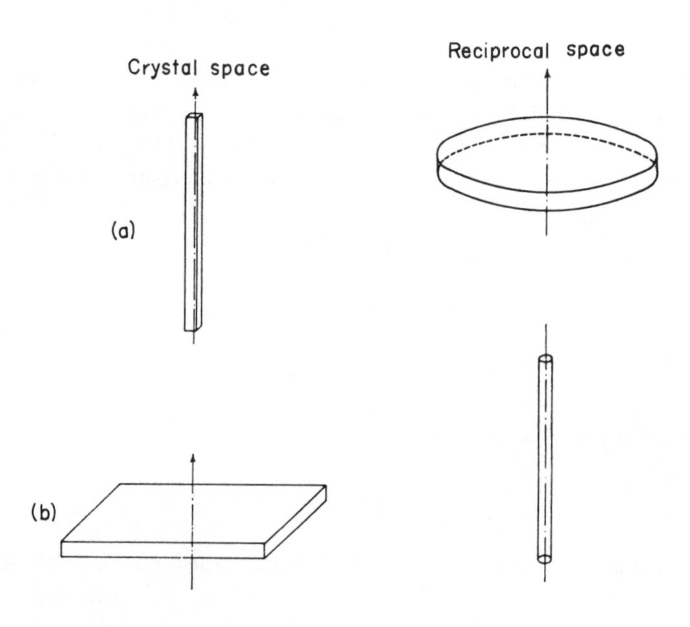

factors of the product appearing in Eq. (*5.9*) is nearly zero for $s_i > 1/N_i a$. However, because of the function $\sin^2 \Phi/\Phi^2$, shown in Figure 5.3, we can expect weak secondary maxima. These have been observed experimentally with very thin crystals obtained by evaporation: the spacing of the secondary maxima gives the thickness of the crystals [18].

If the crystals have the form of thin plates of large area ($N_a$ small, $N_b$ and $N_c$ large), the reflection domain, according to Eq. (*5.9*), is elongated like a needle of length $1/N_a a$ and is perpendicular to the plane of the crystal. Inversely, if the crystals are needle shaped ($N_a$ large, $N_b$ and $N_c$ small), the reflection domain has the form of a plate which is perpendicular to the axis of the needle and whose diameter is of the order of $1/N_b a$. The reciprocity existing between the functions and their Fourier transforms is illustrated in Figure 5.6.

We can show from the general properties of Fourier transforms that, when the crystal has plane faces, the surfaces of equal intensity in the

**FIGURE 5.7.** *Diffraction pattern for a square diaphragm obtained by optical means* (*Françon*).

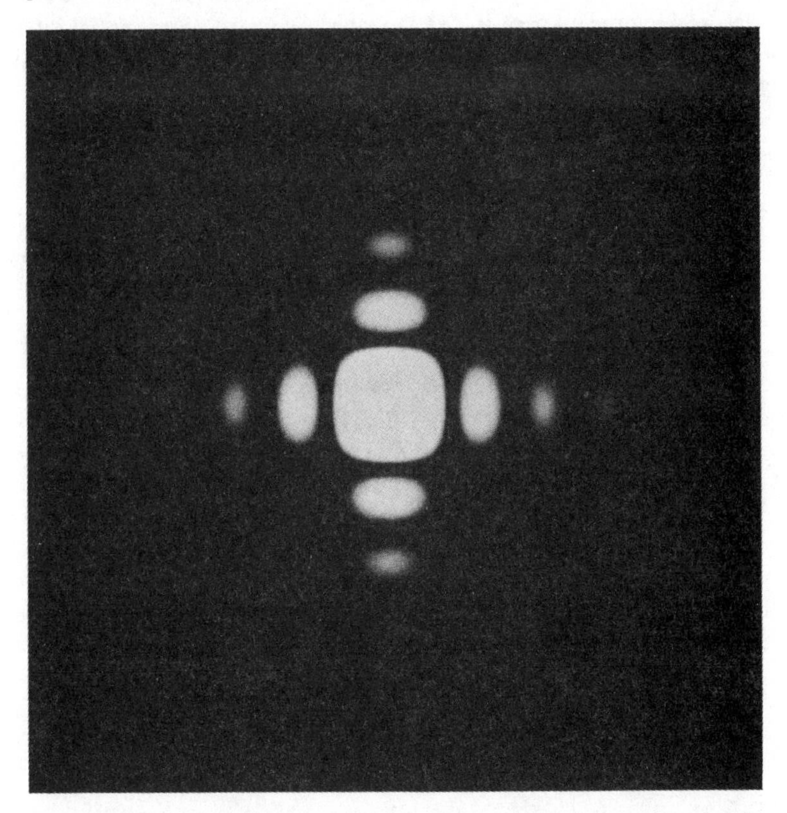

reflection domain have a pair of opposite peaks which are normal to the crystal faces and whose height increases as the area of the faces increases. The domains increase in size in a given direction as the diameter of the crystal decreases in that same direction.

It is interesting to use an optical analogy to find the shapes of the reflection domains [4]. For example, Figure 5.7 shows the diffraction pattern for a square diaphragm: it has a series of secondary maxima in the directions of the smaller diameters of the object. With X-rays, generally, the only observable domain is that limited by the first minimum.

It is not possible experimentally to observe directly the diffraction patterns of an isolated crystal which is small enough that the reflection domain can have an appreciable size, for it would have to be smaller than $0.1\mu$. However, it is in some cases possible to obtain a large number of small crystals all oriented in the same direction, as in the precipitated phase which grows inside a crystal of a metallic solid solution with the "Widmanstätten structure" [5]. The diffraction pattern of such an ensemble is then identical to that of an individual crystal, and intense enough to be observed. If we then obtain successive patterns of the section of reciprocal space in the neighborhood of a node of the lattice for the precipitated phase, it is possible to reconstruct the shape of the reflection domain and thus the size and shape of the grain.

## 5.3. Diffraction Lines for a Crystalline Powder

Small crystals are generally available in the form of a powder with random orientation. Corresponding to the broadened reflection domain, we have broad Debye-Scherrer lines centered exactly on the positions of the lines obtained with a powder composed of larger crystals. This is shown in Figure 5.8.

One can see a priori that it is impossible to determine the shape of the crystals from the powder pattern alone, while this would be possible if we knew the shape of the reflection domains themselves. The problem is to extract from the experimental data on the width and shape of the line a maximum amount of information on the size and shape of the

**FIGURE 5.8.** *Diffraction patterns obtained with nickel.* (A) *Rolled sheet.* (B) *Catalytic nickel powder consisting of very small crystals. The lines are broadened and there is a central spot* (*Chapter 10*).

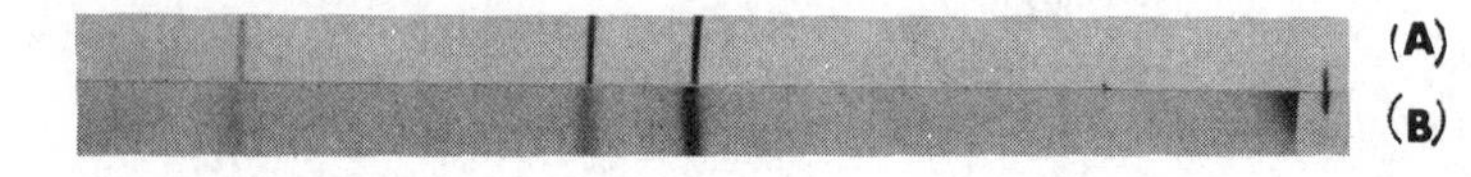

elementary crystals in the powder.

The reciprocal space for the powder is obtained by rotating in all possible directions about the center the reciprocal lattice for a single crystal. *We assume that the powder grains are all identical.* All points of the reciprocal space for the single crystal which are situated at a distance $s$ from the origin correspond to rays which are diffracted by the powder at the same angle $\theta$, defined by $s = (2 \sin \theta)/\lambda$. We shall calculate the shape of the line corresponding to a given node *hkl* surrounded by its reflection domain. If several nodes with different indices are at the same distance from the center, we superpose their effects to account for the observed line.

The distribution of the intensity within the line corresponds to the variation of the integral of $I(\boldsymbol{s})$ over a section of the domain cut by a sphere centered at $O$, of radius $s$, approximately equal to $|\boldsymbol{r}^*_{hkl}|$, as in Figure 5.9. Since the reflection domain is small, we can replace the spherical sections by planes normal to $\boldsymbol{OR}_{hkl}$ or, since the domains around the nodes are all identical, by sections of the domain at the origin such that $s - r^*_{hkl} = s_0$.

Let us take $\theta$ as the Bragg angle corresponding to the node under consideration, and let us take $\Delta(2\theta)$ as the angular distance from the center to a given position on the line. Then

**FIGURE 5.9.** *Calculation of the profiles of the Debye-Scherrer lines.*

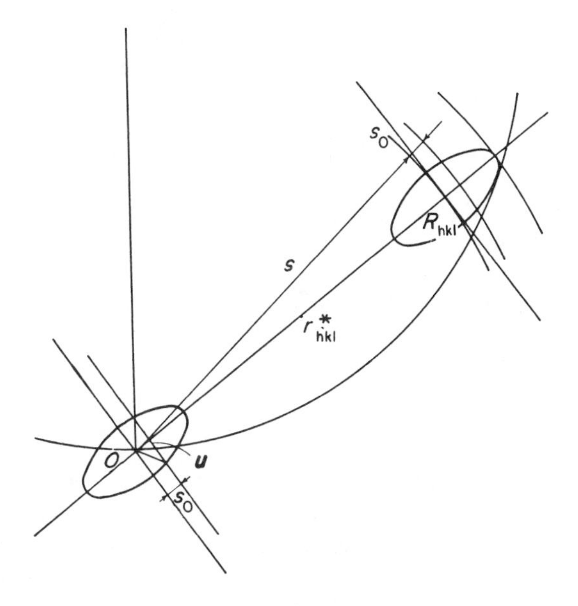

$$s_0 = \frac{\cos\theta}{\lambda} \Delta(2\theta) . \tag{5.10}$$

We also assume that the crystal is sufficiently large that the reflection domains are well separated and do not overlap. If this is not so—that is, if the crystal contains only a very small number of unit cells—the pattern is more easily calculated by the Debye formula given in Eq. (*2.54*).

Since we wish to determine the line shape in the powder pattern, we only require relative values of the scattering power. In the domain surrounding the origin, Eq. (*5.5*) gives

$$I(\boldsymbol{s}) \sim \frac{|\Sigma(\boldsymbol{s})|^2}{V}$$

or, according to Eq. (*2.34*),

$$I(\boldsymbol{s}) \sim \int V(\boldsymbol{x}) \exp(-2\pi i \boldsymbol{s}\cdot\boldsymbol{x})\, dv_x ,$$

where $V(\boldsymbol{x})$ is the function defined in Section 2.4.2. This is the Fourier transform of $|\Sigma(\boldsymbol{s})|^2/V$, and we can therefore write

$$V(\boldsymbol{x}) = \int \frac{|\Sigma(\boldsymbol{s})|^2}{V} \exp(2\pi i \boldsymbol{s}\cdot\boldsymbol{x})\, dv_s . \tag{5.11}$$

We decompose the vector $\boldsymbol{s}$ in the domain around the origin into two components, $\boldsymbol{s}_0$ parallel to $\boldsymbol{OR}_{hkl}$ and $\boldsymbol{u}$ normal to $\boldsymbol{OR}_{hkl}$. In a cross-section of the domain, $\boldsymbol{s}_0$ is a constant (Figure 5.9), and $\boldsymbol{s} = \boldsymbol{s}_0 + \boldsymbol{u}$. Let us consider in object space the vectors $\boldsymbol{t}$ of length $t$ parallel to $\boldsymbol{OR}_{hkl}$, i.e. *normal to the lattice planes of the crystal producing the reflection under consideration.* For these particular vectors $\boldsymbol{t}$, Eq. (*5.11*) can be written as

$$V(t) = \int \frac{|\Sigma(\boldsymbol{s})|^2}{V} \exp(2\pi i s_0 t)\, dv_s . \tag{5.12}$$

because, $\boldsymbol{u}$ being normal to $\boldsymbol{t}$, we have $\boldsymbol{u}\cdot\boldsymbol{t} = 0$ and $\boldsymbol{s}\cdot\boldsymbol{t} = s_0 t$. The above triple integral can be rewritten as

$$V(t) = \iiint \frac{|\Sigma(s_0, \boldsymbol{u})|^2}{V} \exp(2\pi i s_0 t)\, ds_0\, d\boldsymbol{u} . \tag{5.12a}$$

The double integral

$$\iint \frac{|\Sigma(s_0, \boldsymbol{u})|^2}{V}\, d\boldsymbol{u}$$

is none other than the integral of the scattering power over the cross-section of the domain at a distance $s_0$ from the origin. It is therefore a quantity which is proportional to the intensity observed in the line of the powder pattern at an abscissa corresponding to the length $s_0$, or $i(s_0)$.

The quantity $i(s_0)$ therefore gives the profile of the Debye-Scherrer line observed. Therefore, Eq. (*5.12a*) is equivalent to

$$V(t) = \int i(s_0) \exp(2\pi i s_0 t)\, ds_0 \,. \tag{5.13}$$

This equation shows that the function $V(t)$, which depends on a *single variable*, is proportional to the Fourier transform of the function $i(s_0)$. If we invert the above equation, we have

$$i(s_0) = \int V(t) \exp(-2\pi i s_0 t)\, dt \,. \tag{5.14}$$

Let us discuss this important result. The function $i(s_0)$ gives the line profile as a function of the variable

$$s_0 = s - r^*_{hkl} = \frac{2\sin\theta}{\lambda} - \frac{2\sin\theta_0}{\lambda} = \frac{\cos\theta_0}{\lambda}\Delta(2\theta)\,.$$

And from the geometric interpretation for $V(\boldsymbol{x})$ given in Section 2.4.2, we can say that $V(t)$ is equal to the volume which is common to the crystal and to its "double" obtained by a translation $t$ in the direction normal to the reflecting lattice planes, divided by the volume of the crystal. It will be recalled that we consider separately each of the families of planes contributing to a given line. Within a constant factor, the functions $V(t)$ and $i(s_0)$ are Fourier transforms, one of the other. We can calculate the shape of the line for a known powder from Eq. (*5.14*). Inversely, through Eq. (*5.13*), the observed profile of a diffraction line gives $V(t)$, which is a rather complex function of the shape of the grains; it is the only function which is determined by this profile.

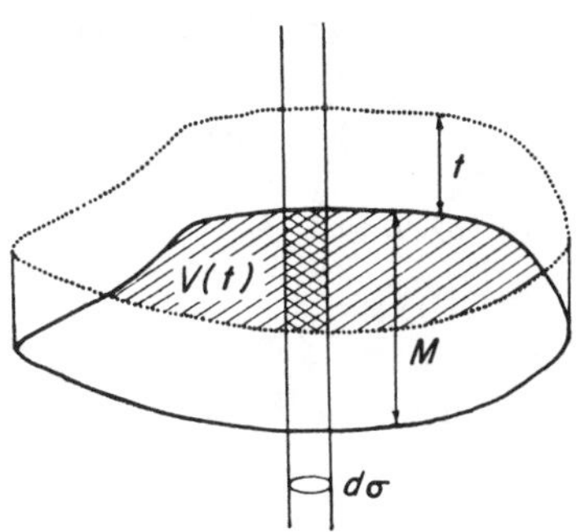


**FIGURE 5.10.** *The function* $V(t)$.

Equation (*5.14*) can also be written in other forms which are sometimes more convenient. Let us divide the crystal into cylindrical columns of cross-section $d\sigma$ normal to the lattice planes $(hkl)$. The height of a column is the thickness of the crystal $M$ at the point considered, as shown in Figure 5.10. In the volume $V(t)$ the column has the length $M - |t|$ for $|t| < M$, and it does not exist when $|t| > M$. Let us set $d\sigma_M$ as the cross-section of the columns whose heights in the crystal lie in the interval between $M$ and $M + dM$. The common volume is

$$VV(t) = \int_{M=0}^{M=\infty} \{M - |t|\} d\sigma_M , \tag{5.15}$$

where $\{M - |t|\}$ is the function shown in Figure 5.11. The above equation can then be rewritten as

$$V(t) = \int_0^\infty \left\{1 - \frac{|t|}{M}\right\} \frac{M d\sigma_M}{V} = \int_0^\infty \left\{1 - \frac{|t|}{M}\right\} g(M) dM , \tag{5.16}$$

where

$$g(M)dM = \frac{M d\sigma_M}{V} \tag{5.17}$$

is the distribution function for the column heights. It represents the fraction of the volume of the grain for which the length of the diameter normal to the $(hkl)$ plane lies between $M$ and $M + dM$. The advantage of this equation is that it can be generalized immediately to the case where the *powder grains are not identical.* The distribution function $g(M)$ applies to the diameters of all the grains, whether they are in the same grain or not.

Since the function $V(t)$ gives the most complete information which can be obtained directly from the pattern, powders composed of grains of widely different shapes and sizes give the same pattern if the function $g(M)$ remains the same. *It is therefore not possible to determine the shape of nonuniform powder grains by using only the diffraction pattern.*

Let us consider the example of spherical grains all having the same diameter, and the particular function $g_1(M)$ representing the distribution of the lengths of the diameters of these spheres. Let us consider also a powder formed of cubes of side $M$ such that we again have the same function $g_1(M)$. The profile for the line (100) is the same in both cases and, as we shall see later on, the differences are small even for the

**FIGURE 5.11.** *The function* $\{M - |t|\}$.

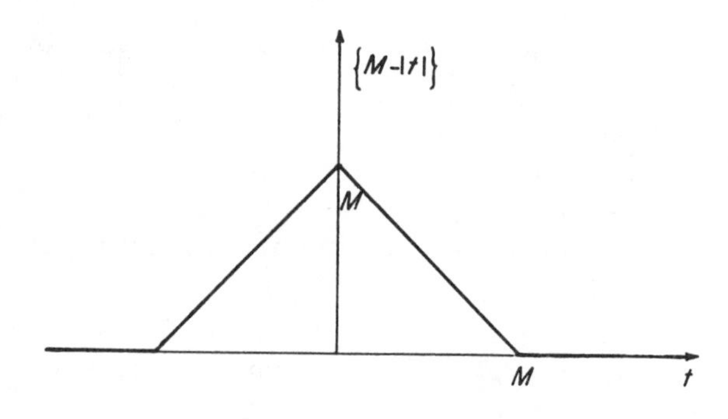

other lines.

We can also find another expression for $i(s_0)$, starting from Eq. (5.16). As will be shown in Appendix A (Table A.1, 10), the function of Figure 5.11,

$$\left\{1 - \frac{|t|}{M}\right\},$$

is the transform of

$$\frac{1}{M}\frac{\sin^2(\pi M s_0)}{(\pi s_0)^2}.$$

Since $i(s_0)$ is the transform of $V(t)$, it is therefore the sum of the transform of the functions

$$\frac{\sin^2(\pi M s_0)}{M(\pi s_0)^2},$$

with the distribution function $g(M)$ of Eq. (5.16). Therefore

$$i(s_0) \sim \int_0^\infty \frac{\sin^2(\pi M s_0)}{M(\pi s_0)^2} g(M)\, dM. \tag{5.18}$$

The function

$$\frac{\sin^2(\pi M s_0)}{M(\pi s_0)^2}$$

is the interference function for an object having the form of a segment of a straight line of length $M$ whose electron density per unit length is uniform and equal to unity. Hence the theorem given by Bertaut [6]:

"The profile of the (*hkl*) line is the same as that of the diffraction pattern for a set of parallel and incoherent segments whose lengths are distributed like the diameters of the grains normal to the reflecting plane (*hkl*)".

We shall now show how these general results can be used to obtain information from the widths and profiles of the lines.

### 5.3.1. Line Width

Let us recall that $\theta_0$ is the Bragg angle corresponding to the reflecting plane (*hkl*) and that it corresponds to the center of a broad line. Setting $\Delta(2\theta)$ as the angular position with this center as origin, the variable $s_0$ is equal to $(\cos\theta_0/\lambda)\Delta(2\theta)$. If the line profile is now drawn as a function of $s_0$, the integral width $\Delta s_0$ is the ratio of $\int i(s_0)\, ds_0$ to the maximum intensity $i(0)$. From Eq. (5.4), the apparent size of a crystal is

$$L = \frac{1}{\Delta(2\theta)}\frac{\lambda}{\cos\theta_0} = \frac{1}{\Delta s_0} = \frac{i(0)}{\int i(s_0)\, ds_0}.$$

According to Eq. (*5.14*),

$$i(0) = \int V(t)\,dt$$

and, from Eq. (*5.13*),

$$V(0) = 1 = \int i(s_0)\,ds_0\ ,$$

whence

$$L = \int V(t)\,dt\ . \tag{5.19}$$

We can also find a relation between $L$ and the diameters $M$. From Eq. (*5.18*),

$$i(0) = \int Mg(M)\,dM = \bar{M}\ ;$$

then

$$L = \bar{M}\ . \tag{5.19a}$$

We shall call the average diameter $\bar{M}$ the apparent size of the crystal. The term "apparent size" now has a rigorous meaning, whereas it represented only an order of magnitude of the crystal size in the elementary calculation.

It is important to specify carefully the exact meaning of the average $\bar{M}$. This is a *volumetric average*,

$$\bar{M} = \frac{\int M\,dv_M}{V}\ ,$$

where $dv_M/V$ is the fraction of the volume of the powder for which the diameter of a grain which is normal to the reflecting plane lies between $M$ and $M + dM$. There are also other possible definitions for an average diameter. In particular, Bertaut [6] has given a completely different expression for the line width which is consistent with the above.

For a powder composed of homogeneous grains of given shape, it is possible to calculate the apparent size $L$ as a function of the geometric parameters defining the grain. Thus, for a spherical grain of diameter $D$, the direction of the reflecting plane is unimportant and we find that

$$L = 0.75D\ .$$

Let us call $l$ the length, equal to

$$V^{1/3} = \left(\frac{\pi D^3}{6}\right)^{1/3}\ .$$

Then $L = 0.93l$. It can be shown, for many fairly regular grain shapes, that $L$ is equal to or somewhat less than $l$, the difference between the two being in general less than 20% [7]. Thus, for a cube whose sides are parallel to the axes of a cubic crystal, $L = l$ for the lines (100) and $L = 0.87l$ for the line (111); for the other indices $L$ takes on intermediate values. If the cubic crystal has the shape of a parallelepiped, the line (100) is the result of the superposition of the three reflections (100), (001), (010), and $L$ is equal to the arithmetic mean of the three sides. For the other lines, $L$ is a very complicated function of the lengths of the sides. It is difficult to find the form of the crystal from the widths of the various lines. But this would be possible for crystals of the hexagonal system having the form of cylinders parallel to the six-fold axis. In this case the lines $(00.l)$ depend only on the height of the cylinder, and the lines $(hk.0)$ depend only on its diameter.

In the case of a powder composed of uniform grains of volume $V$, the apparent size of a crystal as deduced by the Scherrer formula of Eq. (*5.3*) from the integral width of any line is therefore equal to $V^{1/3}$, generally within 20% or better. The volume of the crystal is therefore equal to $L^3$, often within less than 50%. The precise meaning of the apparent size of the grain can in some cases permit a complete determination of the grain geometry if its shape is known and *if the powder is homogeneous*, by comparing the widths of the lines corresponding to different indices.

### 5.3.2. The Line Profile

If we wish to go further, we must use the true line profile, which is determined as described in the next section. We must first check the *symmetry of the line* for, according to Eq. (*5.18*), whatever the type of powder, $i(s_0) = i(-s_0)$. If this is not so, the broadening comes from causes other than the small size of the crystals.

It is not possible to do more than find the distribution of diameters $g(M)$. To do this we must, according to Eq. (*5.18*), decompose the function $i(s_0)$ into a sum of functions

$$\frac{\sin^2 \pi M s_0}{M(\pi s_0)^2} \, .$$

It is mathematically impossible to do this directly, but the desired result is obtained by performing the analysis on the transform of $i(s_0)$, i.e. on $V(t)$ [Equation (*5.16*)]. It will be noticed that

$$V(0) = \int g(M)\, dM = 1 \, ,$$

and

$$\left(\frac{dV}{dt}\right)_{t=0} = -\int \frac{1}{M} g(M)dM = -\frac{1}{\bar{M}_1}. \tag{5.20}$$

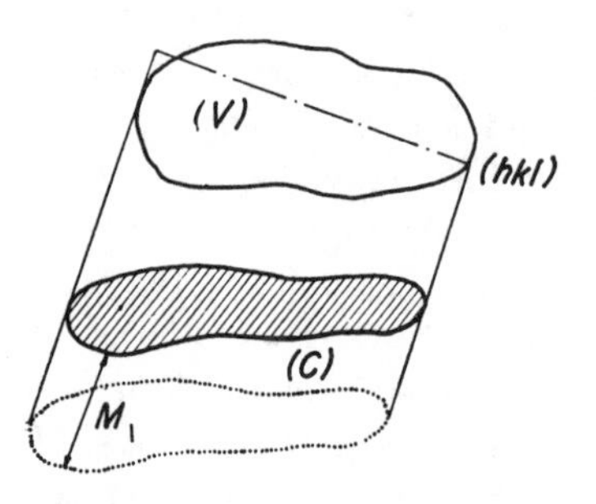


**FIGURE 5.12.** *The average value $\bar{M}_1$ of the diameters normal to (hkl) is such that the volume of the cylinder (C) is equal to V.*

According to Eq. (5.17),

$$\frac{1}{\bar{M}_1} = \int \frac{1}{M} g(M)\, dM = \frac{1}{V} \int d\sigma_M ,$$

where $\int d\sigma_M$ is the total cross-section of the grains normal to the reflecting plane, and $V$ is their total volume. The quantity $\bar{M}_1$ is therefore an "average" value of the thickness of the grain which has the simple geometrical interpretation illustrated in Figure 5.12. It will be noticed that $\bar{M}_1$ is different from the average value $\bar{M}$ calculated previously.

The tangent to the $V(t)$ curve at $t = 0$ intersects the $t$ axis at $t_0$ such that

$$\frac{V(0)}{t_0} = -\left(\frac{dV}{dt}\right)_{t=0},$$

or

$$t_0 = -\frac{1}{(dV/dt)_{t=0}} = \bar{M}_1 .$$

This property also applies to a curve $KV(t)$ and therefore to the trans-

**FIGURE 5.13.** *Transform of the experimental curve i(s) and determination of $\bar{M}_1$.*

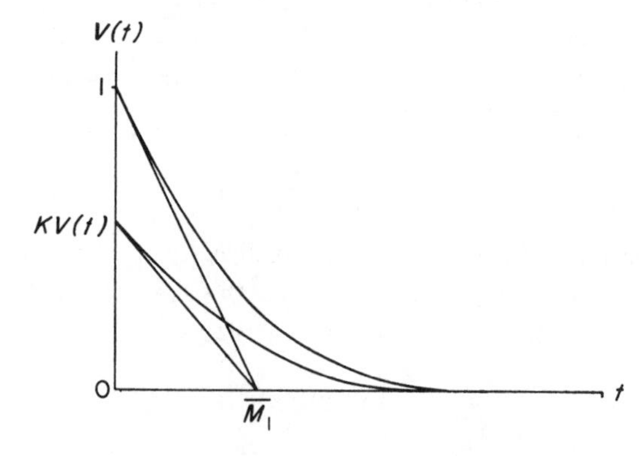

form of the experimental curve $i(s_0)$, which is proportional to $V(t)$ within an unknown factor. This transform is drawn as a function of $t$ on an arbitrary scale, as in Figure 5.13. Then, to obtain $V(t)$, one uses a scale such that the value of the transform at the origin is equal to unity. The intersection of the tangent at $t = 0$ with the axis of $t$ gives $\bar{M}_1$ with sufficient accuracy.

The measurement of the line width also gives another average function of the diameters $\bar{M}$, which we have already defined. The average $\bar{M}_1$ is always smaller than $\bar{M}$, except if the diameters are constant. Thus, for a spherical grain, $\bar{M}_1 = 0.67D = 0.89M$. For a cubical grain, $\bar{M}_1 = \bar{M}$ for the plane (100), but $\bar{M}_1 = 0.66\bar{M}$ for the plane (111). The ratio $\bar{M}_1/\bar{M}$ decreases when the spread in diameters increases.

From Eq. (*5.16*), we have

$$\frac{dV}{dt} = -\int_t^{\infty} \frac{g(M)}{M}\, dM\,,$$

$$\frac{d^2V}{dt^2} = \frac{g(t)}{t}\,, \qquad (5.21)$$

$$g(M) = M\left(\frac{d^2V}{dt^2}\right)_{t=M}.$$

The distribution in diameters is therefore determined by the second derivative of $V(t)$, and thus by the Fourier transform of the line profile. The second derivative of the curve of $V(t)$ is zero for $M = 0$, and is also zero for $M = \infty$. It is always positive.

The graphic determination of the second derivative is unfortunately difficult and, although Eq. (*5.21*) permits in principle the determination of $g(M)$, it is very often of doubtful value in practice. Other properties of the function $V(t)$ will be found in the work of Bertaut [6], as well as methods for deducing from it further information on the distribution of the diameters.

Since this method leads to a single reliable result, namely $\bar{M}_1$, it may appear to be of no advantage since a simple measurement of the line widths leads to $\bar{M}$, which is another average function of the diameters. As we shall see in Section 5.5, it is impossible to arrive at the correct values of the line widths without making corrections which can only be accurate by calculating the Fourier transforms. Since these corrections require a function which is proportional to $V(t)$, it is natural to use this function. Moreover, it is possible to check whether the transform really has zero curvature at the origin, as shown in Figure 5.13, and thus to find out whether the broadening of the line can be attributed exclusively to the small size of the grains. We shall find in Section 7.3.2.1 that broadening arising from other causes corresponds to other types of transforms.

## 5.4. Widths of the Diffraction Spots for a Crystalline Fiber

A crystalline fiber is formed of a set of crystals having a common axis and oriented at random around this axis. Its pattern is analogous to that of a rotating crystal. In such a fiber the elementary crystals are usuallv small cylinders whose axes coincide with that of the fiber and whose lengths are large compared to their diameters. Under these conditions the reflection domains for each node have the shape of flat disks normal to the axis of the fiber (Figure 5.14). Along the equator the spots are elongated and one can deduce the average length of the crystals from their height. In many cases, however, the broadening of the spot will be too small for an accurate measurement. The distribution of the intensity along the equator, when studied like the profile of an ordinary Debye-Scherrer line, gives the diameters of the crystals which are normal to the axis of the fiber.

The extraequatorial spots could not provide further information, but

**FIGURE 5.14.** *Spot shapes for a fiber composed of thin chainlike crystals. Case of a disk-shaped reflection domain giving an elongated spot on the equator.*

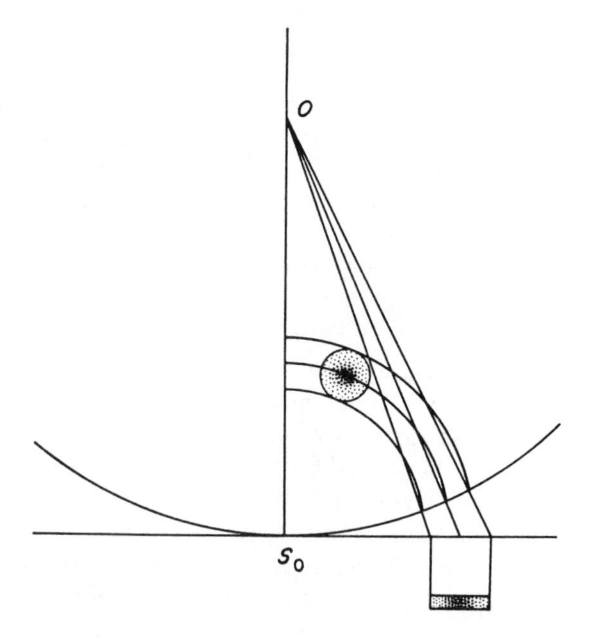

we could not apply the Scherrer formulas to them because their shape is determined by the geometric conditions at the intersection of the reflection domain with the Ewald sphere. Let us consider, for example, a

**FIGURE 5.15.** *Extra-equatorial spots in a fiber pattern.* (a) *The reflection domain around the node penetrates into the Ewald sphere by rotation about the fiber axis.* (b) *The reflection domain grazes the Ewald sphere.*

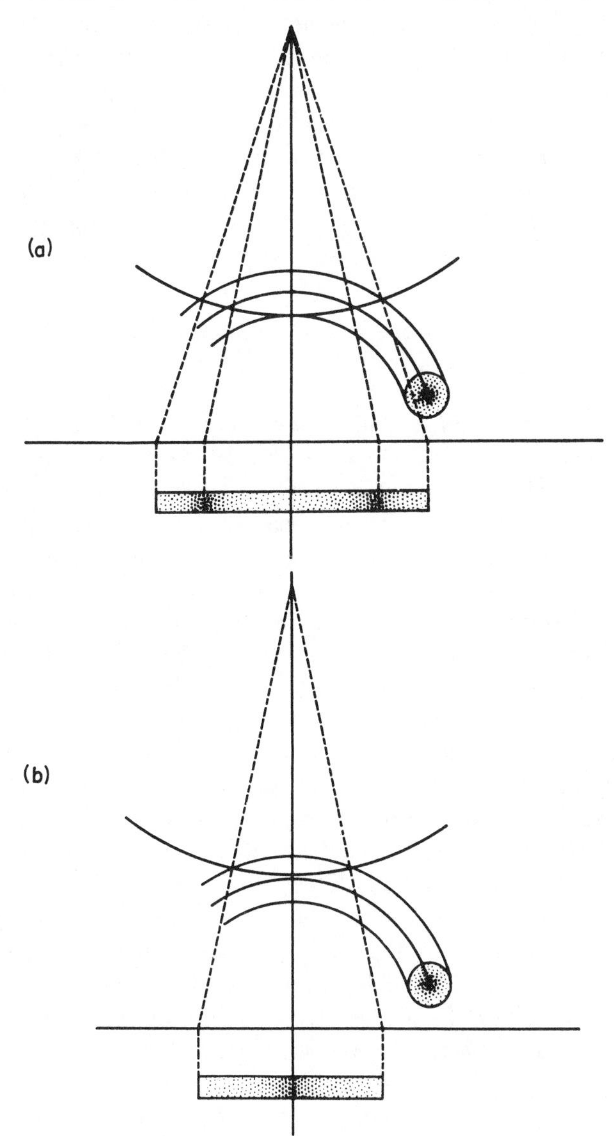

node such that its distance to the axis is about equal to the distance between the axis and the Ewald sphere in the plane of the node. Figure 5.15 shows that even a small domain can produce a long diffuse streak joining the two symmetrical intense spots near the perpendicular bisector when the domain around the lattice point penetrates into the Ewald sphere. When it simply grazes the sphere, we observe only a small diffuse streak on the perpendicular bisector and no true diffraction spots.

Very often the crystal axes are not perfectly well aligned with the axis of the fiber and the diffraction spots are then broadened more or less, depending on the position of the node with respect to the axis of the fiber. The profile of the equatorial spots is, however, not disturbed and still permits the determination of the transverse dimensions of the crystals. To determine the lengths of the elementary crystals we would have to utilize the spots ($0k0$), which can be made to appear by tilting the incident rays with respect to the axis of the fiber.

## 5.5. Experimental Measurement of the Width and of the Profile of a Line

For the determination of grain size by means of the above formulas, one must know either the integral width or the complete profile of the lines. It is unfortunately not possible to use the width and the profile as observed directly on the pattern because, even for a perfectly crystallized powder with fairly large grains of the order of one micron, the lines have a finite width which comes from instrumental factors. It is therefore necessary to correct the experimentally observed values.

The method of correction depends on the manner in which the numerical results are to be used. We shall consider two very different cases: either we require simply an order of magnitude of the crystal dimensions, and this is supplied by the simple formula of Scherrer, or we may wish to obtain the most complete and accurate information possible.

(a) In the first case, a good pattern obtained by the most classical types of instruments is sufficient (Debye-Scherrer camera or counter-type diffractometer). The simplest procedure is to measure the line width at half intensity above the background interpolated from its values on either side of the line. It is necessary to take into account the possible existence of tails on either side at the foot of the line. In the case of two neighboring lines the curves can overlap and the determination of the background in the region between the two lines then becomes inaccurate.

Equations (*5.4*) and (*5.19*) involve the integral width, but the rather lengthy measurement of the area under the curve is not justified for approximate measurement. It is preferable to calculate the apparent size $L$ of the grains from the width at half maximum from Eq. (*5.3*):

$$L = \frac{0.9\lambda}{\cos\theta_0 \Delta'(2\theta)}.$$

We have assumed that the coefficient 0.9 is valid for the general case, as well as for the schematic calculation of Section 5.1.

To obtain the true value of $\Delta'(2\theta)$, we must compare the lines for the sample under consideration with those for a standard powder with large, well-crystallized grains. The two patterns must be recorded in identical experimental conditions, for example by mixing the two powders together. The measurement of the widths of the lines for the standard powder which are close to the line under consideration gives, by interpolation, the width $\Delta_0'$ of a theoretically infinitely narrow line for the Bragg angle $\theta$. If $\Delta_1'$ is the measured width, the corrected value is given by

$$\Delta' = \sqrt{\Delta_1'^2 - \Delta_0'^2} \tag{5.22}$$

with reasonable accuracy [8]. This formula, which we shall justify later on, is preferable to the simpler one $\Delta' = \Delta_1' - \Delta_0'$. In any case, *the correction must be small* if the final result is to be valid.

(b) If the aim is to achieve accuracy and to use the rigorous formulas given in Section 5.3.2, *it is essential to start with very good measurements* made with a carefully designed instrument, since it is obviously absurd to apply complex mathematical calculations to poor data. For this reason it is important to use methods giving *high dispersion* patterns with a minimum of geometric broadening and a minimum of background. A Geiger-counter goniometer, possibly with a Ross filter (see Section 6.3.1), or a focusing camera equipped with a monochromator is particularly appropriate. With the diffractometer one must be careful to use a narrow slit in front of the counter. Point-by-point measurements using pulse counting are preferable to the recording of the counting rate, for this can be inaccurate in the case of narrow lines. In the usual instruments, where the sample is studied by reflection, an important source of line broadening is the penetration of the incident ray into substances with low coefficients of absorption. It is important to use a thin sample, to avoid this broadening, even at the cost of a decrease in intensity. It is also possible to correct for this broadening [9].

The most general method for correcting the observed profile is due to Stokes [10]. Here again we use as reference a powder whose theoretical line width is negligible. Setting $s = (2\sin\theta)/\lambda$, let $i_1(s)$ and $i_0(s)$ be the

**FIGURE 5.16.** *Determination of the true profile $i(s)$ from the observed profile $i_1(s)$ and from that of a line with zero theoretical width, $i_0(s)$.*

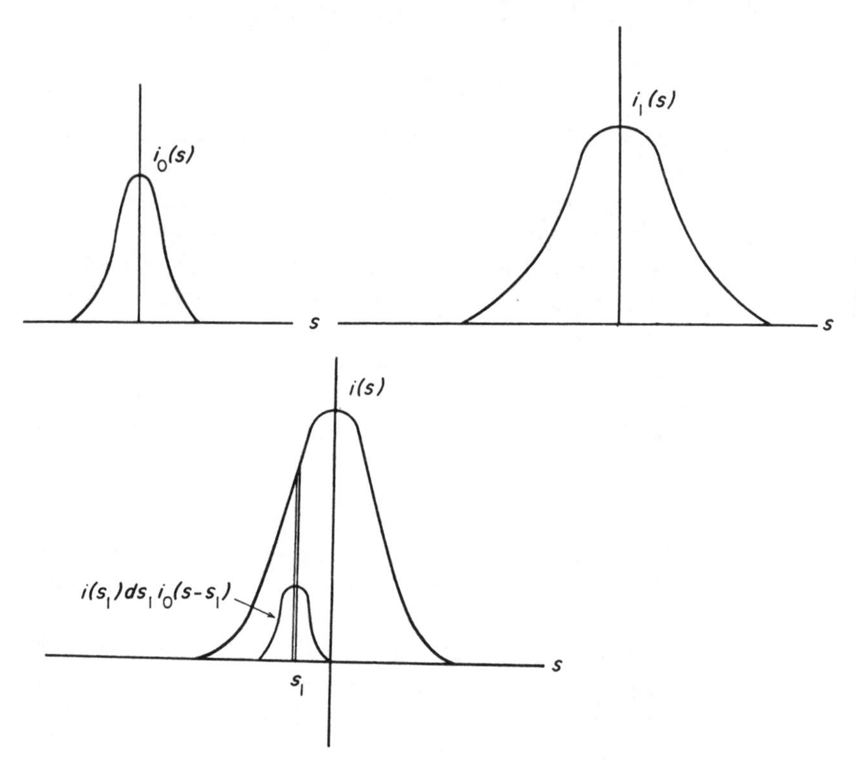


observed profiles for the line under consideration and for the standard line, the Bragg angles being chosen as close together as possible (Figure 5.16). We assume that the intensity scale is chosen in such a way that $i_0(s)$ is normalized:

$$\int i_0(s)ds = 1 .$$

Finally, let $i(s)$ be the true profile of the line.

We wish to determine how this profile is distorted by the experimental conditions. Since an infinitely narrow line is transformed into the broadened line $i_0(s)$ when its total intensity is unity, each element of the line $i(s_1)ds$ gives an element of line of profile $i_0(s - s_1)i(s_1)ds$, The observed total line is therefore

$$i_1(s) = \int i(s_1)i_0(s - s_1)ds , \tag{5.23}$$

where the integral extends from $-\infty$ to $+\infty$, assuming that the line profiles extend to infinity on either side with zero intensity. Then $i_1(s)$ is the faltung of the functions $i(s)$ and $i_0(s)$. According to the theorem of Appendix A.5, relating the Fourier transforms of the three functions, we have

$$\text{transf}\, i_1(s) = \text{transf}\, i(s) \times \text{transf}\, i_0(s)\,. \tag{5.24}$$

Since the functions $i_0(s)$ and $i_1(s)$ are known, their transforms $\int i_0(s) \exp(2\pi i s t)\, ds$ and $\int i_1(s) \exp(2\pi i s t)\, ds$ can be calculated and, from the above equation, their ratio gives directly the Fourier transform of the corrected profile. We have previously called this $V(t)$, in Eq. (*5.13*). This function is fundamental to the rigorous methods of interpretation of line profiles (Section 5.3.2). It is also possible to determine the true profile of the line by inversion of $V(t)$.

Let us apply this theory to the case where both the line under study and the standard line are represented by Gaussian curves of widths $\varepsilon_1$ and $\varepsilon_0$, respectively:

$$i_1(s) = C \exp\left(-\frac{\pi s^2}{\varepsilon_1^2}\right),$$

$$i_0(s) = \frac{1}{\varepsilon_0} \exp\left(-\frac{\pi s^2}{\varepsilon_0^2}\right),$$

where the coefficient $i_0(s)$ is chosen so the condition of normalization is satisfied, since

$$\int_{-\infty}^{+\infty} \exp\left(-\frac{\pi s^2}{\varepsilon^2}\right) ds = \varepsilon\,.$$

According to Table A.1, Appendix A, the Fourier transforms of the functions $i_1(s)$ and $i_0(s)$ are

$$\text{transf}\, i_1(s) \sim \exp(-\pi \varepsilon_1^2 t^2)\,,$$
$$\text{transf}\, i_0(s) \sim \exp(-\pi \varepsilon_0^2 t^2)\,.$$

The transform of the corrected curve is therefore [Eq. (*5.24*)]

$$\text{transf}\, i(s) \sim \exp[-\pi(\varepsilon_1^2 - \varepsilon_0^2) t^2]\,.$$

By inversion, the true line profile is finally given by

$$i(s) \sim \exp\left(-\frac{\pi s^2}{\varepsilon_1^2 - \varepsilon_0^2}\right).$$

This is a Gaussian curve whose width is $(\varepsilon_1^2 - \varepsilon_0^2)^{1/2}$. It is the formula utilized by Warren [8] and which we have given as a simplified method of correction [Eq. (*5.22*)]. It will be seen that it is strictly valid only for lines with Gaussian profiles. There exists no general formula giving the corrected width as a function of the widths of the lines which are compared, for the relation depends on the shapes of the two curves.

Jones [11] has given a graph for the true width as a function of the measured width and of the width of the standard line. He assumes a law of the form $1/(1+h^2s^2)$ for the theoretical profile of the broadened line, the shape of the standard line being measured experimentally. This graph is therefore valid only for a given instrument.

In principle the Stokes method provides a general and exact correction, whatever the shapes of the lines which are compared. Appendix A.7 gives an example of the practical use of this method. It will be seen that, as often happens in using Fourier transforms, the numerical calculations are of sufficient accuracy only in certain special conditions. In practice the method is useful only if *the standard line is narrow compared to the line under investigation*. It is also necessary to know the line profile out to distances, measured from the center, of the order of four times its width. One must remember that the possibility of making a mathematical correction does not eliminate the necessity of reducing the geometric broadening to a minimum by performing the experiment with great care.

There is one source of broadening which is unavoidable with the usual instruments. This is the spectral width of the characteristic radiation [12]. The line width of the standard powder with a geometrically perfect instrument is not infinitely narrow, but corresponds to the spectrum of the radiation. We normally use the $K\alpha$ radiation, and thus the doublet $K\alpha_1\alpha_2$; but even with a single line—$K\beta$, for example—or a monochromator sufficiently well adjusted to isolate the $K\alpha_1$ line, the spectral width $(\Delta\lambda)/\lambda$ is of the order of 1/2000. As a function of the Bragg angle, the minimum width of the standard line is given by

$$\Delta(2\theta) = 2\tan\theta\,\frac{\Delta\lambda}{\lambda} \simeq 10^{-3}\tan\theta\,.$$

The apparent size of crystals giving an equivalent line width is

$$\frac{\lambda}{L\cos\theta} \simeq 10^{-3}\tan\theta$$

or

$$L \simeq \frac{1000\lambda}{\sin\theta}\,.$$

This limiting size decreases with $\theta$: it is $3000\lambda$ for $\theta = 20°$, and $1000\lambda$ for $\theta = 90°$. It is therefore preferable to use the lines diffracted at small angles for the measurement of grain sizes. For example, with $\theta = 20°$ and $\lambda = 1.5$Å, it is impossible to measure crystal sizes larger than 1000–

2000Å with accuracy, even with the best instruments.

For small Bragg angles, the $\alpha_1$ and $\alpha_2$ components are not resolved, even in the narrow standard line. If we wish to analyze lines at large angles $\theta$, the doublet may appear in the narrow standard line, but its two components are not resolved in a broader line.

The Stokes method of correction can again be applied, but a simpler method is available to determine the shape of a line which would be produced by the single $K\alpha_1$ radiation (Rachinger [13]).

It is well known that the relative intensities of the two components of the doublet $\alpha_1$ and $\alpha_2$ are in the ratio 2 to 1, and their difference in wavelength $\Delta\lambda$ is also known. The $\alpha_2$ diffraction line is therefore a copy of the $\alpha_1$ line reduced by a factor of 2 and displaced by an angle

$$\Delta(2\theta) = 2 \tan\theta \frac{\Delta\lambda}{\lambda} .$$

If $i(2\theta)$ is the profile of the line corresponding to a single line $\alpha_1$, i.e. the required profile, the observed profile is

$$I(2\theta) = i(2\theta) + \frac{1}{2}\, i[2\theta + \Delta(2\theta)] . \tag{5.25}$$

This equation can be solved graphically in the manner illustrated in Figure 5.17. We start from the abscissa $2\theta_1$, just at the foot of the line, such that $2\theta_1 + \Delta(2\theta)$ is sufficiently far from the center of the line to make

**FIGURE 5.17.** *The Rachinger method. The solid line shows the observed value of I, and the dotted line the components $\alpha_1$ and $\alpha_2$. We have $PB = PA + PC = PA + (QD/2)$.*

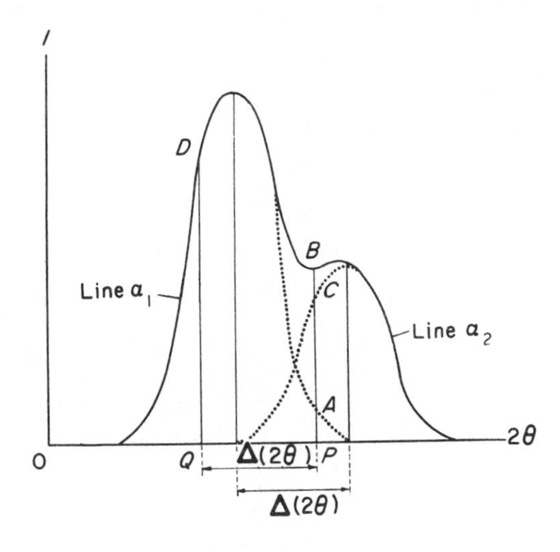

$i[(2\theta_1 + \Delta(2\theta)]$ equal to zero. Then

$$i(2\theta_1) = I(2\theta_1) .$$

We then use Eq. (*5.25*) for the angles $2\theta_1 - \Delta(2\theta)$, $2\theta_1 - 2\Delta(2\theta)$, $\cdots$, and we find successively

$$i[2\theta_1 - \Delta(2\theta)] = I[2\theta_1 - \Delta(2\theta)] - \frac{1}{2}\, i(2\theta_1) ,$$

$$i[2\theta_1 - 2\Delta(2\theta)] = I[2_1\theta - 2\Delta(2\theta)] - \frac{1}{2}\, i[2\theta_1 - \Delta(2\theta)] ,$$

$$\cdots\cdots\cdots\cdots\cdots\cdots ,$$

which gives a set of points on the curve $i(2\theta)$.

A method of "achromatization" has been suggested to eliminate experimentally the effect of the spectral width of the characteristic radiations (Ekstein and Siegel [14]). The principle, illustrated in Figure 5.18, is as follows.

A beam is incident on a crystal so as to reflect the $K\alpha$ radiation and a knife edge is placed at $A$ against the crystal. The various rays of the reflected beam do not have the same wavelength. For example, the $K\alpha_1$ and $K\alpha_2$ lines give rise to the rays $AM_1$ and $AM_2$, forming the Bragg angles $\theta_{\alpha_1}$ and $\theta_{\alpha_2}$ with the lattice planes. We now place the sample on this reflected beam. Let us consider lattice planes with a lattice spacing $d$. The diffraction angle varies with the wavelength in such a way that $\lambda/\sin\theta = 2d$ and the diffracted rays $M_1P$ and $M_2P$ form angles $2\theta'_{\alpha_1}$ and $2\theta'_{\alpha_2}$ with $AM_1$ and $AM_2$, respectively. If the photographic film is placed at the point $P$ where they meet, *the two components of the*

**FIGURE 5.18.** *Achromatization of diffraction lines.*

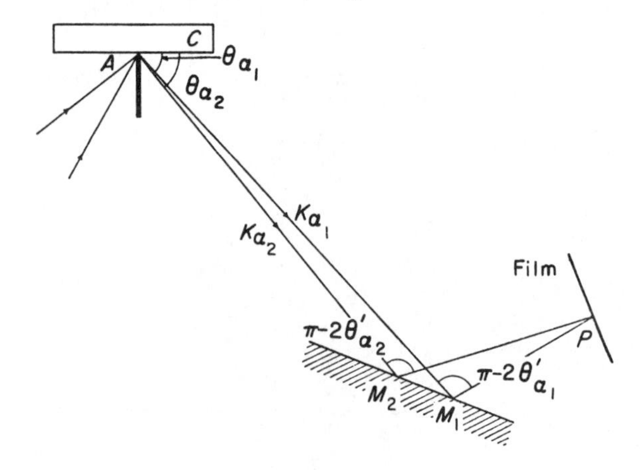

*doublet are superposed.* More precisely, the diffracted rays arriving at each point of the film correspond to the same value of $\lambda/\sin\theta$, whatever the value of $\lambda$ in a narrow band around the wavelength of the $K\alpha$ line. The line profile thus obtained is therefore independent of the band width $\Delta\lambda$.

This principle of achromatization is utilized in the focusing camera associated with a bent-crystal monochromator (Guinier camera [15]). The conditions for chromatic focusing are satisfied theoretically for only one value of $\theta$, which is of the order of 10–30° with the usual monochromators. This instrument has been investigated in detail by Sebilleau [16] for small values of $\theta$, and by Chaulet, Guinier, and Sébilleau for large values of $\theta$, using back reflection [17]. The aim of this method is to eliminate from the diffraction line profiles the broadening which is due to the spectral width of the incident radiation. After everything has been done to reduce the geometric broadening, the spectral width of the incident radiation is the last important source of line broadening.

## References


1. J. Strong, *Concepts of Classical Optics*, Freeman, San Francisco (1958).
2. W. L. Bragg, *The Crystalline State*, Vol. I, Bell, London (1933), p. 189.
3. R. W. James, *The Crystalline State*, Vol. II, Bell, London (1948), p. 4.
4. R. Hosemann, *Naturwiss.*, **19** (1954), 440.
   M. von Laue and Rierve, *Z. Krist.*, *A* **95** (1936), 408.
5. C. S. Barrett, *Structure of Metals*, McGraw-Hill, New York (1952), p. 542.
6. F. Bertaut, *Acta Cryst.*, **3** (1950), 14.
7. A. J. C. Wilson, *X-ray Optics*, Methuen, London (1949).
8. B. E. Warren, *J. Appl. Phys.*, **12** (1941), 375.
9. D. T. Keating and B. E. Warren, *Rev. Sci. Instr.*, **23** (1952), 519.
10. A. R. Stokes, *Proc. Phys. Soc. London*, **61** (1948), 382.
    M. S. Paterson, *Proc. Phys. Soc. London, A*, **63** (1950), 477.
    C. G. Shull, *Phys. Rev.*, **70** (1946), 679.
11. F. W. Jones, *Proc. Roy. Soc. London, A*, **166** (1938), 16.
12. H. Ekstein and S. Siegel, *Acta Cryst.*, **2** (1949), 99.
13. W. A. Rachinger, *J. Sci. Instr.*, **25** (1948), 254.
14. H. Ekstein and S. Siegel, *Phys. Rev.*, **73** (1948), 1207.
15. A. Guinier, *Théorie et Technique de la Radiocristallographie*, Dunod, Paris (1956), p. 196.
16. F. Sébilleau, Achromatisme en rayonnement X. Application à l'étude des raies de diffraction de la solution solide Al-Cu à 4% au cours du durcissement structural. Publication O.N.E.R.A. No. 87, Paris (1957).
17. R. Chaulet, A. Guinier and F. Sébilleau, *Acta Cryst.*, **13** (1960), 332,
18. P. Croce, G. Devant, M. Gandais and A. Marraud, *Acta Cryst.* **15** (1962), 424

CHAPTER

# 6

# DIFFRACTION BY AN IMPERFECT CRYSTAL LATTICE

## 6.1. Crystal Imperfections of the First and Second Types

Let us first define the nature of the crystal imperfections which we shall discuss in this chapter. We have already introduced one concept of imperfection in the theory of X-ray diffraction by crystals, and we have used it to explain the intensity of the reflections. We have used the model of the *mosaic structure*, which is certainly too simplified, but yet satisfactory. According to this model, the crystal is assumed to be formed of small blocks which are slightly rotated or displaced, one with respect to the other. The faults occur only between the blocks, which measure $10^3$ to $10^4$ interatomic distances on the side, and which are assumed to have perfect lattices. This model provides a satisfactory explanation of the observed facts, at least as a first approximation, and it is legitimate to use it as a starting point for the theories and applications discussed in the previous chapters, as well as for all crystal structure determinations.

According to this model, the faults are so rare that, for the vast majority of the atoms, the relative positions of the neighboring atoms is the same as in an ideal lattice out to considerable distances, and the few atoms to which this does not apply have no appreciable effect on the diffraction. However, as the size of the blocks decreases, the diffraction patterns change, as we have seen in Chapter 5. Although this model of a real crystal has been used successfully in crystallography and crystallochemistry, it has proved to be inadequate for many aspects of the physics of solids [1]. One must consider the nodes of the lattice to be only the approximate positions of the atoms, *on the whole,* the atoms being slightly displaced away from their theoretical positions. The imperfections are thus distributed throughout the mass of the crystal, and not localized in distinct faults. One example of this is the thermal agitation of the atoms about their lattice points; the theory of the specific heat of solids is based on this phenomenon. Another example is the distortion following plastic deformation, which brings about important changes in the physical and mechanical properties of metals during work hardening. A third example is that of metal alloys whose properties are modified by introducing imperfections in the lattice during the age hardening process.

We now consider the particular type of imperfections which *alters the rigorous periodicity* of the crystal lattice. We shall investigate their effect on diffraction phenomena and we shall find that the diffraction patterns of real crystals cannot be explained by the simple Bragg-Laue theory when they are recorded with the most refined methods. In certain cases these anomalies can be important and we shall examine the methods whereby one can arrive at the real structure of a crystal by analyzing the diffraction patterns.

This subject is progressing rapidly at the present time. It is of considerable interest because crystal imperfections often play the *most important role* in determining the physical properties of solids. Unfortunately for the solid-state physicist, these imperfections have only secondary effects in diffraction phenomena. X-rays exaggerate the perfection of crystals, and imperfections are in general difficult to detect, and even more difficult to measure. This fundamental fact has been most favorable to the development of crystallochemistry since it has been possible to establish the ideal structure of crystals using real imperfect crystals. It is, however, an unfortunate handicap for solid-state physics, which must even now be based on structures that experimentally are not sufficiently well known.

We first require a mathematical definition for the imperfect crystal which is derived from the perfect lattice. The general formulas will then show the difference between the diffraction pattern and that of a perfect crystal.

Two different methods are used. With the first method, we start with a perfect crystal and displace the atoms away from their theoretical positions, the displacements being small compared to the interatomic distances. This type of imperfection is characterized by the fact that there exist throughout the solid an *ideal average lattice* which, at all points, is not much different from the real lattice. Besides this *displacement disorder*, there can also be a *substitution disorder*, certain atoms of a particular type being replaced by atoms of other types in given proportions. We shall assume that the imperfect crystal is homogeneous—that all the unit cells exhibit the same statistical distribution of their atoms. An example of displacement disorder is that of thermal agitation, while a disordered mixed crystal is an example of substitution disorder.

Such imperfections, which have been called imperfections of the first type, introduce fluctuations in the distances between corresponding atoms, but preserve a long-range order which is no more disturbed than the short-range order. Thus, for the one-dimensional arrangement shown schematically in Figure 6.1b, the distance between the $n$th neighbors $A_0A_n$ remains approximately equal to the ideal distance $na$, and the fluctuation $(A_0A_n - na)$ has an amplitude which is independent of $n$.

**FIGURE 6.1.** (a) *Ideal linear lattice.* (b) *Linear lattice with perturbations of the first type. The amplitude of the fluctuations is independent of position.* (c) *Linear lattice with perturbations of the second type. The amplitude of the fluctuations increases with distance.*

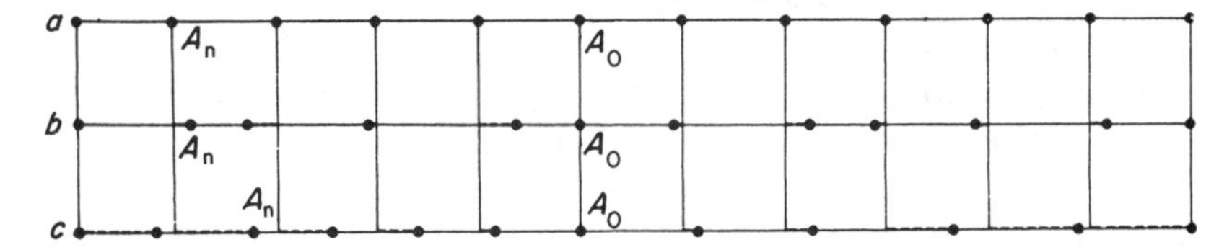


In the second type of imperfection there is, on the contrary, no long-range order. One can imagine that an ideal crystal is built up by positioning atoms successively according to a rigid plan. If this plan is followed only approximately, the distances between close neighbors fluctuate slightly, but the fluctuations increase in amplitude as the number of intermediate atoms increases. In the one-dimensional lattice of Figure 6.1c $(A_0A_n - na)$ shows fluctuations which increase with $n$, while the fluctuation in the distance between two close neighbors is of the same order as in the previous models. If the plan for positioning close neighbors is very indefinite, we have an amorphous body. On the other hand, if it is quite well defined but nevertheless not perfect, we have a crystal which is poorly crystallized. This seems to be the case for many organic compounds composed of large molecules which dispose themselves in an orderly fashion because of Van der Waals' forces, which are weak compared to the interatomic forces which produce a rigorous order in inorganic crystals. The two types of imperfect crystals give diffraction patterns which are very distinct and which we shall study separately.

## 6.2. General Formulas for the Diffraction by an Imperfect Crystal Referred to an Average Lattice[1]

Let us consider a periodic three-dimensional lattice in object space whose nodes are defined by the vectors $\boldsymbol{x}_{n_a n_b n_c}$ or $\boldsymbol{x}_n$, where $n$ indicates the group of three whole numbers $n_a$, $n_b$, $n_c$. Corresponding to this lattice we have a reciprocal lattice whose nodes are at the extremities of the vectors $\boldsymbol{r}^*_{hkl}$.

For the perfect crystal each node of the lattice corresponds to a unit cell which, by definition, repeats itself identically from node to node. This unit cell appears in the diffraction calculations through its structure factor:

$$F(\boldsymbol{s}) = \sum f_r \exp(-2\pi i \boldsymbol{s} \cdot \boldsymbol{u}_r) ,$$

for $\boldsymbol{s} = \boldsymbol{r}^*_{hkl}$. The summation is extended to all the atoms of the unit cell, $f_r$ is the scattering factor, and $\boldsymbol{u}_r$ the vector joining the lattice node to the $r$th atom.

For an imperfect crystal derived from the same lattice, the unit cell corresponding to a given node has a position and a content which depend on the node considered. The structure factor is a function of the coordinates of the unit cell:

$$F_n(\boldsymbol{s}) = \sum f_{nr} \exp(-2\pi i \boldsymbol{s} \cdot \boldsymbol{u}_{nr}) .$$

In fact, corresponding atoms can be of different types ($f_{nr} \neq f_{n'r}$) or they can be displaced ($\boldsymbol{u}_{nr} \neq \boldsymbol{u}_{n'r}$). Thus substitution disorder and displacement disorder are both identified mathematically by a variation of $F_n$ with $n$. In one case it is the *modulus* of the terms of the above sum which varies and, in the other, it is the *argument*.

If the crystal has a single atom per unit cell, and if the atoms occupy their ideal positions, we have a pure substitution disorder; $F_n$ remains real and it has a value which varies from one unit cell to another. If, on the contrary, the atoms are all identical and simply displaced, $F_n$ has a constant modulus and its phase alone varies:

$$F_n = f \exp(-2\pi i \boldsymbol{s} \cdot \Delta \boldsymbol{x}_n) , \qquad (6.1)$$

where $\Delta \boldsymbol{x}_n$ is the displacement of the $n$th atom from the corresponding

[1] Reference [2] at the end of this chapter gives the list of books and articles on which this section is based.

node of the average lattice, chosen so that $\sum_n \Delta \boldsymbol{x}_n = 0$ for all the atoms of the crystal.

Once the structure factors for each unit cell are defined, the scattering power for a crystal of volume $V$ containing $N$ unit cells is easily calculated from Eq. (2.7):

$$I_N(\boldsymbol{s}) = \sum_n \sum_{n'} F_n F^*_{n'} \exp[-2\pi i \boldsymbol{s}\cdot(\boldsymbol{x}_n - \boldsymbol{x}_{n'})] \tag{6.2}$$

If we group in the double sum the terms containing the same vector $\boldsymbol{x}_m = \boldsymbol{x}_{n'} - \boldsymbol{x}_n$ where $m = n' - n$, or $m_a = n'_a - n_a$, $m_b = n'_b - n_b$, $m_c = n'_c - n_c$, then we can write Eq. (6.2) as

$$I_N(\boldsymbol{s}) = \sum_m (\sum_n F_n F^*_{n+m}) \exp(2\pi i \boldsymbol{s}\cdot\boldsymbol{x}_m) \tag{6.2a}$$

We now introduce the homogeneity of the structure so as to use the average value of $F_n F^*_{n+m}$. The sum $\sum F_n F^*_{n+m}$ for a given and sufficiently large number $p$ of unit cells is constant, whatever the particular nodes considered. We set

$$\overline{F_n F^*_{n+m}} = \frac{1}{p} \sum_{p \text{ unit cells}} F_n F^*_{n+m} = y_m\,. \tag{6.3}$$

In Eq. (6.2a) the sum $\sum_n F_n F^*_{n+m}$ contains a number of terms, equal to the number of nodes of the lattice in the finite crystal (of volume V), such that the node $n+m$ is also in the crystal. Let us now displace the volume $V$ by a distance $\boldsymbol{x}_m$. The volume common to the two solids has already been called $VV(\boldsymbol{x}_m)$ in Section 2.4.2. The number of terms required is therefore $VV(\boldsymbol{x}_m)/v_1$, where $v_1$ is the volume of the unit cell of the average lattice, and Eq. 6.2 can be rewrittten as

$$I_N(\boldsymbol{s}) = \frac{V}{v_1} \sum_m V(\boldsymbol{x}_m) y_m \exp(2\pi i \boldsymbol{s}\cdot\boldsymbol{x}_m)\,. \tag{6.4}$$

The unit scattering power referred to one unit cell in the crystal is

$$I(\boldsymbol{s}) = \frac{I_N(\boldsymbol{s})}{N} = \frac{I_N(\boldsymbol{s}) v_1}{V} = \sum_m V(\boldsymbol{x}_m) y_m \exp(2\pi i \boldsymbol{s}\cdot\boldsymbol{x}_m)\,. \tag{6.5}$$

This equation shows that, *if the coefficients $y_m$ are independent of* $\boldsymbol{s}$, the intensity is periodic in reciprocal space, the periods being those of the reciprocal lattice of the average lattice. Thus, if we add to $\boldsymbol{s}$ any vector $\boldsymbol{r}^*_{hkl}$ of this lattice, the product $\boldsymbol{s}\cdot\boldsymbol{x}_m$ remains unchanged since, for any vector $\boldsymbol{x}_m$ of the crystal lattice, $\boldsymbol{r}^*_{hkl}\cdot\boldsymbol{x}_m$ is a whole number. The intensity diffracted by such an imperfect crystal reproduces itself identically from one unit cell to another in reciprocal space. However, this result is not valid if the coefficients $y_m$ depend on $\boldsymbol{s}$. This is the usual case because the structure factors involve the atomic scattering coefficients which are themselves functions of $|\boldsymbol{s}|$; moreover, whenever there

**FIGURE 6.2.** *Variation of the scattered intensity from one unit cell to the next in reciprocal space (schematic).*

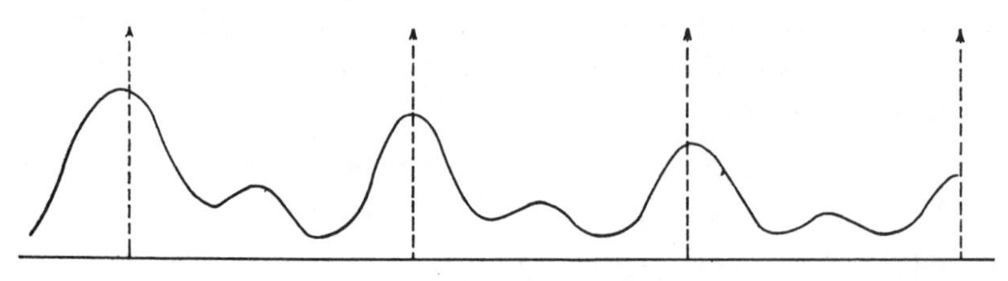

is a displacement disorder, the phase of the factors $F_n$ are also functions of $\boldsymbol{s}$ [see Eq. (*6.1*)]. In general, these variations of $y_m$ with $\boldsymbol{s}$ are slow on the scale of the unit cell of the reciprocal lattice. The quantity $I(\boldsymbol{s})$ is thus quasi-periodic, in the sense that its fluctuations repeat themselves in neighboring unit cells, altered only by a factor which varies little from one end of the unit cell to the other (Figure 6.2). More precisely, if $y_m$ can be written under the form $y_1(\boldsymbol{s})y_2(m)$, the second factor being independent of $\boldsymbol{s}$, it is $I(\boldsymbol{s})/y_1(\boldsymbol{s})$ which is periodic in reciprocal space.

Equation (*6.5*) shows the intensity diffracted by an imperfect crystal in a form which is analogous to that for the intensity diffracted by any homogeneous subject [Eq. (*2.33*)]. The quantity $I(\boldsymbol{s})$ is represented by a Fourier series instead of an integral, each term being the product of two factors, one of which depends on the size of the crystal and the other on its internal structure. Let us apply the faltung theorem [Eq. (*2.26*)] to Eq. (*6.5*). The coefficients $y_m$ correspond to a function $Y(\boldsymbol{s})$ in reciprocal space defined by the series

$$Y(\boldsymbol{s}) = \sum y_m \exp(2\pi i \boldsymbol{s}\cdot\boldsymbol{x}_m)\,. \qquad (6.6)$$

As to $V(\boldsymbol{x}_m)$, it is the value of $V(\boldsymbol{x})$ for the particular points $\boldsymbol{x}_m$ given in Section 2.4.2. According to Eq. (*2.34*), its transform is $|\Sigma(\boldsymbol{s})|^2/V$, where $\Sigma(\boldsymbol{s})$ is the transform of the form factor for the crystal. Finally, by analogy with Eq. (*2.35*), we can write

$$I(\boldsymbol{s}) = \frac{|\Sigma(\boldsymbol{s})|^2}{V} * Y(\boldsymbol{s})\,. \qquad (6.7)$$

Let $y(\boldsymbol{x})$ be a function which is regular and such that $y = y_m$ for $\boldsymbol{x} = \boldsymbol{x}_m$. The summation of Eq. (*6.6*) can then be replaced, by using the delta function, by the integral

$$Y(\boldsymbol{s}) = \sum y_m \exp(2\pi i \boldsymbol{s}\cdot\boldsymbol{x}) = \int \sum \delta(\boldsymbol{x} - \boldsymbol{x}_m) y(\boldsymbol{x}) \exp(2\pi i \boldsymbol{s}\cdot\boldsymbol{x})\, dv_x\,. \qquad (6.8)$$

The function $Y(\boldsymbol{s})$ is then the Fourier transform of the product of the functions $\sum \delta(\boldsymbol{x} - \boldsymbol{x}_m)$ and $y(\boldsymbol{x})$. We call $Y_1(\boldsymbol{s})$ the transform of $y(\boldsymbol{x})$ and, according to appendix A.4, the transform of $\sum \delta(\boldsymbol{x} - \boldsymbol{x}_m)$ is $(1/v_1) \sum \delta(\boldsymbol{s} - \boldsymbol{r}^*_{hkl})$. The faltung theorem then gives

$$Y(\boldsymbol{s}) = \frac{1}{v_1} \sum_{hkl} \delta(\boldsymbol{s} - \boldsymbol{r}^*_{hkl}) * Y_1(\boldsymbol{s}) = \frac{1}{v_1} \sum_{hkl} Y_1(\boldsymbol{s} - \boldsymbol{r}^*_{hkl}) \,. \qquad (6.9)$$

The function $y(\boldsymbol{x})$ is not completely defined, since we only know its values $y_m$ at the points $\boldsymbol{x} = \boldsymbol{x}_m$. Similarly, $Y_1(\boldsymbol{s})$ is not uniquely defined, but the sum $\sum Y_1(\boldsymbol{s} - \boldsymbol{r}^*_{hkl})$ is independent of the choice of $y(\boldsymbol{x})$. If we choose a function $y(\boldsymbol{x})$ which is as uniform as possible, $Y_1(\boldsymbol{s})$ is sharply limited around the origin of the reciprocal lattice and we then have only one important term in the summation of Eq. (*6.9*)—that relative to the nearest node.

In order to interpret Eq. (*6.7*) which gives the scattering power per unit cell of the perturbed lattice, we shall utilize the results of Section 2.4.2 on the faltung of a function by $|\Sigma(\boldsymbol{s})|^2$. If $Y(\boldsymbol{s})$ has an extremely sharp peak at $\boldsymbol{s} = \boldsymbol{s}_0$, the intensity will have at this point a peak whose profile will be determined by $|\Sigma(\boldsymbol{s} - \boldsymbol{s}_0)|^2$. This is what happens when the crystal imperfections are nonexistent or negligible. Under these conditions, according to Eqs. (*6.1*) and (*6.3*), the coefficients $y_m$ are all equal to $|F|^2$, and $Y(s)$ has a series of peaks at the nodes of the reciprocal lattice. We then have the case of the small crystal studied in Chapter 5.

When the crystal contains imperfections and the function $Y(\boldsymbol{s})$ does not have sharp peaks, and when, on the other hand, the crystal is not too small, $I(\boldsymbol{s})$ is simply proportional to $Y(\boldsymbol{s})$ and, since

$$\frac{1}{V} \int |\Sigma(\boldsymbol{s})|^2 \, dv_s = 1 \,,$$

by Eq. (*2.38*), then

$$I(\boldsymbol{s}) = Y(\boldsymbol{s}) = \sum y_m \exp(2\pi i \boldsymbol{s} \cdot \boldsymbol{x}_m) \,. \qquad (6.10)$$

A complex case is that where the function $Y(\boldsymbol{s})$ has peaks whose width is of the same order as the function $|\Sigma(\boldsymbol{s})|^2$, i.e. of the width of the domain due to the small size of the crystals. Then the faltung cannot be written in a simple form.

We can rewrite Eq. (*6.10*) in another way if we separate the real and imaginary parts of the coefficient $y_m$. Equation (*6.3*) shows that for equal and opposite values of $m$ the values of $y_m$ are conjugate. There fore, setting

$$y_m = u_m + iv_m \,,$$

we have

$$y_{-m} = u_m - iv_m \, .$$

On the other hand, $y_0 = \overline{F_n^2}$. Therefore

$$Y(\boldsymbol{s}) = \sum_{-\infty}^{+\infty} y_m \exp(2\pi i\boldsymbol{s}\cdot\boldsymbol{x}_m)$$

$$= \overline{F_n^2} + 2\sum{}' u_m \cos(2\pi\boldsymbol{s}\cdot\boldsymbol{x}_m) - 2\sum{}' v_m \sin(2\pi\boldsymbol{s}\cdot\boldsymbol{x}_m) \, . \qquad (6.10a)$$

The summations $\sum'$ are extended to half the nodes of the lattice, the origin being excluded.

The function $Y(\boldsymbol{s})$, which represents the diffracted intensity in the many cases where Eq. (*6.10*) is valid, is the transform in reciprocal space of an object space constituted by the nodes of the crystal lattice to which are related coefficients $y_m$, which vary from one node to the other. It will be recalled that, in the case of a crystal in which the atoms are not localized at the lattice points but rather distributed in the unit cell, there corresponds a reciprocal lattice whose nodes have variable coefficients, namely the structure factors of the unit cell [Eq. *4.29*)]. Inversely, because of the perfect reciprocity of the Fourier transformation, a lattice of points with variable coefficients corresponds to a transform $Y(\boldsymbol{s})$ which has the same period as the reciprocal lattice, but which can have *nonzero values* in the unit cell outside the nodes. If the coefficients $y_m$ vary little, i.e. if the perturbations in the crystal are small, the nodes of the reciprocal lattice are again points of high intensity for $Y(\boldsymbol{s})$.

It is possible to foresee in a very general way the effects of perturbations in the crystal lattice on the diffraction pattern. *An imperfect crystal produces diffracted radiation which is not well localized, as is that produced by a perfect crystal.* There are still strong diffractions according to the Bragg law for the average ideal lattice, but in other directions *the wavelets emitted by the individual atoms do not cancel perfectly by interference* because cancellation is the direct consequence of the perfect periodicity of the diffracting medium.

The diffraction can be appreciable in directions close to those of the perfect average crystal and then, experimentally, the diffraction lines or spots are broad. We have already seen an example of this in our study of small crystals, but irregularities in large crystals can have similar effects.

The diffracted intensity can also be appreciable in all directions, even in directions remote from those of a perfect crystal. This effect is called *diffuse scattering*, the term *diffraction* being reserved for the lines or spots which can be either sharp or broad, but whose positions are given by the Bragg law. In general, diffuse scattering is very weak compared to that of normal diffraction.

Let us start with a perfect crystal and displace the atoms, while preserving the same average lattice. From Section 2.3, the integral of $I(\boldsymbol{s})$ over the unit cell of the reciprocal lattice remains constant. In the case of the perfect crystal, the intensity is all concentrated at the node of the reciprocal lattice. In the case of the imperfect crystal, a fraction of this intensity is removed from the node and distributed in the unit cell. Diffuse scattering therefore results in a *decrease in the intensity of the selective reflections.* However, if the imperfections have only the effect of broadening the reflection domain around the node, and if we measure correctly the intensity of the line or of the spot taking into account the complete intensity curve, we find a total intensity which is the same as that for the unperturbed lattice.

### 6.2.1. The Case where the Imperfections Produce a Broadening of the Reflection Domain. The Profile of the Powder Pattern Line

Let us consider crystals whose structure is perturbed in such a way that the coefficients $y_m = \overline{F_n F^*_{n+m}}$ decrease slowly as the indices $m$ increase. A function $y(\boldsymbol{x})$, which has the value $y_m$ at the nodes $\boldsymbol{x}_m$ is maximum for $\boldsymbol{x} = 0$ and decreases with increasing $x$. The Fourier transform $Y_1(\boldsymbol{s})$ of a function of this type shows a peak at the center for $\boldsymbol{s} = 0$, which becomes narrow when $y(\boldsymbol{x})$ is a slowly decreasing function. A few examples of this are given in Appendix A.4. Under these conditions, it is certain that the function $Y(\boldsymbol{s})$ of Eq. (*6.9*) is formed of the peak of $Y_1(\boldsymbol{s})$ repeated around each node of the reciprocal lattice. Since these peaks are narrow and do not overlap, Eq. (*6.9*) becomes

$$Y(\boldsymbol{s}) = \frac{1}{v_1} Y_1(\boldsymbol{s}'),$$

where $\boldsymbol{s}'$ is the vector joining the point in reciprocal space defined by $\boldsymbol{s}$ and the nearest node in the reciprocal lattice.

From Eq. (*6.10*), the scattering power around each node per unit cell of the crystal is given by

$$\begin{aligned} I(\boldsymbol{s}') &= \frac{1}{v_1} Y_1(\boldsymbol{s}') \\ &= \frac{1}{v_1} \int y(\boldsymbol{x}) \exp(2\pi i \boldsymbol{s}' \cdot \boldsymbol{x})\, dv_x . \end{aligned} \tag{6.11}$$

If, however, the crystal is either not much perturbed or sufficiently small, we must use the general Eq. (*6.7*), which involves the form factor

$$I(\boldsymbol{s}') = \frac{1}{Vv_1} [\, \Sigma(\boldsymbol{s}')\,|^2 * Y_1(\boldsymbol{s}')$$

$$= \frac{1}{v_1} \int V(\boldsymbol{x}) y(\boldsymbol{x}) \exp(2\pi i \boldsymbol{s}' \cdot \boldsymbol{x})\, dv\,. \tag{6.12}$$

The approximation of Eq. (*6.11*) can be legitimate only if the error introduced in replacing $V(\boldsymbol{x})$ by unity remains reasonable up to the point where $y(\boldsymbol{x})$ has become negligible.

In general, the coefficients $y_m$ depend on $\boldsymbol{s}$. The reflection domains around the different nodes therefore depend on their indices. Since the domain is restricted to the immediate neighborhood of the node, we can replace $\boldsymbol{s}$ by $\boldsymbol{r}^*_{hkl}$, which is the value of the vector $\boldsymbol{s}$ for the node considered, in the expression for $y_m$.

In a powder pattern the domains surrounding the nodes give rise to broad Debye-Scherrer lines. We have seen in our study of small crystals (Section 5.3) that when the domains are sufficiently small the line profile can be represented on an arbitrary scale by the integral of Eq. (*5.14*):

$$i(s_0) = \int V(t) \exp(-\,2\pi i s_0 t)\, dt\,.$$

The line profile is expressed as a function of the abscissa $s_0 = (2\sin\theta)/\lambda - (2\sin\theta_0)/\lambda_0$, the origin being the position of the line for the unperturbed, perfect, and infinite lattice. The quantity $t$ is the modulus of the vector $\boldsymbol{t}$, which remains normal to the reflecting lattice planes for the line under consideration.

Starting from Eq. (*6.12*), it is possible to show in exactly the same way that, for distorted crystals, the line profile is given by

$$i(s_0) = \int V(t) y(t) \exp(-\,2\pi i s_0 t)\, dt\,. \tag{6.13}$$

When it is possible to neglect the effect of the size of the crystal on the width of the line, the formula simplifies to

$$i(s_0) = \int y(t) \exp(-\,2\pi i s_0 t)\, dt\,. \tag{6.14}$$

The Fourier transform of the broad line therefore represents the function $V(t)y(t)$, or simply $y(t)$ if the size effect is unimportant. The function $y(t)$ is the average value in the crystal of the product of the structure factors for the two unit cells situated at the extremities of a vector of length $t$ normal to the reflecting planes. The quantity $V(t)$ is equal to the volume common to the object and to its double displaced by the vector $\boldsymbol{t}$, divided by the volume of the object. If several nodes give rise to the same line, we must add their effects, calculated separately according to Eqs. (*6.13*) or (*6.14*).

If we measure the integral width of the line and if we apply the Scherrer formula given in Eq. (*5.4*), we have

$$\Delta(2\theta) = \frac{\lambda}{L\cos\theta} .$$

Each line therefore has a parameter $L$ having the dimensions of a length which we have called the *apparent size of the crystal*, when the broadening of the line is due exclusively to the size effect.

Let us examine the meaning of the parameter $L$ when the broadening is due to lattice distortions. According to Eq. (*5.10*), the width of $i(s_0)$, expressed as a function of the abscissa $s_0$, is given by

$$\Delta s_0 = \frac{\cos\theta}{\lambda}\Delta(2\theta) = \frac{1}{L} = \frac{\int i(s_0)\,ds_0}{i(0)} ,$$

and, from Eq. (*6.14*),

$$L = \frac{\int y(t)\,dt}{y(0)} \tag{6.15}$$

When there are both size and distortion effects [Eq. (*6.13*)],

$$L = \frac{\int V(t)y(t)\,dt}{y(0)}$$

and we have again Eq. (*5.19*) if there is no distortion, i.e. if $y(t)$ is a constant.

*The essential difference between size and distortion effects is that, in the first case, the apparent size is independent of the order of the reflection, while in the second case it is not independent.* It will be recalled that $y(t)$ depends on the modulus of $\mathbf{s}$, but $V(t)$ depends only on its direction (see Section 5.3).

One particular case which is often used in the calculation of lattice distortions is that where the function $y(t)$ is of the form

$$y(t) \sim \exp(-\gamma|t|) .$$

Under these conditions, from Eq. (*6.14*), the shape of the line is given by

$$i(s_0) \sim \int \exp(-\gamma|t|)\exp(-2\pi i s_0 t)\,dt ,$$

or, according to Table A.1, Appendix A,

$$i(s_0) \sim \frac{2\gamma}{\gamma^2 + (2\pi s_0)^2}, \tag{6.16}$$

and the apparent size of the crystal is

$$L = \frac{\int_{-\infty}^{+\infty} y(t)\, dt}{y(0)} = 2\int_0^{\infty} \exp(-\gamma t)\, dt = \frac{2}{\gamma}.$$

If we use Eq. (*6.16*), we also find (Section 5.3.1) that

$$L = \frac{i(0)}{\int i(s_0)\, ds_0} = \frac{2/\gamma}{\int \frac{2\gamma}{\gamma^2 + (2\pi s_0)^2}\, ds_0} = \frac{2}{\gamma},$$

and the integral width of the line expressed as a function of the angle $2\theta$ is

$$\Delta(2\theta) = \frac{\lambda}{L\cos\theta} = \frac{\lambda\gamma}{2\cos\theta}.$$

In the case where a broadened node corresponds to a broadened line in the powder pattern, it is possible to relate both the total intensity and the profile of a line to the scattering power in the reciprocal space of the single crystal. This is what Warren has called the Powder Pattern Power Theorem [3].

Let us consider a small imperfect crystal of $N$ unit cells whose reciprocal space contains a node $m$, in the vicinity of which the scattering power is given by $I(\mathbf{s})$ or by $I(hkl)$, where $h, k, l$ are the coordinates with respect to the axes of the reciprocal lattice: $\mathbf{s} = h\mathbf{a}^* + k\mathbf{b}^* + l\mathbf{c}^*$. We assume that the sample contains $M$ identical crystals having random orientations. Corresponding to the node $m$ we have a ring situated at an average diffraction angle of $\theta$. We now wish to calculate the total energy $\Phi$ scattered in this ring. We have already performed such a calculation for a perfect crystal, in which case the node was very small. This calculation is valid in the present case because we assume that the line broadening is relatively small. According to Eq. (*4.47*), for a single node ($n = 1$) and with no temperature effect ($D = 1$), with the volume of the diffracting material given by $dV = MNV_c$, we have

$$\Phi = I\, 2\pi r \sin 2\theta = I_0 r_e^2 \frac{1 + \cos^2 2\theta}{2} \frac{1}{4\sin\theta} \lambda^3 F_{hkl}^2 \frac{1}{V_c^2} MNV_c. \tag{6.17}$$

From Eq. (*4.28a*), the integral of the scattering power for a perfect crystal of $N$ unit cells is given by

$$\int I(\mathbf{s})\, dv_s = \frac{NF_{hkl}^2}{V_c}.$$

Since $1/V_c$ is the volume of a unit cell in the reciprocal lattice,

$dv_s = (1/V_c)\, dh\, dk\, dl$ , and

$$\iiint I(hkl)\, dh\, dk\, dl = V_c \int I(\boldsymbol{s})\, dv_s = NF_{hkl}^2 .$$

Thus

$$\Phi = I_0 r_e^2 \frac{1 + \cos^2 2\theta}{2} \frac{M\lambda^3}{4 V_c \sin\theta} \iiint I(hkl)\, dh\, dk\, dl . \qquad (6.18)$$

The integral is extended to the reflection domain surrounding a node. This expression remains valid when this domain is broadened. However, if the broadening is such that $\theta$ varies appreciably within the reflection domain, the correct equation is rather

$$\Phi = I_0 r_e^2 \frac{1 + \cos^2 2\theta}{2} \frac{M\lambda^3}{4 V_c} \iiint \frac{I(hkl)}{\sin\theta}\, dh\, dk\, dl . \qquad (6.19)$$

The line profile, i.e. the distribution of the intensity within the line, is given by $\Phi_{2\theta}$ so that

$$\Phi = \int \Phi_{2\theta} d(2\theta) . \qquad (6.20)$$

We assume that the axes $\boldsymbol{a}^*$, $\boldsymbol{b}^*$, $\boldsymbol{c}^*$ of the reciprocal lattice are orthogonal and that the node considered lies on the $\boldsymbol{c}^*$ axis. We perform the integration in Eq. (*6.18*) around the $00l$ node, assuming that the reflection domain is small enough that a section of reciprocal space for constant $\theta$ is approximately a plane cross-section of the domain in the direction normal to the [001] axis. Thus

$$s = \frac{2 \sin\theta}{\lambda} = lc^*$$

and

$$dl = \frac{\cos\theta\, d(2\theta)}{\lambda c^*} .$$

The differential $dl$ in Eq. (*6.19*) can be expressed as a function of $d(2\theta)$ and, from Eq. (*6.20*),

$$\Phi_{2\theta} = I_0 r_e^2 \frac{1 + \cos^2 2\theta}{2} \frac{M\lambda^2}{4 V_c c^* \tan\theta} \iint I(hkl)\, dh\, dk . \qquad (6.21)$$

Experimentally, one measures the energy diffracted per unit length in the diffraction ring at a distance $r$ from the object, or the quantity

$$i_{2\theta} = I_0 r_e^2 \frac{1 + \cos^2 2\theta}{2} \frac{M\lambda^2}{16\pi V_c c^* r \sin^2\theta} \iint I(hkl)\, dh\, dk . \qquad (6.22)$$

### 6.2.2. The Scattering Intensity

In the case where we have both diffuse scattering and narrow diffraction spots, it is important to isolate the two corresponding terms in the complete formula given in Eq. (*6.5*). We shall indicate later the conditions necessary for this to happen. Let us call $\bar{F}$ the average value of the structure factor of the unit cell for the perturbed lattice:

$$\bar{F} = \frac{1}{N} \Sigma F_n \, .$$

We set

$$F_n = \bar{F} + \varphi_n \, ,$$

$$y_m = \overline{F_n F^*_{n+m}} = \overline{(\bar{F} + \varphi_n)(\bar{F}^* + \varphi^*_{n+m})} \qquad (6.23)$$

$$= \bar{F}\bar{F}^* + \bar{F}\overline{\varphi^*_{n+m}} + \bar{F}^*\bar{\varphi}_n + \overline{\varphi_n \varphi^*_{n+m}} \, .$$

By definition,

$$\overline{\varphi_n} = \overline{\varphi_{n+m}} = 0 \, ,$$

and therefore

$$y_m = |\bar{F}|^2 + \overline{\varphi_n \varphi^*_{n+m}} = |\bar{F}|^2 + \Phi_m \, . \qquad (6.24)$$

Substituting this expression in Eq. (*6.5*), we separate $I(\boldsymbol{s})$ into two terms, $I_1$ and $I_2$:

$$I(\boldsymbol{s}) = I_1(\boldsymbol{s}) + I_2(\boldsymbol{s})$$
$$= |\bar{F}|^2 \, \Sigma \, V(\boldsymbol{x}_m) \exp(2\pi i \boldsymbol{s} \cdot \boldsymbol{x}_m) + \Sigma \, V(\boldsymbol{x}_m) \Phi_m \exp(2\pi i \boldsymbol{s} \cdot \boldsymbol{x}_m) \, . \qquad (6.25)$$

The term $I_1$ represents the intensity diffracted by a perfect crystal having the volume $V$ of the real crystal, in which all the unit cells have a structure factor $\bar{F}$ equal to the average factor of the real crystal. Therefore, according to Eq. (*4.28*),

$$I_1(\boldsymbol{s}) = \frac{|\bar{F}|^2}{V v_1} | \, \Sigma \, (\boldsymbol{s} - \boldsymbol{r}^*_{hkl}) \, |^2 \, .$$

It is the second term of Eq. (*6.25*) which represents the diffuse scattering. The size effect is unimportant in this term or, in other words, $\Phi_m$ rapidly becomes equal to zero when $m$ increases, so it is possible to set $V(\boldsymbol{x}_m)$ equal to unity for all the terms which are not negligible. Therefore

$$I_2(\boldsymbol{s}) = \Sigma \, \Phi_m \exp(2\pi i \boldsymbol{s} \cdot \boldsymbol{x}_m) \, . \qquad (6.26)$$

According to the definition of $\Phi_m$ given in Eq. (*6.24*) we can see that $\Phi_m = \Phi^*_{-m}$, and $\Phi_0$ is real. Separating the real and imaginary parts, $\Phi_m = J_m + iK_m$ and, grouping the terms $m$ and $-m$ two by two, the above summation can be rewritten in a form analogous to that of Eq. (*6.10a*):

$$I_2(\boldsymbol{s}) = \Phi_0 + 2 \, \Sigma' \, J_m \cos(2\pi \boldsymbol{s} \cdot \boldsymbol{x}_m) - 2 \, \Sigma' \, K_m \sin(2\pi \boldsymbol{s} \cdot \boldsymbol{x}_m) \qquad (6.27)$$

or

$$I_2(\mathbf{s}) = \sum_{-\infty}^{+\infty} J_m \cos(2\pi\mathbf{s}\cdot\mathbf{x}_m) - \sum_{-\infty}^{+\infty} K_m \sin(2\pi\mathbf{s}\cdot\mathbf{x}_m)\,. \tag{6.28}$$

As indicated previously, it is possible to replace these summations by integrals if we can define regular functions $\Phi(\mathbf{x})$, $J(\mathbf{x})$, $K(\mathbf{x})$, which have the values $\Phi_m$, $J_m$, $K_m$ for $\mathbf{x} = \mathbf{x}_m$:

$$I_2(\mathbf{s}) = \sum \Phi_m \exp(2\pi i\mathbf{s}\cdot\mathbf{x}_m) = \frac{1}{v_1}\int \Phi(\mathbf{x}) \exp(2\pi i\mathbf{s}\cdot\mathbf{x})\, dv_x\,,$$

$$= \frac{1}{v_1}\left[\int J(\mathbf{x}) \cos(2\pi\mathbf{s}\cdot\mathbf{x})\, dv_x - \int K(\mathbf{x}) \sin(2\pi\mathbf{s}\cdot\mathbf{x})\, dv_x\right]. \tag{6.29}$$

Let us find the value of $\Phi_0$. Using the definition (*6.24*),

$$y_0 = \overline{F_n F_n^*} = \overline{|F_n|^2} = |\bar{F}|^2 + \Phi_0\,.$$

$$\Phi_0 = \overline{|F_n|^2} - |\bar{F}|^2. \tag{6.30}$$

This is the difference between the average of the squares of the moduli of $F_n$ and the square of the modulus of the average of $F_n$. This first term in the summation for $I_2$ is particularly interesting because the coefficients of all the other terms depend not only on the fluctuations of the structure factor from unit cell to unit cell, but also on the *correlation between these fluctuations* in two unit cells separated by a given vector. If there is no correlation between the perturbations existing in these pairs of unit cells, the average of the product $\varphi_n \varphi_{n+m}^*$ is the product of the averages $\bar{\varphi}_n \overline{\varphi_{n+m}^*}$ and these averages are zero. Therefore, for $m \neq 0$, $\Phi_m = 0$ and the scattering intensity reduces to

$$I_2 = \overline{|F_n|^2} - |\bar{F}|^2\,. \tag{6.31}$$

We have here an application of a very general result established in Section 2.7. Since the vector $\mathbf{s}$ does not appear explicitly in Eq. (*6.31*), we have a very diffuse scattering which is superposed over the diffraction by the perfect average lattice. In general, this diffuse scattering is weak, even if the local imperfections are large but perfectly incoherent. It shows up experimentally as a general haze between the diffraction spots or diffraction lines, and it cannot be distinguished from the parasitic scattering which is found on all patterns, unless special precautions are taken. Therefore, as a first approximation, the diffraction pattern of the perturbed lattice is identical to that of a perfect crystal. This will justify our remarks at the beginning of this chapter: X-rays exaggerate the regularities in structure, and crystal imperfections produce only second-order effects.

The scattering becomes more intense and *modulated* in the unit cell of the reciprocal lattice if the coefficients $\Phi_m$ are not all zero. The coefficient $\Phi_m$, which is the average of the product $\varphi_n \varphi_{n+m}^*$, is different from zero only if there exists a correlation between $\varphi_n$ and $\varphi_{n+m}$, i.e. between the fluctuations of the structure factor in the unit cells $n$ and $n+m$, or between the atomic displacements or substitutions in these two unit cells. In a way, the function $\Phi_m$ is a measure of this correlation at the distance $\boldsymbol{x}_m$. In general, correlation exists between neighboring unit cells but decreases as the distance between the unit cells increases. This is why the coefficients $\Phi_m$ decrease with $m$. The approximation which we have used in establishing Eq. (*6.26*), by setting $V(\boldsymbol{x}_m)$ equal to unity, is therefore justified.

### 6.2.3. General Interpretation of Scattering Measurements on Imperfect Crystals

The scattering intensity is shown as a Fourier series in Eq. (*6.26*). If the coefficient $\Phi_m$ is independent of $\boldsymbol{s}$, then $I_2(\boldsymbol{s})$ is a periodic function of $\boldsymbol{s}$, its periodicities being those of the reciprocal lattice. We can then perform an inversion to calculate the coefficients $\Phi_m$ [Eq. (*6.29*)]:

$$\Phi_m = v_1 \int_{\text{unit cell}} I_2(\boldsymbol{s}) \exp(-2\pi i \boldsymbol{s} \cdot \boldsymbol{x}_m)\, dv_s\,. \tag{6.32}$$

If the function $I_2(\boldsymbol{s})$ is not periodic, this means that $\Phi_m$ depends on $\boldsymbol{s}$. Let us suppose that $I_2$ fluctuates in two neighboring unit cells in a similar but not identical manner. Then the above equation is still approximately true, since the integral gives the value of $\Phi_m$ for a value of $\boldsymbol{s}$ corresponding to the center of the chosen unit cell. The calculation can be made accurate if we can find a function $\Phi_1(\boldsymbol{s})$ such that $I_2(\boldsymbol{s})/\Phi_1(\boldsymbol{s})$ is periodic within a good approximation. We then invert this ratio and the coefficients thus found are multiplied by $\Phi_1(\boldsymbol{s})$ to obtain the required coefficients $\Phi_m$. In general, $\Phi_1(\boldsymbol{s})$ must contain the factor which shows the decrease of the atomic scattering factors with $|\boldsymbol{s}|$.

Under these conditions, the scattering power $I_2$ per unit cell for the diffuse scattering in reciprocal space is calculated according to the fol-fowing theorem. The quantity $I_2(\boldsymbol{s})$ is the Fourier transform of the pattern formed by the nodes $\boldsymbol{x}_m$ of the average crystal lattice which have the coefficients $\Phi_m$, where $\Phi_m$ is the difference between the average of the product of the form factors $\overline{F_n F_{n+m}^*}$ and the squared average $|\bar{F}|^2$.

This theorem leads us to the following conclusions, some of which we have already found.

(a) When there is no correlation between the perturbations existing in the unit cells even when they are neighbors, only $\Phi_0$ is different from zero and $\Phi_m$ is represented by a delta function $\delta(\boldsymbol{x})$, whose transform is a constant. This is expressed by Eq. (*6.31*).

(b) If there exists a strong correlation between the perturbations in the various unit cells, i.e. if the coefficients $\Phi_m$ decrease slowly with $m$, the scattered intensity is concentrated around the nodes of the reciprocal lattice. This is the case which was studied in Section 6.1.1, where we simply have a broadening of the nodes or of the lines in a powder pattern.

(c) We can find these results again and generalize them by interpreting Eq. (*6.32*) in the following manner. The function $\Phi_m$ can be decomposed as a sum of plane waves of amplitude $I_2(\mathbf{s})\,dv_s$ whose wave fronts are normal to $\mathbf{s}$ and whose wavelength is $1/|\mathbf{s}|$. The vector $\mathbf{s}$ can be replaced by the vector $\mathbf{s}'$ joining the extremity of $\mathbf{s}$ to any node of the reciprocal lattice, since $\exp(-2\pi i\mathbf{s}\cdot\mathbf{x}_m)$ of Eq. (*6.32*) is equal to $\exp(-2\pi i\mathbf{s}'\cdot\mathbf{x}_m)$, if $\mathbf{s}'=\mathbf{s}+\mathbf{r}^*_{hkl}$. We can therefore represent all the waves by vectors $\mathbf{s}'$ within a single unit cell of reciprocal space, drawn in such a way that the nodes are at its center.

The points which are close to the nodes of the reciprocal lattice correspond to long wavelengths, and the external regions of the unit cells correspond to the shorter wavelengths. The regions of reciprocal space where the scattering is important are those which correspond to the predominant components of $\Phi_m$.

If $\Phi_m$ is periodic and has the form of a plane wave proportional to $\cos(2\pi\mathbf{k}\cdot\mathbf{x}_m)$, the summation of Eq. (*6.26*) is zero except for equal and opposite values of $\mathbf{s}'$, $\pm\mathbf{k}$. The scattered intensity is then concentrated in two points which are symmetrical with respect to the nodes of the reciprocal lattice at the distance $\mathbf{k}$, which is the propagation vector for the perturbation, its direction being normal to the wave fronts and its modulus being the inverse of the wavelength of the perturbation. This is illustrated in Figure 6.3. If all the short wavelength components of $\Phi_m$ have coefficients which are very small compared to those of larger wavelengths, the scattering is concentrated around the node. This is the case which we have discussed previously in Section 6.1.1, where the distortion in the lattice produces only a broadening of the lines and negligible diffuse scattering.

On the contrary, if the components all have the same amplitude, whatever their wavelength, there is no correlation in the disorder from one unit cell to another, and the scattered intensity is then constant.

It is obvious that all the intermediate cases are possible where we have at the same time broad diffraction spots and diffuse scattering which is more or less intense and more or less concentrated.

Experimental measurements of the scattered intensity cannot give

**FIGURE 6.3.** *The coefficients $\Phi_m$ in object space and the corresponding distribution of the scattering power in reciprocal space (schematic representation for one dimension).* (a) *$\Phi_m(x)$ has a period $\Lambda$. In this case there are two satellites on either side of each node of the reciprocal lattice, at a distance of $1/\Lambda$.* (b) *$\Phi_m(x)$ is a delta function. The scattering is then uniform in reciprocal space.* (c) *$\Phi_m(x)$ decreases from a maximum at $x = 0$. The scattering power is then concentrated near the nodes of the reciprocal lattice, the concentration being greatest when the $\Phi_m(x)$ curve is wide.*

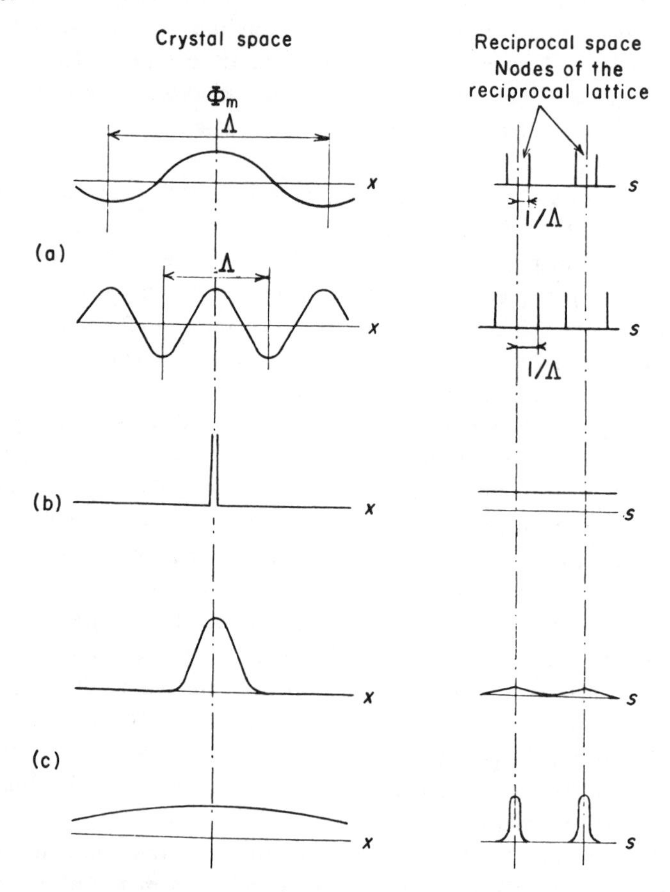


more than the coefficients $\Phi_m$ given by Eq. (*6.32*) without introducing a priori hypotheses on the crystal imperfections.

We obtain in this way an average value for the product of the factors $\varphi_n$ shown in Eqs. (*6.23*), which we may call the disorder factors, for two unit cells, as a function of the distance between them. This is the form taken by the general rule for the diffraction pattern produced by a sample of undefined structure (Section 2.2) in the case of perturbed crystals: the experiments give the average value of the product of the electron densities at two points as a function of their distance. The function $\Phi_m$ is far from adequate to give a proper description of an imperfect crystal in terms of a statistical structure. Neverthless, we shall study cases where one can arrive more or less easily at a model for the atomic distribution. We have already encountered such cases in discussing amorphous bodies.

In the study of crystal imperfections we therefore have not only experimental difficulties arising from the weak intensity of the characteristic scattered radiation, but also fundamental theoretical difficulties which arise from the fact that the diffraction patterns are in general insufficient to determine the structure of the object.

### 6.2.4. Particular Case of Planar and Linear Disorders

There are some very important cases where the disorder is essentially anisotropic in the crystal.

Let us first consider *planar disorder*. The lattice planes of a certain family $(\pi)$ then preserve their accurate periodicity and remain parallel between themselves, but they are not arranged regularly, because the distance from one to the other is not constant, or because they are displaced parallel to themselves in some irregular fashion, or, finally, because they contain different atoms. In general, $\varphi_n = \varphi_{n+m}$ if $\boldsymbol{x}_m$ is equal to any vector $\boldsymbol{x}_p$ of the plane $(\pi)$. Therefore $\Phi_m$ is constant in all the lattice planes which are parallel to $(\pi)$. Then the integral of Eq. (*6.32*) must remain unchanged if we add to $\boldsymbol{x}_m$ *any* vector $\boldsymbol{x}_p$, and $I_2(\boldsymbol{s})$ must be zero except if $\boldsymbol{s}$ is normal to the plane $(\pi)$, since $\boldsymbol{s}\cdot\boldsymbol{x}_p$ is then equal to zero. *Planar disorder is therefore characterized by the fact that scattering is limited to the rows of the reciprocal lattice which are normal to the lattice planes whose structure is intact* (Figure 6.4).

This result, which we shall often use, can be demonstrated as follows. Let us consider the plane $(\pi)$. Corresponding to this two-dimensional lattice we have a reciprocal space formed of a series of straight lines perpendicular to $(\pi)$ and passing through the nodes of the plane lattice. If we consider a collection of $N$ identical parallel planes arranged in a random manner, the diffracted intensity is simply $N$ times the intensity diffracted by a single plane (gaseous diffraction). Along these straight lines in reciprocal space the intensity is proportional to the structure factor.

If, on the contrary, the $N$ planes are arranged to form a regular space

**FIGURE 6.4.** *Planar disorder. The planes* ($\pi$) *are identical and parallel, but arranged irregularly. Scattering in reciprocal space is then limited to the rows normal to* ($\pi$).

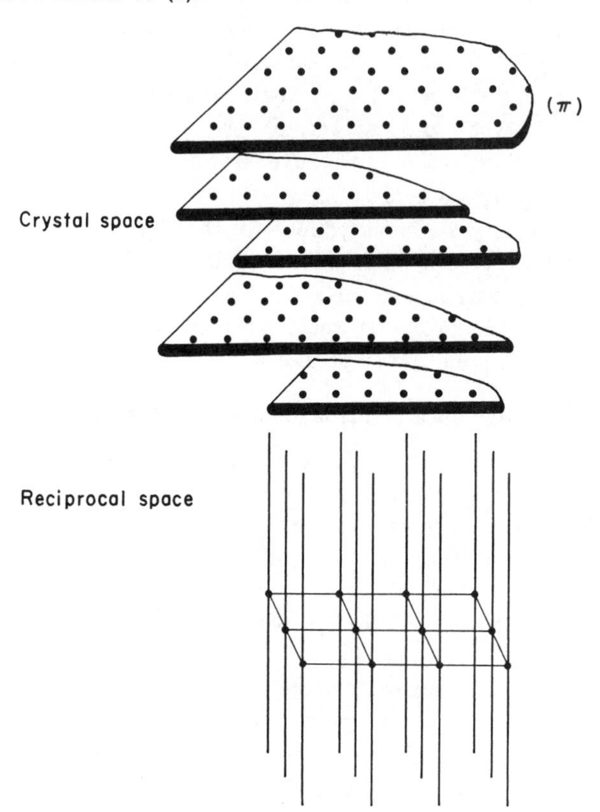


lattice, the intensity along the rows becomes zero by interference, except at the nodes of the space lattice. When the planes are slightly perturbed with respect to the arrangement in a perfect crystal, the nodes on the rows again give intense diffractions corresponding to the average lattice but, along the rows between these nodes, the intensity is not completely reduced to zero. This is the origin of the scattering mentioned above.

Another particular case is that of the *linear disorder* illustrated in Figure 6.5. In this case, the periodicity of the lattice is preserved in only one direction. The crystal is then composed of perfectly periodic, identical, and parallel rows, but these are not arranged regularly, one with respect to the other, or some of them are occupied by different types of atoms.

**FIGURE 6.5.** *Linear disorder. The rows (D) are identical and parallel but arranged irregularly. Scattering in reciprocal space is then limited to planes normal to (D).*

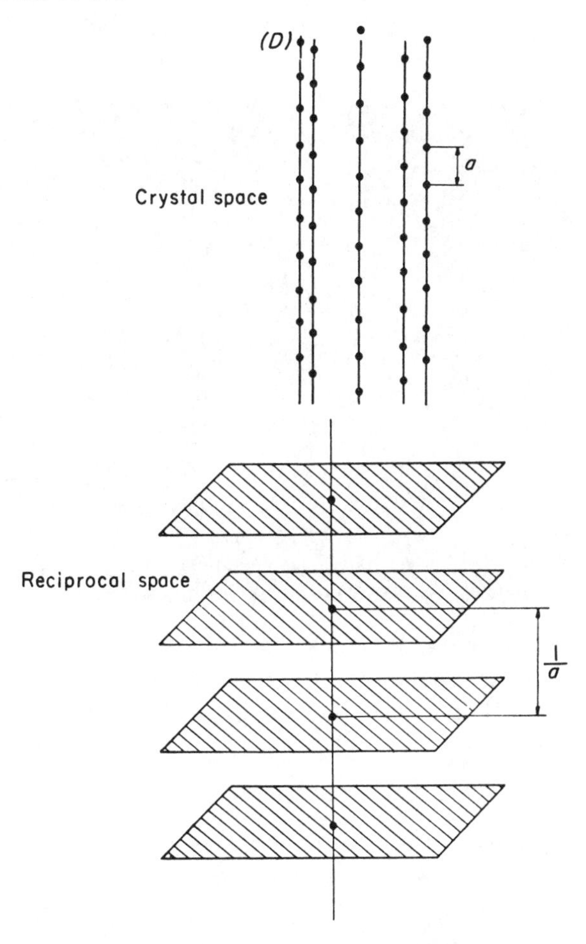


It is possible to show, by means of an argument which is similar to the preceding one, that *the scattering is then concentrated on a family of parallel planes, normal to the rows, and spaced by a distance $1/a$, $a$* being the period of the linear lattice. It is this set of planes which forms the reciprocal space of the linear lattice. The interference function would be constant along the planes if the disorder in the arrangement of the rows were complete (Figure 6.6). If it is not, the distribution of the scattering in these planes depends on the disorder in the arrangement of the rows.

**FIGURE 6.6.** *Al Mg Ge single monocrystal during age hardening as an illustration of linear disorder. The pattern was obtained with the crystal fixed in position and with monochromatic Mo Kα radiation. The scattering planes parallel to {100} intersect the Ewald sphere along circles corresponding to very slightly curved hyperbolas on the pattern (Lambot [21]).*

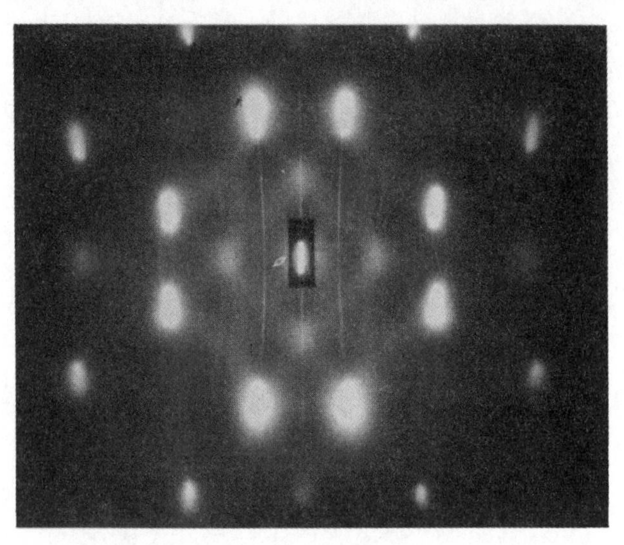

In both of these examples, and especially in the first, the scattering is concentrated in narrow regions instead of being distributed throughout the unit cell; it is thus more intense and easier to detect.

## 6.3. Scattering Experiments

The general theory developed in the preceding chapters relates the structure of an imperfect crystal to the distribution of the scattering power in reciprocal space. The task of the experimenter is therefore to determine this distribution. We have already discussed the particular case where this scattering is limited to the immediate neighborhood of the nodes of the reciprocal lattice. This involves the experimental study of the width and of the profile of the diffraction spots or diffraction lines (Section 5.5). We shall now discuss the measurement of the scattered intensity distributed over the unit cells of the reciprocal lattice.

We consider a monochromatic beam of X-rays of wavelength $\lambda$ incident on a small fixed crystal along the direction $\boldsymbol{S}_0$. If the direction of observation is $\boldsymbol{S}$, the vector $\boldsymbol{s}$ is equal to $(\boldsymbol{S} - \boldsymbol{S}_0)/\lambda$. From the orientation of the crystal with respect to $\boldsymbol{S}_0$ and $\boldsymbol{S}$, we can situate the vector

$\boldsymbol{s}$ in reciprocal space, and therefore identify the point in reciprocal space which corresponds to each direction of observation. The theoretical formulas give the value of $I(\boldsymbol{s})$, which is the scattering power expressed in electrons per unit cell:

$$I(\boldsymbol{s}) = \frac{\text{scattered intensity (experimental)}}{\text{intensity scattered by a single electron (theoretical)} \times \text{effective number of unit cells.}} \quad (6.33)$$

The scattering by a free electron situated at the position of the object is given by the Thomson formula [Eqs. (*1.21*) to (*1.23*)], all other experimental conditions remaining the same.

The *effective number* of unit cells in the object is that which is contained in a volume which would produce the same scattering as the object if there were no absorption, the true absorption of both the incident and scattered rays being taken into account. This number is $(V/V_c)A$, where $V$ is the volume of the object, $V_c$ is the volume of the average unit cell, and $A$ is the absorption factor as defined for the measurement of diffracted intensity. The method used for calculating $A$ is given in the *International Tables for Crystallography* [4], which also gives its values for various shapes of objects. We shall give an example later, in Section 6.3.3.

If it is not possible to measure the incident intensity, but if instead it is kept constant, the distribution $I(\boldsymbol{s})$ as a function of $\boldsymbol{s}$ is expressed in arbitrary units.

The detectors can be a photographic film, a Geiger counter, or an ionization chamber. We shall first discuss the experimental requirements for scattering measurements and then examine the properties of the various instruments.

### 6.3.1. Experimental Requirements

(a) Except in particular cases, it is necessary to use a *single crystal.* With a crystalline powder, one can only determine the average value of $I(\boldsymbol{s})$ for a given modulus $s$ of the vector $\boldsymbol{s}$ and, with a complex distribution of the scattering in reciprocal space, this average value cannot be interpreted at all. Also, if the scattered radiation is localized, as for example on certain rows, the average value is very small even if the scattering is locally of appreciable intensity. It is therefore impossible to detect anomalous scattering on powder patterns. It was only after studying isolated crystals of Al Cu alloys that it became possible to observe the scattering which is characteristic of the solid solution during age hardening [5]. It was also the study of single crystals of Cu Si [6] and of $AuCu_3$ [7] which contributed most to our present knowledge on stacking faults for the planes (111) in cubic metals, and on order and disorder phenomena, respectively. Unfortunately, substances are in some cases available only in the form of microscopic or submicroscopic grains of irregular shape and random orientation (catalysts), or they may be composed of fibers for which only one axis has a known direction. It is then impossible to determine completely the reciprocal lattice, and the analysis of the imperfections is therefore limited in scope. Nevertheless, it is possible in certain special cases to reconstitute the reciprocal lattice (disordered layer structures, Mering [2]).

(b) It is highly recommended, and often essential, that the *incident beam be strictly monochromatic.* Let us consider a crystal containing imperfections. Its reciprocal space is that of the average perfect crystal, plus an anomalous scattering which is, in general, of relatively low intensity. If the line used is accompanied by a continuous spectrum, the Laue pattern of the average perfect crystal is superposed over the scattering pattern, and the former can be much more intense than the latter [8].

For example, there is a very strong Laue reflection at the intersection of any axis of the reciprocal lattice with the Ewald sphere [9], and it is impossible to measure the scattered intensity along these axes. In certain cases, such as that of age-hardened Al Cu [10], it is just this scattered intensity along the axes {100} which is the most significant.

The continuous spectrum also exhibits the undesirable effect of increasing the background due to stray scattering in the air and to fluorescence radiation.

In general, therefore, it is necesssary to use monochromatic radiation for the incident beam, *especially for quantitative measurements.* For best accuracy one should use a crystal monochromator, but if a counter is employed as detector, the double filter technique of Ross [11] is simpler and in some cases adequate.

The Ross method consists in making two successive measurements with two different filters, whose absorption discontinuities are on either side of the wavelength $K\alpha$ which is characteristic of the anode. The filters are designed to have the same coefficient of transmission, except between the two absorption discontinuities. Thus, to isolate Cu $K\alpha$, we utilize nickel and cobalt filters, and the curves of Figure 6.7 show that a cobalt filter with mass $p$ per unit area is matched with a nickel filter whose mass per unit area is $0.9p$. The difference in the intensities measured after filtering with one and then with the other corresponds to the contribution of the Cu $K\alpha$ line, and also of that band of the

**FIGURE 6.7.** *Absorption coefficients of cobalt and nickel as a function of wavelength.*

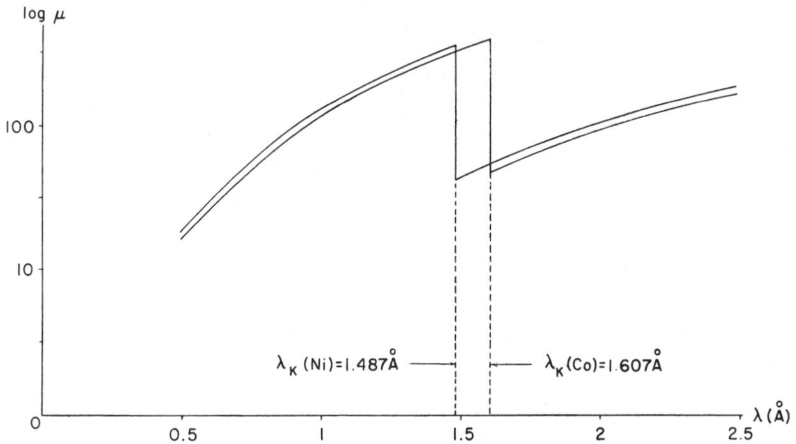


continuous spectrum lying between 1.487 Å and 1.607 Å. This band is narrow enough to be negligible. It can be shown that, with a pair of filters of $6.7 \times 10^{-3}$ g/cm$^2$ of cobalt and $6.0 \times 10^{3}$ g/cm$^2$ of nickel (optimum pair [12]), the difference between the measurements corresponds to 0.63 times the incident intensity of the Cu K$\alpha$ line. This double filter method of Ross has the advantage of providing a higher intensity than reflection from a crystal.

It is important to remember that a crystal monochromator does not only give the required radiation $\lambda$, but also narrow harmonic bands $\lambda/2$, $\lambda/3$, $\cdots$ reflected by the crystal in the second, third, $\cdots$ orders. For certain types of work this parasitic radiation can be highly undesirable, even when its intensity is relatively low. Thus, for a cubic face-centered crystal, the $\lambda/2$ radiation produces Bragg reflections at points which are exactly superlattice nodes for $\lambda$ and are therefore particularly interesting positions for studies of order.

It is then necessary to separate the main radiation from its harmonics in the scattered beam. The best method is to use a scintillation or a proportional counter, together with an electronic discriminator to eliminate pulses corresponding to the $\lambda/2$ photons. With a Geiger counter one can also use the double filter method of Ross, and, with a photographic plate, appropriate filters can be utilized. For example, with the Mo K$\alpha$ radiation, two different films are used on either side of a 0.2 mm copper foil. This stops the Mo K$\alpha$ radiation completely, but has little effect on the $\lambda/2 = 0.35$ Å radiation, and even less on the $\lambda/3$ radiation. Whatever appears on the second film is then subtracted from the first, since it is not due to Mo K$\alpha$.

(c) The main problem in scattering measurements is the weak intensity of the scattered radiation. It is quite possible that anomalous scattering will not be detected if one does not use a very intense source, especially with photographic methods. It is advantageous to use a rotating anode X-ray tube for this work and to adjust its focus with great care. Bent-crystal monochromators should be used since they provide the most intense beams[13]. With an ordinary tube and with a bent-quartz-crystal monochromator, the scattering patterns of alloys in the process of age hardening, or with partial order, require exposure times of the order of 4 hours; with a rotating anode the time is only of the order of one half hour.

Cylindrical monochromators have the disadvantage of giving a linear beam, and the vector $\boldsymbol{s}$ then changes from one point to another in the sample. One then measures an average value of the scattering in a region of reciprocal space which can be large enough to spoil the accuracy of the measurements, especially at very small angles. It would be interesting to use a point monochromator in such cases [14]. Warren [15] has devised a monochromator which gives a very intense beam, permitting scattering measurements for a well-defined value of the vector $\boldsymbol{s}$; he has used it for measurements of temperature-diffuse scattering.

(d) Anomalous scattering can be observed only if parasitic scattering is carefully eliminated. The incident rays must therefore not encounter objects which can either scatter or diffract rays into the detector. In particular, the air along the path of the incident beam, especially near the detector, can produce more background radiation than the anomalous scattering. Operation in a vacuum improves the results considerably, especially with photographic plates. With a Geiger counter it is possible to take zero readings by eliminating the radiation scattered by the sample.

It is essential to *eliminate the fluorescence radiation* by selecting the incident radiation appropriately. One can use either a long wavelength which cannot excite fluorescence, or a short wavelength with an appropriate filter in front of the detector so that the fluorescence radiation can be stopped without affecting the useful radiation to any great extent. For example, with a copper sample one can use either the Cu K$\alpha$ line or the Mo K$\alpha$ line with a 0.2 mm aluminum filter.

(e) Certain types of scattered radiation are inevitable, namely *Compton radiation* and *thermal agitation scattering*. Both are minimum for small scattering angles, where weak scattered radiation is therefore most easily detected.

The intensity of Compton radiation can be calculated as a function of the angle and of the atomic composition of the sample. It can thus be subtracted from the measured intensity. Recently more accurate calculations [16] have appreciably changed the formerly adopted values listed in Compton and Allison [17], and these new values seem to be in good agreement with the experimental measurements [18]. The correction for Compton radiation is very important when the scattering to be measured is of the same order of magnitude. But Compton scattering often corresponds to a scattering power per atom which is of the order of only one-tenth of the scattering due to the crystal imperfections.

As to thermal agitation scattering, the measurements made to date on pure crystals make it possible to apply a satisfactory correction [19]. It would seem appropriate to reduce this scattering by lowering the temperature of the sample, but this is experimentally difficult and it cannot be eliminated entirely because of zero point energy (Section 7.1.2). In any case, a correction can be made by comparing measurements made at two different temperatures.

### 6.3.2. Measurements with Photographic Plates

In this case the beam emerging from the monochromator is incident on a fixed monocrystal. The scattered radiation is recorded on a plane film which is perpendicular to the incident beam or, preferably, on a photographic film supported on a cylindrical surface whose axis passes through the sample (Figure 6.8). If the observations can be limited to angles smaller than 60°, which is often satisfactory with the Mo K$\alpha$ radiation, the best type of sample is a small plate whose thickness is of the order of $1/\mu\rho$ [20] and through which the diffracted rays are transmitted. The preparation of such crystals is quite delicate because the lattice must be left undistorted. Chemical or electrolytic methods are therefore preferable to mechanical methods. On the other hand, the transmission technique cannot be used in practice with strongly absorbing samples. In such cases the measurements must be made by reflection. The beam reflected by a bent-crystal monochromator converges at an angle of 1–2°, but the resulting inaccuracy is in general unimportant because the diffuse nature of the scattered radiation makes accurate focusing useless.

A more serious difficulty is that the beam has the shape of a small straight line at its focus. The pattern which is recorded is deduced from the theoretical pattern by sweeping in the vertical direction over a height which is equal to that of the beam, i.e. a few millimeters. This reduces the accuracy of the measurements in the direction of the height of the

**FIGURE 6.8.** *Photographic camera for diffuse scattering studies*: A, *rotating anode X-ray tube*; B, *monochromator*; C, *movable sample holder*; D, *slits for reducing background. The film is placed along a cylinder whose axis passes through the sample at* C.

photographic film, and this also distorts the diffuse spots. Thus, for a certain crystal of Al Mg Si [21] the theoretical scattering pattern shows thin horizontal and vertical lines. The horizontal lines give wide bands, but the vertical ones remain sharp and are much more intense, since their energy is concentrated over a much smaller surface. In fact, it is only the vertical lines which can be observed. This difficulty can be avoided [22] but at the cost of a considerable increase in exposure time.

A special problem of considerable difficulty is the study of the diffuse scattering in the immediate vicinity of a node of the reciprocal lattice. The normal spot has a finite size depending upon the geometry of the incident beam; furthermore, it is so intense that the photographic film is fogged out to a few millimeters from the center of the spot. Manenc [23] has obtained interesting results by a double focusing method with a monochromator and with the sample placed in an adequate position.

A section of reciprocal space passing through the Ewald sphere corresponds to a pattern recorded with the sample fixed in position. If the orientation of the crystal with respect to the average incident ray is known, it is possible to situate the surface explored with respect to the reciprocal lattice of the crystal; each point on the film then corresponds to a point on the sphere whose position can be calculated from the coordinates measured on the film. This calculation is rather elaborate, but it can be simplified by the use of appropriate nomograms [24]. One can explore in this way all the reciprocal space of the crystal by recording a series of patterns for various orientations of the crystal. The positions and the shapes of the regions of scattering are thus determined qualitatively, the variations in the film blackening representing variations in the scattering power $I(\mathbf{s})$. One can compare two successive patterns if they are standardized—for example by recording the Debye-Scherrer rings for a given thin metallic foil. A more accurate determination of $I(\mathbf{s})$ would require many corrections, analogous to the ones described in the following paragraph, but they are rarely used because the photographic method is mostly qualitative.

### 6.3.3. Measurements with a Geiger Counter

Geiger counters are especially desirable because of their sensitivity. In a few minutes they can measure radiation which, with photographic film, would require exposure times of the order of 100 hours. Geiger counters are also more accurate and easier to use.

The monochromatic beam is incident on the sample, supported on the axis of a goniometer analogous to those used for powder patterns. The sample can be oriented at will. Except for very small scattering angles, which require transmission methods, one can use a large crystal having one plane face on which the incident radiation is reflected. This eliminates difficulties in the preparation of the sample. In the setup of Figure 6.9, monochromatic radiation converges on the sample: there is no point in seeking focusing conditions as in the case of powder patterns, since the scattered intensity generally depends not only on the scattering angle $2\theta$ but also on the angle between the primary ray and the crystal. The extremity of the vector $\mathbf{s} = (\mathbf{S} - \mathbf{S}_0)/\lambda$, for a given position of the

**FIGURE 6.9.** *Camera for single crystals. A Geiger counter is used.*

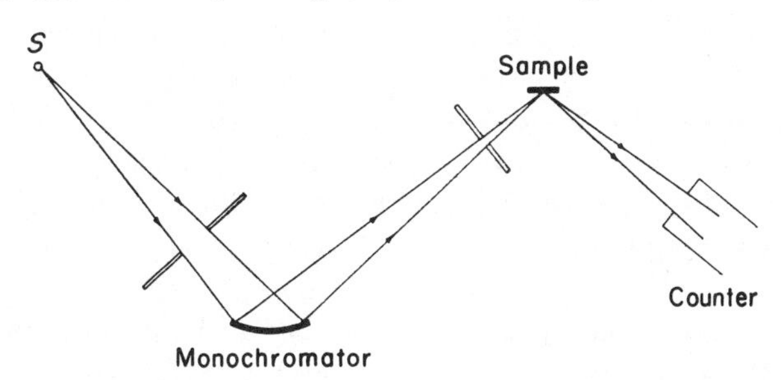

counter, lies in a small volume of reciprocal space. If the function $I(\boldsymbol{s})$ has a weak or zero curvature in the region considered, the measurement gives the correct value of $I(\boldsymbol{s})$ for the center of the volume. In other words, one can define the vector $\boldsymbol{s}$ from the average incident and scattered rays. However, if $I(\boldsymbol{s})$ varies rapidly and has a strong curvature—as, for example, near a node in the case of thermal scattering—corrections become necessary. Olmer [25] has described an accurate method of correction, and Curien [26] has used a simplified one. The detailed exploration of a large region of the reciprocal lattice is lengthy and requires very many measurements for different positions of the detector and different orientations of the crystal. For example, Cowley [7] made 450 measurements to cover 1/32 of the unit cell of the reciprocal lattice of $AuCu_3$, the remaining volume being deduced by symmetry [27], [28].

To calculate $I(\boldsymbol{s})$ according to Eq. (*6.33*), we require the energy which would be received by the counter if a free electron were at the position of the sample from the Thompson formula, [Eq. (1.22)]; this is

$$I_e = r_e^2 I_0 \left( \frac{1 + \cos^2 2\theta}{2} \right) d\omega \qquad (6.34)$$

if the primary beam is unpolarized. The quantity $d\omega$ is the solid angle subtended by the window of the counter at the sample, $2\theta$ is the scattering angle, $I_0$ is the energy per unit cross-section of the primary beam, and $r_e$ is a constant called the classical radius of the electron ($r_e^2 = 7.90 \times 10^{-26}\ \mathrm{cm}^2$). But, after reflection on the crystal monochromator at an angle $\alpha$, the monochromatic beam is partially polarized and, according to Eq. (*1.21*), we must rather use

$$I_e = r_e^2 I_0 \left( \frac{1 + \cos^2 2\alpha \cos^2 2\theta}{1 + \cos^2 2\alpha} \right) d\omega \,. \qquad (6.35)$$

One must also calculate the "effective" volume of the sample. We shall give an example of this type of calculation for the case of a plane sample of infinite thickness. Let $\alpha$ and $\beta$ be the angles formed with the surface by the incident and scattered rays, $S$ the cross-section of the incident beam (Figure 6.10), $\mu$ and $\rho$ the mass absorption coefficient and the density of the sample. Consider the layer of thickness $dz$ at the depth $z$, irradiated by the primary beam; it emits the same scattered energy as would the volume

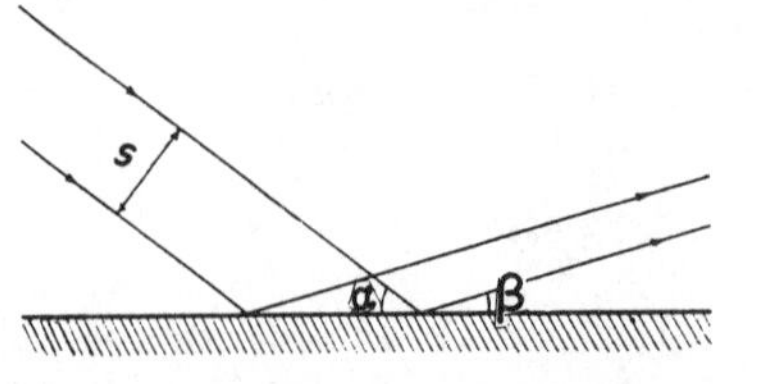


**FIGURE 6.10.** *Absorption correction for scattering measurements by reflection.*

$$S\frac{dz}{\sin\alpha}\exp\left[-\mu\rho z\left(\frac{1}{\sin\alpha}+\frac{1}{\sin\beta}\right)\right] \tag{6.36}$$

if there were no absorption. By integrating from $z=0$ to $z=\infty$, one finds that the "effective" volume is equal to

$$\frac{S}{\mu\rho\,(1+\sin\alpha/\sin\beta)}\,. \tag{6.37}$$

Setting $i(\mathbf{s})$ to be the power as recorded by the counter, we have for the scattering power per unit cell, from Eqs. (6.33) and (6.34),

$$I(\mathbf{s})=\frac{2i(\mathbf{s})\mu\rho[1+(\sin\alpha)/(\sin\beta)]v_1}{r_e^2E_0(1+\cos^2 2\theta)\,d\omega}\,, \tag{6.38}$$

where $v_1$ represents the volume of the unit cell of the crystal and $E_0 = I_0S$, the total power in the incident beam.

If we only require relative values for $I(\mathbf{s})$, it is sufficient to keep the energy of the incident beam constant or to use a monitor counter. If the angle of incidence is kept equal to the angle of emergence, the effective volume of the sample also remains constant.

### 6.3.4. Absolute Measurements of the Scattered Intensity

It is necessary to measure $E_0$ to obtain the absolute value of $I(\mathbf{s})$ from Eq. (6.38). The difficulty comes from the disproportion between the scattered and incident energies, which can be in the ratio of about 1 to $10^5$. Since the counter cannot be used for measuring directly such high intensities, the following devices have been used.

(a) The incident beam is attenuated by absorption through a screen of known nature and thickness. The absorber must be chosen to absorb the harmonic radiation $\lambda/2$ more than the main radiation (Cu for Cu K$\alpha$, Ag for Mo K$\alpha$). The measurement is always quite inaccurate because it is difficult to know the product $\mu p$ within 2%, $p$ being the mass/centimeter$^2$. Thus, if the attenuation $C=\exp(-\mu p)$ is $10^{-5}$, the error from this source will be

$$\frac{dC}{C}=\frac{d(\mu p)}{\mu p}lnC\sim 20\%\,.$$

One must also take into account the possible presence of the harmonic $\lambda/3$. A second filter is added and, from the attenuation, one either deduces that the radiation has the absorption coefficient corresponding to $\lambda$, or finds an appropriate relation to calculate the proportion of $\lambda/3$.

(b) The sample can be replaced by a scatterer whose scattering power is known and is of the same order of magnitude as that of the sample.

Nitrogen has been used for this purpose because it behaves like a perfect gas. Paraffin can also be used at a scattering angle which makes $(\sin \theta)/\lambda$ large enough so that the scattering is entirely due to the Compton effect. The coherent scattering factors of carbon and of hydrogen for Mo $K\alpha$ and $2\theta = 90°$, for instance, are so weak that the scattering can be calculated as if all the electrons were free [*29*].

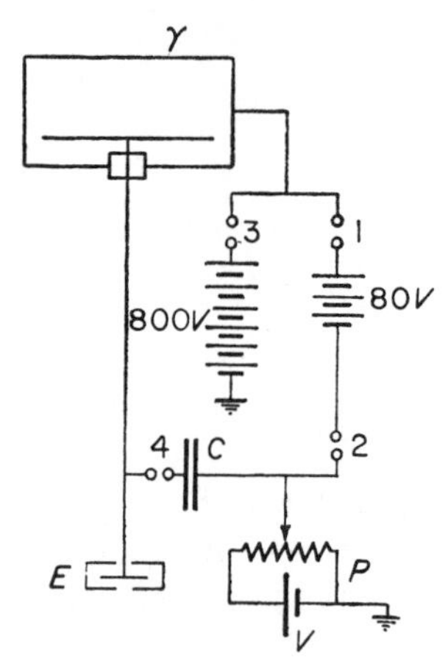


**FIGURE 6.11.** *Ionization chamber and electrometer. For measuring weak intensities, switches 1 and 2 are closed, while 3 and 4 are open. For strong intensities, switches 3 and 4 are closed, 1 and 2 open.*

(c) The most accurate instrument for absolute measurements is the ionization chamber for, even with intense radiation fluxes, the ionization current remains proportional to the incident energy so long as recombination is avoided by using a sufficiently high voltage. Laval and his coworkers [*29*], [26] have used the instrument illustrated in Figure 6.11. For the measurement of weak radiation, the applied voltage is 80 volts. It is possible to compensate for the charges collected by the central electrode by varying the potential on the chamber with the potentiometer *P*. If $\gamma$ is the capacitance of the chamber, and if the potentiometer must be readjusted by $a$ divisions in $t$ seconds to keep the needle of the Lindemann electrometer at zero, the scattered flux is $i = K\gamma \; a/t$ . To measure the direct flux, one uses 800 volts on the chamber and the compensating charges are induced on the needle by means of the capacitor C. The direct flux $E_0 = KCa'/t'$. Therefore

$$\frac{i}{E_0} = \frac{\gamma}{C} \frac{a/t}{a'/t'} .$$

It is possible to measure the ratio $\gamma/C$ with accuracy. This can be quite large: for example [26], $\gamma = 5.61\,pF$ and $C = 0.2\,\mu F$ . Thus, with the same ionization chamber and the same potentiometer, one can measure the absolute value of the ratios $i/E_0$ , which are of the order of $10^5$.

### 6.3.5. Comparison between the Photographic and Counter Methods

It is obvious that the Geiger counter and the ionization chamber have the advantage of *precision* and *sensitivity*. They alone permit a quantitative study of scattering. However, point by point measurements in reciprocal space are lengthy, while the photographic method gives directly a picture of the distribution of the scattering in a single experiment over a large surface of reciprocal space. The resolving power of the film is excellent. Thus, sharply defined scattering regions, for example on rows or on planes of the reciprocal lattice, show up very clearly on the photographic film but are difficult to detect with counter measurements.

Let us note also that the counter is "blind." In many cases inaccuracies in the setup produce stray scattering which is easy to detect on the photographic film but which can spoil a series of measurements made with a counter.

The two methods are therefore complementary and it would not be advisable to neglect one in favor of the other. The photographic method is admirably suited to the *qualitative exploration of an unknown pattern*, since one can then find totally unexpected phenomena. The counter is necessary for *quantitative measurements* on a pattern which is already known qualitively.

## References


1. W. Shockley (Ed.), *Imperfections in Nearly Perfect Crystals*, Wiley, New York (1952).
2. W. H. Zachariasen, *Theory of X-Ray Diffraction in Crystals*, Wiley, New York (1945).
   R. W. James, *The Optical Principles of X-Ray Diffraction*, Bell, London (1948).
   A. J. C. Wilson, *X-Ray Optics*, Methuen, London (1949).
   R. Hosemann, *Ergeb. Exact. Naturwiss.*, **20** (1951), 142.
   R. Hosemann and S. N. Bagchi, *Acta Cryst.*, **5** (1952), 612.
   J. Mering and J. L. Escard, *J. Chimie Phys.*, **51** (1954), 416.
3. B. E. Warren, *Progress in Metal Physics*, Vol. 18, Pergamon, London (1959), p. 148.
4. *International Tables of Crystallography*, Vol. II, Kynoch, (1959) p. 291.
5. A. Guinier, "Heterogeneities in solid solutions," in *Solid State Physics*, Vol. IX, Academic, New York (1958), p. 336.
6. C. S. Barrett, *Trans. AIME*, **188** (1950), 123.
7. J. M. Cowley, *J. Appl. Phys.*, **21** (1950), 24.
8. A. H. Geisler and J. K. Hill, *Acta Cryst.*, **1** (1948), 238.
9. A. Guinier, *Théorie et Technique de la Radiocristallographie*, Dunod, Paris (1956), p. 276.
10. H. K. Hardy and T. J. Heal, *Progress in Metal Physics*, Vol. 5, Pergamon, London (1954), p. 143.
11. P. Kirpatrick, *Rev. Sci. Instr.*, **10** (1939), 186; **15** (1944), 223.
12. A. Guinier, *Théorie et Technique de la Radiocristallographie*, Dunod, Paris (1956), p. 21.
13. G. Hägg and N. Karlsson, *Acta Cryst.*, **6** (1952), 728.
14. D. W. Berreman, J. W. M. DuMond, and P. E. Marmier, *Rev. Sci., Instr.*, **25** (1954), 1219.

B. E. WARREN, *J. Appl. Phys.*, **25** (1954), 814.
15. B. E. WARREN, *J. Appl. Phys.*, **25** (1954), 814.
D. R. CHIPMAN, *Rev. Sci. Instr.*, **27** (1956), 164.
16. *International Tables of Crystallography*, Vol. III, Kynoch, Press (1962).
17. A. H. COMPTON and J. K. ALLISON, *X-Rays in Theory and Experiment*, Macmillan, London (1935), p. 782.
18. J. CORBEAU, *C. R. Acad. Sci. Paris*, **253** (1961), 1553.
19. B. E. WARREN, *Acta Cryst.*, **6** (1953), 803.
20. A. GUINIER, *Théorie et Technique de la Radiocristallographie*, Dunod, Paris (1956), p. 237.
21. H. LAMBOT, *Rev. Met.*, **47** (1950), 709.
22. R. GRAF and A. GUINIER, *Acta Cryst.*, **5** (1952), 150.
23. J. MANENC, *Acta Cryst.*, **10** (1957), 259.
24. J. HOERNI and W. A. WOOSTER, *Acta Cryst.*, **5** (1952), 626.
25. P. OLMER, *Bul. Soc. Franç. Minér. Cryst.*, **71** (1948), 144.
26. H. CURIEN, *Bul Soc. Franç. Minér. Cryst.*, **75** (1952), 343.
27. E. H. JACOBSEN, *Phys. Rev.*, **97** (1955), 654.
28. H. COLE and B. E. WARREN, *J. Appl. Phys.*, **23** (1952), 335.
29. J. LAVAL, *Bull. Soc. Franç. Minér. Cryst.*, **62** (1939), 137.

CHAPTER

# 7

# DISPLACEMENT DISORDER IN CRYSTALS

We shall now study imperfections arising solely from atoms which are left unchanged, but displaced from their normal positions in the crystal lattice.

From the general formulas established in the preceding chapter, let us first draw a fundamental conclusion concerning the effects of such a displacement disorder on the diffraction pattern. The distribution of the scattering power for a unit cell of the crystal in reciprocal space is given by Eq. (*6.5*),

$$I(\boldsymbol{s}) = \sum_m V(\boldsymbol{x}_m) y_m \exp(2\pi i \boldsymbol{s} \cdot \boldsymbol{x}_m) ,$$

where, according to Eq. (*6.3*),

$$y_m = \overline{F_n F_{n+m}^*} ,$$

$F_n$ being the structure factor of the $n$th cell. In the case of displacement disorder, we always have

$$F_n = F \exp(-2\pi i \boldsymbol{s} \cdot \Delta \boldsymbol{x}_n) ,$$

where $F$ is the structure factor of the unit cell before displacement, and where $\Delta \boldsymbol{x}_n$ is the displacement of the $n$th unit cell. The modulus of $F_n$ is therefore constant and equal to $|F|$. We also have

$$|\overline{F_n}|^2 = |F|^2 .$$

On the other hand, when $s$ tends to zero, all the $F_n$ tend to $F$ so long as the displacements remain small. All the coefficients $y_m$ also tend to $|F|^2$, which means that the $\Phi_m$ [Eq. (*6.24*)] tend to zero. Then, from Eq.

(*6.26*), the scattered intensity becomes zero near the center of reciprocal space. However, the scattering increases in intensity as the distance to the center increases, since the functions $\Phi_m$ in general increase with $s$. *This absence of scattering near the center is a general characteristic of displacement disorder, and it can often be used to identify immediately the cause of anomalous scattering.* Let us recall that we have assumed the displacement to be small, which means that the local density of atoms does not fluctuate strongly and that there are no holes nor appreciable accumulations of atoms.

The first example of displacement disorder which we shall study in detail is derived from a very general phenomenon which is found in all crystal, under all circumstances—thermal agitation.

## 7.1. Thermal Agitation in Crystals

The atoms of a solid vibrate with an amplitude which increases with temperature and which does not become zero at the absolute zero of temperature because of the zero point energy. Rigorously speaking, it is therefore never possible to assume that the atoms are situated at the nodes of a regular and rigid lattice.

Thermal agitation has two effects on the diffraction of X-rays by a crystal: it produces a *general background of scattering* in all directions and, as a result, the intensity of the selective reflections is decreased. The elementary formulas for the intensity of a diffraction line or spot must therefore be corrected. We shall also find that the distribution of the scattering can provide information on the elastic properties of crystals and on the interatomic forces.

### 7.1.1. The Effect of Thermal Agitation on the Intensity Diffracted by Crystals. The Debye-Waller Factor

Let us consider a crystal which has a simple lattice containing a single atom per unit cell, with a scattering factor $f$, the nodes of the regular lattice being the positions of equilibrium around which the atoms can vibrate. The period of the vibration is short compared to the duration of an experiment in which the intensity is measured, but very long compared to the period of the X-rays. We can therefore assume that the diffraction is produced by atoms which are fixed in position, the observed intensity being the average of the diffracted intensities for all the possible configurations. Each atom moves according to a single statistical law and, at a given moment, the displacements of the various atoms are statistically identical to those of a given atom taken at $N$

arbitary times separated by intervals which are large with respect to the period of vibration. Let us call $\Delta \boldsymbol{x}_n$ the displacement of the $n$th atom from its node. Then the average values of $\Delta \boldsymbol{x}_n$ for all the atoms at a given instant are identical to the average values of $\Delta \boldsymbol{x}_n$ in time, for a given atom.

The structure factor for the $n$th unit cell is

$$F_n = f \exp(-2\pi i \boldsymbol{s} \cdot \Delta \boldsymbol{x}_n) . \tag{7.1}$$

According to Section 6.2.2, the diffraction pattern for the real crystal is that of a crystal in which each unit cell has a structure factor which is equal to the average value of $F_n$, or $\bar{F}$. Let us calculate $\bar{F}$ on the assumption that the displacements are small. Then the above exponential function can be expanded as a series,

$$\begin{aligned}\exp(&-2\pi i \boldsymbol{s} \cdot \Delta \boldsymbol{x}_n) \\ &= 1 - 2\pi i \boldsymbol{s} \cdot \Delta \boldsymbol{x}_n - 2\pi^2(\boldsymbol{s} \cdot \Delta \boldsymbol{x}_n)^2 + K_3(\boldsymbol{s} \cdot \Delta \boldsymbol{x}_n)^3 + K_4(\boldsymbol{s} \cdot \Delta \boldsymbol{x}_n)^4 + \cdots ,\end{aligned} \tag{7.2}$$

and the average of this quantity is the sum of the averages of the successive terms. Furthermore,

$$\overline{\boldsymbol{s} \cdot \Delta \boldsymbol{x}_n} = \boldsymbol{s} \cdot \overline{\Delta \boldsymbol{x}_n} = 0 ,$$

since the origin of the vector $\Delta \boldsymbol{x}_n$ is the average position of the atom. Then, neglecting higher order terms,

$$\overline{\exp(-2\pi i \boldsymbol{s} \cdot \Delta \boldsymbol{x}_n)} = 1 - 2\pi^2 \overline{(\boldsymbol{s} \cdot \Delta \boldsymbol{x}_n)^2} = 1 - 2\pi^2 s^2 \overline{\Delta x_{1n}^2} . \tag{7.3}$$

We have written the scalar product $\boldsymbol{s} \cdot \Delta \boldsymbol{x}_n$ as a function of the modulus of the vector $\boldsymbol{s}[s = (2 \sin \theta)/\lambda]$ and $\Delta x_{1n}$ is the projection of $\Delta \boldsymbol{x}_n$ on $\boldsymbol{s}$. Now let $\Delta x_{2n}$ and $\Delta x_{3n}$ be the projections of $\Delta \boldsymbol{x}_n$ on two other axes such that the three axes are mutually orthogonal. We then have

$$\Delta x_n^2 = \Delta x_{1n}^2 + \Delta x_{2n}^2 + \Delta x_{3n}^2 .$$

If the displacements of the atom are perfectly isotropic around the node, it is evident that the averages $\overline{\Delta x_{1n}^2}$, $\overline{\Delta x_{2n}^2}$, $\overline{\Delta x_{3n}^2}$ are equal and that

$$3\overline{\Delta x_{1n}^2} = \overline{\Delta x_n{}^2} = \Delta X^2 , \tag{7.4}$$

if we set $\Delta X$ as the quadratic average of $|\Delta \boldsymbol{x}_n|$. We therefore have

$$\bar{F} = f\left(1 - \frac{2\pi^2 s^2}{3} \Delta X^2\right) , \tag{7.5}$$

which is generally used in the practically equivalent form

$$\bar{F} = f \exp(-M) ,$$

where

$$M = \frac{2\pi^2 s^2 \Delta X^2}{3} = \frac{8\pi^2 \sin^2\theta}{\lambda^2} \frac{\Delta X^2}{3}. \tag{7.6}$$

The above two values of $\bar{F}$ are equivalent within the approximations used. With thermal agitation the diffraction pattern of a crystal is therefore the same as that of a perfect crystal in which the atoms have a scattering factor $f\exp(-M)$, and *the intensity of the diffraction is reduced in the ratio*

$$D = \exp(-2M) = \exp\left(-\frac{16\pi^2 \sin^2\theta}{\lambda^2} \frac{\Delta X^2}{3}\right), \tag{7.7}$$

where *D is the Debye, or Debye-Waller factor* [1], [2]. We have already indicated briefly, in Section 4.5.3, the necessity of correcting the intensity formulas for the temperature effect.

According to Eq. (*7.7*) the factor $D$ increases with the diffraction angle, and the diffraction lines at large angles are therefore more affected than the others. This result is readily understood if we write $M$ in the form

$$M = 2\pi^2 \left(\frac{\Delta X/\sqrt{3}}{d}\right)^2, \tag{7.8}$$

where $d = 1/s = \lambda/(2\sin\theta)$ is the reticular spacing for the planes giving rise to the reflection under consideration, and $\Delta X/\sqrt{3}$ is the quadratic average of the atomic displacement from the ideal lattice plane, according to Eq. (*7.4*). Thermal agitation therefore has the effect of making the lattice planes appear diffuse, the apparent thickness being measured by the factor $\Delta X/\sqrt{3}$. The effect on the reflected intensity depends on the ratio of this "thickness" to the reticular spacing. The decrease in intensity is therefore greater for planes which are closer together and which correspond to large Bragg angles.

In the case of KCl, which has nearly the simple crystal lattice which we have assumed at the beginning of this section—since the ions $K^+$ and $Cl^-$ have approximately equal scattering factors—it has been found from experimental measurements of the reflecting power [3] that $\Delta X$ is equal to 0.25 Å at room temperature and to 0.43 Å at 900°K. The displacements are therefore quite large, since they are respectively 8% and 13% of the average distance between neighboring atoms. The values of $D = \exp(-2M)$ are given in Table 7.1 for the first intense line and

TABLE 7.1. The Temperature Factor $D$ for KCl

| Temperature | 300°K | 900°K |
|---|---|---|
| 200 line ($d = 3.14$ Å) | 0.92 | 0.74 |
| 800 line ($d = 0.785$ Å) | 0.27 | 0.01 |

for the last visible one on the pattern observed with Cu K$\alpha$ radiation. The table shows that the Debye factor is so large for the last lines that the approximations used in Eqs. (*7.6*) and (*7.7*) are quite rough.

The intensities of the diffraction lines give the Debye temperature from Eqs. (*4.47*) and (*7.15*). The result is generally in agreement with that obtained by other methods [4] except that, in some cases, the Debye temperature found by X-ray methods varies slightly with temperature [5].

We shall show in the following section that the intensity subtracted from the diffraction lines is found in the background, which is always weak with respect to the maximum intensity of the line, even in its immediate neighborhood. The width of the peak, measured at half intensity of the maximum, is unaffected by the temperature effect (Figure 7.1). *Thermal agitation therefore does not broaden the diffraction lines.*

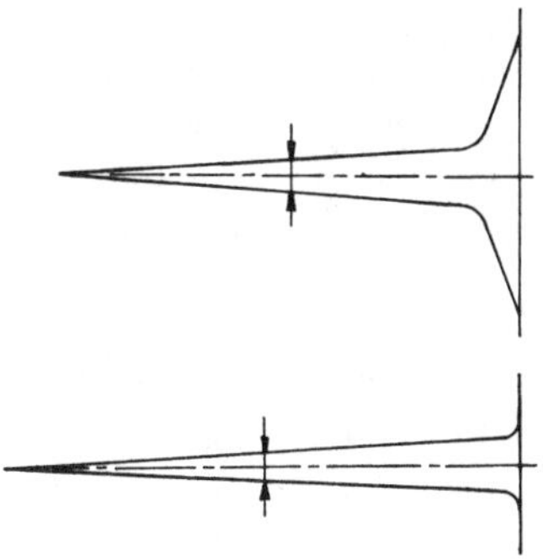

**FIGURE 7.1.** *The profile of a diffraction line for a crystal.* (a) *Without thermal agitation.* (b) *With thermal agitation.*

Equation (*7.6*) applies only if the atomic displacements are isotropic in the crystal. If, on the contrary, the vibrations are anisotropic, Eq. (*7.8*) applies only if we replace $\Delta X/\sqrt{3}$ by the quadratic average of the projections of the atomic displacements on the normal to the reflection plane, and this average varies with the indices corresponding to the reflection [6]. The lines whose intensities are most greatly reduced are therefore those which correspond to the planes which are normal to the directions in which the amplitude of the vibrations is largest.

### 7.1.2. Calculation of the Debye Factor

To calculate the Debye factor a priori, we must know the quadratic average of the displacement of the atoms as a function of the temperature. This calculation is given by the theory of the specific heat of solids. We shall not perform this calculation in detail, but rather refer the reader to Kittel [7].

The thermal vibrations of an atom can be considered to be the result of the superposition of waves propagating in the crystal. In the case of a simple lattice containing $N$ atoms in a volume $V$, there are $3N$ such waves. An elementary wave of frequency $\nu$ corresponds to a displacement $\xi_\nu = A_\nu \cos 2\pi\nu t$, and the quadratic average of the displacement is

$$\overline{\xi_\nu^2} = \frac{A_\nu^2}{2}.$$

The total average displacement is the sum of the $3N$ vibrations

$$\Delta X^2 = \int \frac{A_\nu^2}{2} Q_\nu d\nu \,, \tag{7.9}$$

where $Q_\nu d\nu$ is the number of vibrations whose frequencies lie between $\nu$ and $\nu + d\nu$. We must therefore calculate $Q_\nu$ and $A_\nu$. In the first-approximation theory of Debye, the $3N$ waves have a frequency which lies between zero and a maximum frequency $\nu_m$ such that

$$h\nu_m = k\Theta \,, \tag{7.10}$$

where $h$ is Planck's constant and $k$ is Boltzmann's constant. The quantity $\Theta$ is called the Debye characteristic temperature and, within this interval, the frequencies are distributed so that

$$Q_\nu = \frac{9\nu^2 N}{\nu_m^3}. \tag{7.11}$$

The value of $A_\nu$ can be found by calculating the total energy of the elementary wave $E_\nu$ in two different ways. If $m$ is the mass of the atom, then the energy of the harmonic oscillation of frequency $\nu$ and amplitude $A_\nu$ is $2\pi^2\nu^2 m A_\nu^2$. Since the $N$ atoms execute identical vibrations,

$$E_\nu = 2\pi^2 N\nu^2 m A_\nu^2 . \tag{7.12}$$

We know from quantum mechanics that, at temperature $T$°K,

$$E_\nu = h_\nu \left\{ \frac{1}{2} + \frac{1}{[\exp(h\nu/kT)] - 1} \right\}. \tag{7.13}$$

Thus, at a sufficiently high temperature $kT \gg h\nu$, or $T \gg \Theta$, the energy corresponding to this degree of freedom is $kT$. At low temperature the energy increases slowly with $T$, starting from the value $h\nu/2$ at absolute zero.

Using the value of $A_\nu$ calculated from Eq. (*7.12*) and (*7.13*), and using Eqs. (*7.11*) and (*7.9*), we find that

$$\Delta X^2 = \frac{9h^2}{4\pi^2 mk\Theta}\left[\frac{1}{4} + \frac{T}{\Theta}\varphi\left(\frac{\Theta}{T}\right)\right] = \frac{9h^2}{4\pi^2 mk\Theta}\,\frac{T}{\Theta}\left[\frac{1}{4}\,\frac{\Theta}{T} + \varphi\left(\frac{\Theta}{T}\right)\right],$$

where

$$\varphi\left(\frac{\Theta}{T}\right) = \frac{T}{\Theta}\int_0^{\Theta/T} \frac{y\,dy}{(\exp y) - 1}. \tag{7.14}$$

Numerically, if $M_a$ is the atomic mass and $\Delta X$ is expressed in Ångströms,

$$\Delta X^2 = \frac{424}{M_a\Theta}\frac{T}{\Theta}\left[\frac{1}{4}\frac{\Theta}{T}+\varphi\left(\frac{\Theta}{T}\right)\right].$$

Substituting in Eq. (7.6), we have

$$M = \frac{\sin^2\theta}{\lambda^2}\frac{6h^2}{mk\Theta}\left[\frac{1}{4}+\frac{T}{\Theta}\varphi\left(\frac{\Theta}{T}\right)\right] = \frac{\sin^2\theta}{\lambda^2}\frac{6h^2}{mk\Theta}\frac{T}{\Theta}\left[\frac{1}{4}\frac{\Theta}{T}+\varphi\left(\frac{\Theta}{T}\right)\right] \quad (7.15)$$

or, numerically, if we express $\lambda$ in Ångströms,

$$M = \frac{1.14\times 10^4}{M_a\Theta}\frac{\sin^2\theta}{\lambda^2}\left[\frac{1}{4}+\frac{T}{\Theta}\varphi\left(\frac{\Theta}{T}\right)\right].$$

Above the Debye critical temperature ($T>\Theta$), it is preferable to use the second of the two expressions for $M$, and similary $\Delta X^2$ since, as shown in Table 7.2, the function

$$\left[\frac{1}{4}\frac{\Theta}{T}+\varphi\left(\frac{\Theta}{T}\right)\right]$$

is equal to unity within 3%.

TABLE 7.2. Function $\left[\frac{1}{4}\frac{\Theta}{T}+\varphi\left(\frac{\Theta}{T}\right)\right]$ vs $\frac{\Theta}{T}$

| $\frac{\Theta}{T}$ | 0.0 | 0.2 | 0.4 | 0.6 | 0.8 | 1.0 | 1.2 | 1.4 | 1.6 | 1.8 | 2.0 | 2.5 |
|---|---|---|---|---|---|---|---|---|---|---|---|---|
| $\frac{1}{4}\frac{\Theta}{T}+\varphi\left(\frac{\Theta}{T}\right)$ | 1.000 | 1.001 | 1.004 | 1.010 | 1.018 | 1.028 | 1.040 | 1.054 | 1.069 | 1.087 | 1.107 | 1.164 |

Figure 7.2 shows the expression

$$\frac{1}{4}+\frac{T}{\Theta}\varphi\left(\frac{\Theta}{T}\right)$$

as a function of $T/\Theta$ at low temperatures. This curve can be used for any crystal, whenever the Debye temperature is known. As the temperature approaches absolute zero, the limiting values of $\Delta X^2$ and $M$ are given by the expressions

$$\Delta X^2_{T\to 0} = \frac{9h^2}{16\pi^2 mk\Theta} = \frac{106}{M_a\Theta}\ (\text{Å}^2),$$
$$M_{T\to 0} = \frac{\sin^2\theta}{\lambda^2}\frac{3h^2}{2mk\Theta} = \frac{2.8\times 10^3}{M_a\Theta}\frac{\sin^2\theta}{\lambda^2}. \quad (7.16)$$

**FIGURE 7.2.** *The quantity* $(1/4)+[T/\Theta]\varphi(\Theta/T)$ *as a function of* $T/\Theta$. *The curve is asymptotic to the line passing through the origin and whose slope is equal to unity.*

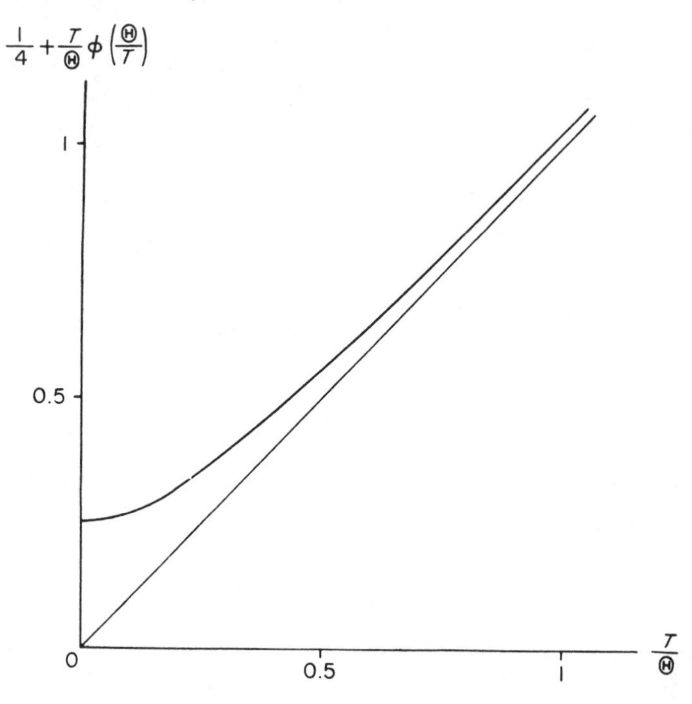


Table 7.3 gives the values of $\Delta X^2$ at room temperature ($T = 293°K$) and at absolute zero for a few simple substances. It also gives the value of the Debye factor [$D = \exp(-2M)$] at room temperature for the last back-reflection line ($\sin\theta = 1$) of a pattern using Cu K$\alpha$, as well as the ratio

TABLE 7.3. Values of $\Delta X^2$ and of the Debye Factor at Room Temperature and at Absolute Zero

| Substance | Characteristic Temperature (°K) | $\Delta X^2$ (293°K) (Å$^2$) | $D = \exp(-2M)$ (293°K) | $\frac{M_{T=0}}{M_{T=293}}$ |
|---|---|---|---|---|
| Lead | 88 | 0.074 | 0.18 | 0.075 |
| Silver | 215 | 0.026 | 0.57 | 0.18 |
| Tungsten | 280 | 0.009 | 0.82 | 0.22 |
| Copper | 315 | 0.020 | 0.64 | 0.26 |
| Aluminum | 398 | 0.030 | 0.49 | 0.32 |
| Iron | 453 | 0.012 | 0.77 | 0.36 |
| Diamond | 1,860 | 0.005 | 0.88 | 0.85 |

$M_{T=0}/M_{T=293}$. This shows the importance of the decrease of the diffraction intensity under ordinary experimental conditions and *the impossibility of completely eliminating thermal agitation by cooling the object, even in the most radical manner.*

When the crystal contains several atoms per unit cell, the analysis of the vibrations in terms of elementary waves is much more complex. We then distinguish between *acoustical and optical waves.* We shall return to these in a later section (7.1.7). For the moment, we shall only say that the optical waves have higher frequencies and, from Eqs. (*7.12*) and (*7.13*), the amplitude of these vibrations is smaller and they therefore have less influence on the diffraction pattern. As an approximation, one may assume that the atoms have average displacements which are approximately equal to a common value $\Delta X$, and use a temperature factor to modify the structure factor for the unit cell $\bar{F} = F\exp(-M)$, with

$$M = \frac{2\pi^2 s^2}{3} \Delta X^2 .$$

It is more accurate to correct for the temperature effect each one of the atomic scattering factors entering into the expression for the structure factor of the unit cell. If $\Delta X_r$ is the average quadratic displacement of the atom $r$,

$$\bar{F} = \sum f_r \exp\left(-\frac{2\pi^2 s^2}{3} \Delta X_r^2\right) \exp\left(-2\pi i \boldsymbol{s} \cdot \boldsymbol{x}_r\right)$$

Thus, for NaCl, there are two types of lattice planes with structure factors $f_{Na} + f_{Cl}$ and $f_{Na} - f_{Cl}$. The corrected factors are $f_{Na}\exp(-M_{Na}) \pm f_{Cl}\exp(-M_{Cl})$. It is possible to determine experimentally the average displacement of the chlorine and sodium ions and it has been found that, at room temperature, these are 0.22 Å and 0.24 Å, respectively [8].

### 7.1.3. Temperature-diffuse Scattering

The energy which is missing in the diffraction lines is found in the general background in the other directions. In his first theory Debye [1] had assumed that the atomic vibrations were perfectly independent, or that there was no correlation between the instantaneous values of the displacements of the atoms in two unit cells, even when they were neighbors. With this hypothesis, Eq. (*6.31*) gives for the *scattered* intensity per atom

$$I_2 = \overline{|F|^2} - |\bar{F}|^2 .$$

We already know that $|\bar{F}| = f\exp(-M)$. As to the modulus of $F_n$, it is equal to $f$ from Eq. (*7.1*), whatever the value of the displacement $\Delta \boldsymbol{x}_n$

**FIGURE 7.3.** *Temperature-diffuse scattering pattern for a KCl crystal 0.5 mm thick. Mo Kα radiation was used with a bent-crystal monochromator. The plane (100) was normal to the incident beam as well as to the plane film.*

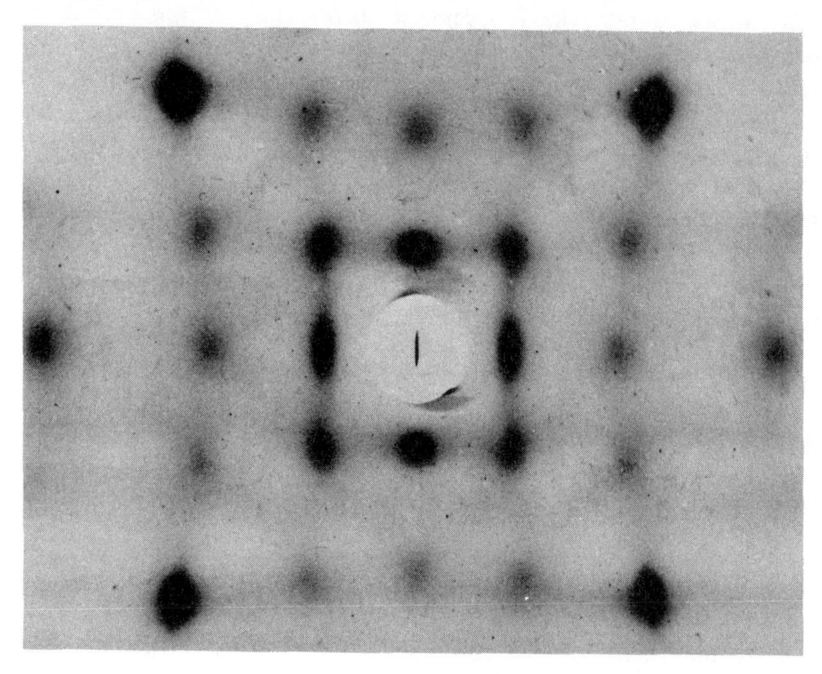

of the atom in the unit cell. Therefore

$$I_2 = f^2\Big[1 - \exp(-2M)\Big] = f^2\Big[1 - \exp\Big(-\frac{4\pi^2 s^2}{3}\,\Delta X^2\Big)\Big] = f^2 4\pi^2 s^2 \frac{\Delta X^2}{3}\,. \tag{7.17}$$

According to this formula, the intensity depends only on $|\mathbf{s}|$, so there is spherical symmetry in reciprocal space, and the intensity increases as the square of the distance from the center of this space. Preliminary experiments [9] appeared at first to verify this equation, but very accurate measurements by Laval [10] showed for the first time that the distribution of thermal scattering in reciprocal space is much more complex and, in particular, highly anisotropic.

Figure 7.3 shows a typical example of the phenomenon. It was taken with a small crystal of sylvite (KCl) under the following experimental conditions. The sample had the form of a plate 0.5 mm in thickness,

**FIGURE 7.4.** *Ewald construction corresponding to Figure 7.3: reciprocal cubic lattice of parameter* $1/3.14\,\text{Å}^{-1}$ *and* $1/\lambda = 1/0.704\,\text{Å}^{-1}$; *incident ray along the [010] axis, with the [001] axis normal to the plane of the figure.*

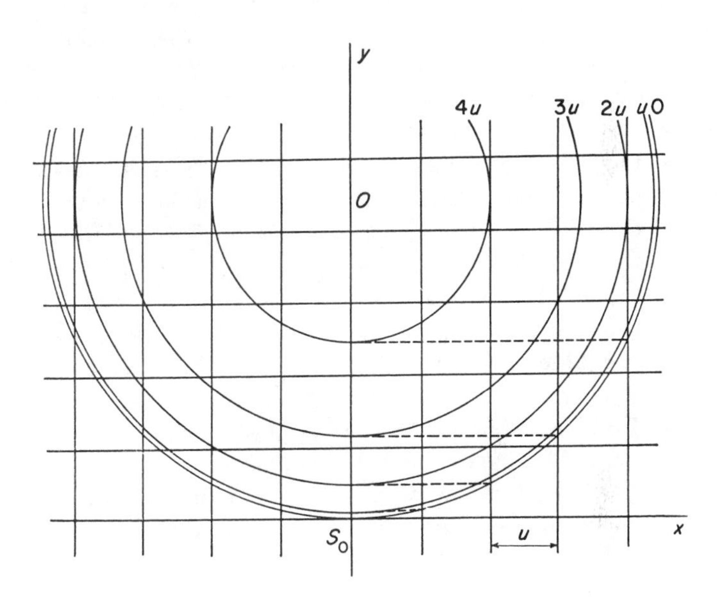


and rigorously monochromatic Mo K$\alpha$ radiation was directed along the axis [100]. The axis [010] was vertical and the crystal was at room temperature. As the temperature was increased or decreased, the pattern became more or less intense, without much other change. Since the crystal was fixed during the exposure, the photographic plate shows a section of reciprocal space through the sphere of reflection, or the variations in the scattering intensity from one point to another of this surface. To interpret this figure, we must consider the position of the sphere of reflection with respect to the reciprocal lattice of the crystal. This is shown in Figure 7.4, which corresponds to the position of the KCl crystal in Figure 7.3. We have shown, on the same scale, a projection of the reciprocal lattice of the crystal on the planes (001) and the sphere of reflection of radius $1/\lambda = (1/0.71)\,\text{Å}^{-1}$. Since the $K^+$ and $Cl^-$

ions have equal numbers of electrons, the KCl lattice can be considered to be a simple cubic lattice with a parameter of 3.14 Å. We have shown separately the circles at heights $u$, $2u$ and $3u$. It will be seen that only the four nodes 122, $1\bar{2}2, \cdots$ are on the surface of the sphere. These correspond to the four Bragg spots which are shown overexposed in Figure 7.3. Near the the center of the figure the scattering is very weak. *There is a diffuse spot corresponding to each node, outside the reflection sphere but close to its surface.* Starting from the center, we first see the spots close to the nodes 100, 110 and then 200, 201. The diffuse spots are more intense if the nodes are close to the surface of the sphere. The 100 spots have an elongated shape in the direction perpendicular to the radius drawn from the center of the pattern to the spot. Aside from the spots related to the nodes, other scattering regions can be seen in Figure 7.3: these join consecutive nodes on the rows parallel to the axes $\langle 100 \rangle$. Others, which are less intense, are directed in the direction which is parallel to the $\langle 110 \rangle$ rows. The contrast between strong and weak scattering regions decreases with the distance from the center of the pattern and, at large angles, no clear-cut pattern can be observed.

Equation (*7.17*) can in no way explain such observations because it predicts only a slowly varying scattering intensity which is isotropic in reciprocal space. The hypothesis that the atomic displacements are independent must therefore be rejected. This is not unexpected from what we have already said concerning the analysis of thermal vibrations: interatomic forces are such that the displacement of one atom necessarily acts on those of its neighbors.

The result of the calculations on the dynamics of the crystal lattice is that the vibrations of a given atom can be considered to be the superposition of displacement waves crossing the crystal [11]. The fundamental phenomenon in thermal agitation is not therefore the vibration of an isolated atom, but a harmonic wave traversing the whole crystal and displacing all the atoms. The theoretical study of temperature diffuse scattering ]2], [10], [12] must therefore start with the calculation of the scattering by a crystal traversed by a single wave of given frequency, wavelength, direction of propagation, and amplitude. We then add the results for the various waves, using the theory of lattice dynamics to obtain the distribution of the various waves in the crystal.

Let us first explain what is meant by a displacement wave in a crystal. The atom situated at the node $\boldsymbol{x}_n$ is displaced from this node by a distance

$$\Delta \boldsymbol{x}_n = \boldsymbol{A} \cos 2\pi(\nu t - \boldsymbol{k} \cdot \boldsymbol{x}_n) . \qquad (7.18)$$

For a given atom, $\Delta \boldsymbol{x}_n$ oscillates at a frequency $\nu$ with an amplitude $\boldsymbol{A}$.

The odes at which the vibrations are in phase are distributed in planes such that $\boldsymbol{k} \cdot \boldsymbol{x}_n = \text{constant}$. The wave fronts are therefore normal to the vector $\boldsymbol{k}$ and the wavelength, or minimum distance between two planes where the phase is the same, is given by $\Lambda = 1/|\boldsymbol{k}|$, where $\boldsymbol{k}$ is the *propagation vector*.

At two different times the structure of the perturbed crystal is the same, except for a general translation which does not affect the diffraction. Experimentally the fact that the diffracting atoms are in motion has no appreciable effect, since the frequency is very low compared to that of the X-radiation. Theoretically the motion of the atoms produces a change in the frequency of the diffracted X-rays which is equal to the frequency of the thermal agitation wave, so the relative variation is of the order of $10^{-7}$.

We can therefore calculate the diffraction by a crystal in which the atoms occupy fixed positions defined by the displacement vectors

$$\Delta \boldsymbol{x}_n = \boldsymbol{A} \cos 2\pi \boldsymbol{k} \cdot \boldsymbol{x}_n \,. \tag{7.19}$$

According to the general theory we must calculate the coefficients $\Phi_m = \overline{\varphi_n \varphi_{n+m}^*}$ to obtain the scattering intensity, where $\varphi_n$ is the difference between the structure factor for the $n$th unit cell and its average value. Then

$$f_n = f \exp(-2\pi i \boldsymbol{s} \cdot \Delta \boldsymbol{x}_n)$$

and, according to Eq. (7.3),

$$\bar{f} = f\left[1 - 2\pi^2 \frac{(\boldsymbol{s} \cdot \boldsymbol{A})^2}{2}\right] \tag{7.20}$$

since the average value of $\overline{\Delta x_n}$, according to Eq. (7.19), is such the $\overline{\Delta x_n^2} = A^2/2$. Whence

$$\begin{aligned}\varphi \varphi_{n+m}^* = f^2 &\left[\exp(-2\pi i \boldsymbol{s} \cdot \Delta \boldsymbol{x}_n) - 1 + 2\pi^2 \frac{(\boldsymbol{s} \cdot \boldsymbol{A})^2}{2}\right] \\ \times &\left[\exp(2\pi i \boldsymbol{s} \cdot \Delta \boldsymbol{x}_{n+m}) - 1 + 2\pi^2 \frac{(\boldsymbol{s} \cdot \boldsymbol{A})^2}{2}\right].\end{aligned} \tag{7.21}$$

As a first approximation, we shall assume that the amplitude of the vibration and the absolute value of the reciprocal lattice vector are sufficiently small that we can neglect the higher powers of $\boldsymbol{s} \cdot \boldsymbol{A}$. Expanding the above exponentials and calling $\alpha$ the angle between the vectors $\boldsymbol{A}$ and $\boldsymbol{s}$, we find that

$$\varphi_n \varphi_{n+m}^* = f^2[-2\pi i s A \cos\alpha \cos(2\pi \boldsymbol{k} \cdot \boldsymbol{x}_n)][2\pi i s A \cos\alpha \cos(2\pi \boldsymbol{k} \cdot \boldsymbol{x}_{n+m})] \,.$$

If we set $\boldsymbol{x}_{n+m} = \boldsymbol{x}_n + \boldsymbol{x}_m$, with $\boldsymbol{x}_m$ a vector in the crystal lattice,

$$\varphi_n \varphi_{n+m}^* = 2\pi^2 f^2 (As\cos\alpha)^2 \{\cos[2\pi \boldsymbol{k} \cdot (2\boldsymbol{x}_n + \boldsymbol{x}_m] + \cos[2\pi \boldsymbol{k} \cdot \boldsymbol{x}_m]\} \,. \tag{7.22}$$

Now $\Phi_m$ is the average value of $\varphi_m \varphi^*_{n+m}$ for all possible values of $n$ [Eq. (*6.24*)]. The first term in the summation is zero and we are left with the second term, which is independent of $n$:

$$\Phi_m = 2\pi^2 f^2 (As \cos \alpha)^2 \cos [2\pi \boldsymbol{k} \cdot \boldsymbol{x}_m] \,. \tag{7.23}$$

We have already seen in Section 6.2.3 that, when $\Phi_m$ is sinusoidal with a propagation vector $\boldsymbol{k}$, the scattered intensity is concentrated at the extremities of the vectors $\pm\boldsymbol{k}$ originating at the nodes of the reciprocal lattice. We can consider this scattering to be due to Bragg reflections on the wave fronts of the perturbing wave (Laval [10]) since the diffracted ray corresponding to one of these points ($P$ of Figure 7.5) is the ray reflected by a series of equidistant planes which are normal to $\boldsymbol{OP}$ and whose spacing $\Lambda = 1/OP$. These are the wave fronts with propagation vector $\boldsymbol{OP}$. This wave is none other than that whose propagation vector is $\boldsymbol{R}_{hkl}\boldsymbol{P}$, since one may show that, because of the periodic structure of the crystal lattice, two propagation vectors whose difference is a reciprocal lattice vector represent the same wave [13]. The smaller of the two possible propagation vectors is called the fundamental propagation vector and the set of the fundamental vectors having their origin at a given node is contained within a region called the first Brillouin zone. The points in this zone are closer to the node considered than to any other node.

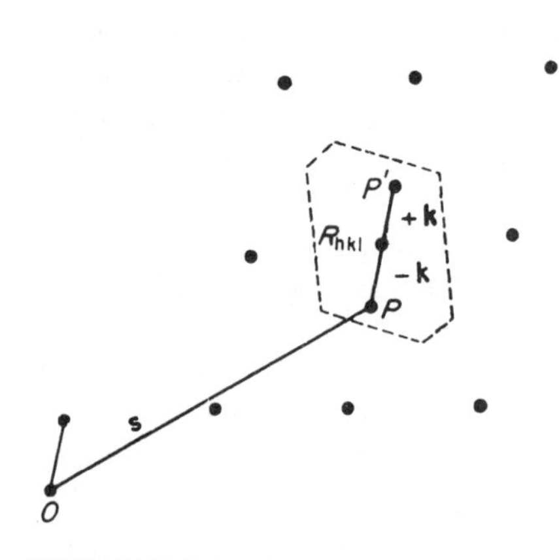


**FIGURE 7.5.** *Satellites of the lattice point $R_{hkl}$ corresponding to the wave whose propagation vector is* $\boldsymbol{k}$. *The first zone of Brillouin surrounding $R_{hkl}$ is shown with a dashed line.*

It is not possible to use Eq. (*6.26*) to find the scattered intensity, since this equation was obtained on the hypothesis that the scattered intensity is a slowly varying function. We therefore have to start from the initial Eq. (*6.25*), in which we substitute the value of $\Phi_m$ from Eq. (*7.23*):

$$\begin{aligned} I(\boldsymbol{s}) &= f^2 \sum V(\boldsymbol{x}_m) \exp (2\pi i \boldsymbol{s} \cdot \boldsymbol{x}_m) \\ &\quad + 2\pi^2 f^2 (As \cos \alpha)^2 \sum V(\boldsymbol{x}_m) \cos [2\pi \boldsymbol{k} \cdot \boldsymbol{x}_m] \exp (2\pi i \boldsymbol{s} \cdot \boldsymbol{x}_m) \\ &= f^2 \sum V(\boldsymbol{x}_m) \exp (2\pi i \boldsymbol{s} \cdot \boldsymbol{x}_m) \\ &\quad + \pi^2 f^2 (As \cos \alpha)^2 \sum V(\boldsymbol{x}_m) \exp [2\pi i(\boldsymbol{s} + \boldsymbol{k}) \cdot \boldsymbol{x}_m] \\ &\quad + \pi^2 f^2 (As \cos \alpha)^2 \sum V(\boldsymbol{x}_m) \exp [2\pi i(\boldsymbol{s} - \boldsymbol{k}) \cdot \boldsymbol{x}_m] \,. \end{aligned}$$

Under this form the intensity appears as the sum of three series of crystal reflections. The first is the normal diffraction at the nodes of the reciprocal lattice. The other two are satellite reflections at points which are deduced from the nodes by the translations $\pm \boldsymbol{k}$. The term $\pi^2 f^2(As \cos \alpha)^2$ now plays the role of a structure factor and, according to the general Eq. (*4.39*), the scattering power associated with the satellites, per unit cell of the crystal, is given by

$$I_{\text{satellite}} = \frac{1}{V_c} \pi^2 f^2 (As \cos \alpha)^2 . \qquad (7.24)$$

The average structure factor for the cells of the crystal perturbed by the wave is [Equation (*7.20*)]

$$\bar{f} = [1 - \pi^2 (As \cos \alpha)^2] = f \exp(-M) .$$

The intensity of the diffraction is reduced by the temperature factor $\exp(-2M)$. This means that the integrated diffracting power associated with the Bragg diffraction spot, instead of being $f^2/V_c$, [Eq. (*4.28a*)], is rather

$$\frac{f^2}{V_c} [1 - 2\pi^2 (As \cos \alpha)^2] . \qquad (7.25)$$

If we now compare Eqs. (*7.25*) and (*7.24*), it becomes evident that the energy lacking in the main diffraction spot is exactly equal to that in the two satellite points. This simple example demonstrates the complementarity between the decrease in intensity of the normal spots and the appearance of anomalous scattering.

The above calculation is not rigorous; an exact calculation [14] shows that aside from the main satellites there are also harmonic satellites at distances $\pm 2\boldsymbol{k}$, $\pm 3\boldsymbol{k}, \cdots$ from each node, but the intensity associated with these harmonic satellites is relatively small.

In conclusion, the fundamental result of the calculation is that the perturbation arising from a single displacement wave produces scattering at only two points of the unit cell of the reciprocal lattice. Inversely, at these two points, any wave having a different propagation vector has no effect. The X-rays provide a sort of harmonic analysis of the thermal agitation in the crystal, and one can expect that a measure of the in tensity scattered at a given point of reciprocal space could permit

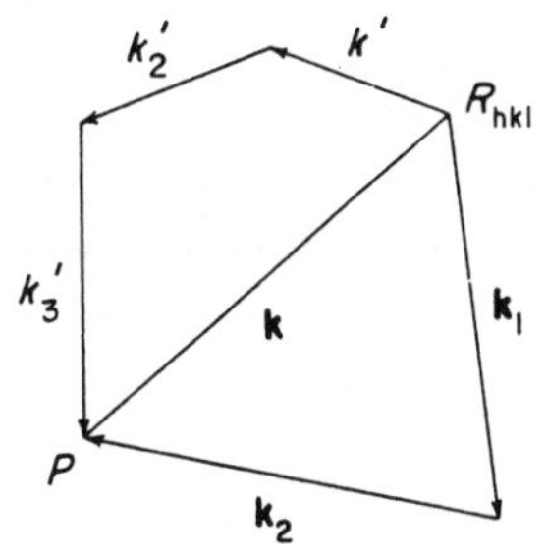


**FIGURE 7.6.** *The propagation vector* **k** *giving second order* ($k_1$, $k_2$) *and third order* ($k_1'$, $k_2'$, $k_3'$) *scattering.*

the determination of the amplitude of a wave of known propagation vector. This is why X-ray scattering is a powerful means of investigation in the field of the dynamics of crystals.

In fact, this is only a first approximation, for Laval [15] has shown that at a given point of reciprocal space (Figure 7.6) we must not only consider the direct wave of propagation vector $\boldsymbol{R}_{hkl}\boldsymbol{P}$, but also second, third,$\cdots$ order scattering by two, three,$\cdots$ elastic waves whose propagation vectors are such that their sum is equal to $\boldsymbol{R}_{hkl}\boldsymbol{P}$. These higher-order scatterings are difficult to calculate, and they are not necessarily negligible compared to first-order scattering [16]. Their influence increases with distance from the center of reciprocal space and their effect is to render thermal scattering more uniform at large scattering angles.

### 7.1.4. Theoretical Expression for the Intensity Scattered by a Crystal

The problem here is to determine the elastic wave or waves having a given propagation vector. This is a rather delicate matter, for we must consider the dynamics of a crystal having the form of a parallelepiped of volume $V$ cut out from an infinite crystal and containing $N$ atoms. Calculations are simplified by the assumption that the displacements of the atoms repeat themselves periodically outside the volume $V$, as if this volume were a giant unit cell of a crystal. Indeed, the results which can be verified experimentally are found to be independent of the choice of $V$. This justifies the "cyclic conditions of Born," which seem at first sight to be rather arbitrary ]17].

There are $3N$ waves corresponding to $N$ vectors $\boldsymbol{k}$, whose extremities form an extremely tight, regular lattice in reciprocal space; the $N$ points fill the unit cell of the reciprocal lattice with a uniform density $N/(1/V_c)$ per unit volume, the volume of the unit cell in the reciprocal lattice being $1/V_c$. We therefore have one point per $1/(NV_c) = 1/V$. Theoretically we therefore have scattering only if the vector $\boldsymbol{s}$ coincides with one of the points of this array. In practice, even under the best experimental conditions, it is only possible to measure the average scattered intensity over a volume $dw$ of reciprocal space which is large compared to $1/V$ and which therefore contains a large number $Vdw$ of points $\boldsymbol{k}$ and the scattering appears continuous. The scattering power related to a point of reciprocal space $I_2(\boldsymbol{s})$ is the measured scattering power divided by the volume $dw$ of the region involved in the measurement or the product of $V$ by the scattering power $I_K$ related to a point $\boldsymbol{k}$ at the center of the region of measurement:

$$I_2(\boldsymbol{s}) = VI_K = NV_c I_K \,. \tag{7.26}$$

We shall use the following results of the dynamics of crystals without demonstration. For a given propagation vector, i.e. for a given point $\boldsymbol{k}$, we have *three independent waves*. One of these is longitudinal, or approximately so, and its amplitude vector $\boldsymbol{A}_1$ is directed along the vector $\boldsymbol{k}$ or slightly inclined with respect to it. The other two are mainly transverse waves, and, in any case, the amplitudes of the three waves are mutually orthogonal. The amplitudes of the three waves are given by Eq. (7.12):

$$A_i^2 = \frac{E_{\nu_i}}{2\pi^2 N m \nu_i^2} \qquad (i = 1, 2, 3)\,, \tag{7.27}$$

where $\nu_1, \nu_2, \nu_3$ are the frequencies of the three waves.

The quantity $E_\nu$ is given as a function of $\nu$ by the rigorous Eq. (7.13). As a first approximation which is valid for sufficiently high temperatures, $E_\nu$ is approximately equal to $kT$ and is the same for all the waves. The scattering intensities due to the three elastic waves simply add because the three scattered radiations are incoherent.

Let us note that the elastic waves associated with the point $-\boldsymbol{k}$ are identical to those associated with $\boldsymbol{k}$ and that they contribute equally to the intensity measured at $\boldsymbol{k}$. Finally, from Eqs. (7.24) and (7.27), the scattering power related to a point of the lattice of points $\boldsymbol{k}$ is

$$I_K = 2\frac{f^2\pi^2}{V_c} s^2 \textstyle\sum_i A_i^2 \cos^2\alpha_i = \dfrac{f^2 s^2}{N V_c m}\left(\textstyle\sum_i \dfrac{E_{\nu_i}\cos^2\alpha_i}{\nu_i^2}\right).$$

The experimentally measured scattering power, from Eq. (7.26), is given by

$$I_2(\boldsymbol{s}) = \frac{f^2 s^2}{m}\left(\frac{E_{\nu_1}\cos^2\alpha_1}{\nu_1^2} + \frac{E_{\nu_2}\cos^2\alpha_2}{\nu_2^2} + \frac{E_{\nu_3}\cos^2\alpha_3}{\nu_3^2}\right), \tag{7.28}$$

where $f$ is the scattering factor of the atom of mass $m$, there being only one atom per unit cell; $s = (2\sin\theta)/\lambda$; $\alpha_i$ is the cosine of the angle between $\boldsymbol{s}$ and the amplitude vector $\boldsymbol{A}_i$ of the three vibrations of frequency $\nu_i$ and of energy $E_{\nu_i}$ asscoiated with the point under consideration in the reciprocal lattice; and $I_2$ is the scattering power per unit cell of the crystal.

We can use the velocities of propagation of the elastic waves $v_i$ and their common wavelengths $\Lambda = 1/|\boldsymbol{k}|$, where $\Lambda$ is the inverse of the distance of the point $\boldsymbol{s}$ to the nearest node in the reciprocal lattice:

$$\Lambda = \frac{v_i}{\nu_i}\,.$$

Replacing $E_\nu$ by $kT$, as is often justified, we have

$$I_2 = \frac{s^2 f^2}{m} \Lambda^2 kT \sum_i \frac{\cos^2 \alpha_i}{v_i^2} . \tag{7.29}$$

Finally, if we assume that the velocities of propagation are equal to an average value $v_m$, we find the approximate relation

$$I_2 \cong \frac{s^2 f^2 \Lambda^2 kT}{m v_m^2} . \tag{7.30}$$

We shall use this equation to find the order of magnitude of the scattering.

It will be seen from Figure 7.5 that, for a point $P$ of reciprocal space,

$$\Lambda = \frac{1}{R_{hkl}P} \quad \text{and} \quad s = OP ,$$

Near the center, the vector $\boldsymbol{k}$ coincides with $\boldsymbol{s}$, and therefore $\Lambda s = 1$ and Eq. (*7.30*) shows that the scattering power is approximately constant near the center:

$$I_0 = \frac{f^2 kT}{m v_m^2} ,$$

where $v_m$ is an average velocity of propagation for the elastic waves. Numerically, we find the following values at room temperature ($T = 300°K$):

| | |
|---|---|
| Aluminum | $I_0 = 0.75$ |
| Copper | $I_0 = 3.38$ |
| Zinc | $I_0 = 3.12$ |

A scattering power of the order of one electron per atom is very weak and can be detected with certainty only with the best experimental techniques. From Eq. (*7.30*), the order of magnitude of the scattering power at $P$ is

$$I_2(P) = I_0 \frac{OP^2}{R_{hkl}P^2} . \tag{7.31}$$

It therefore increases with distance from the center of the reciprocal lattice, and in a given unit cell it increases considerably near the nodes. Thus, at the point (2.1, 0, 0), it is $[2^2/(0.1)^2]I_0 = 400\, I_0$. This is easy to detect. Just as for normal diffraction, thermal scattering is reduced by the Debye factor $D = \exp(-2M)$ [Eq. (*7.7*)]. Since this factor decreases with $s$ while $I_2$ of Eq. (*7.30*) increases with $s$, the intensity of thermal scattering goes through a maximum and then decreases at large values of $s$.

### 7.1.5. Theoretical Estimate of the Temperature-diffuse Scattering Pattern

We shall use Eq. (*7.28*) in two different ways. First, we can calculate, at least for simple lattices, the amplitudes and the frequencies of the elastic waves as functions of their propagation vectors, starting from the elastic coefficients of the crystal. We can therefore calculate a priori the thermal scattering pattern for the crystal and verify the theoretical predictions experimentally. This has been done in the case of sodium, for example [18]. We shall simply indicate a few general and simple facts.

The approximate Eq. (*7.31*) has shown that the temperature-diffuse scattering is mostly concentrated near the nodes of the reciprocal lattice, and this is surely the most striking characteristic of a pattern such as that of sylvite (Figure 7.3). However, the surfaces of equal scattering are not spheres centered on the nodes. Let us consider two points $P$ and $Q$ at equal distances from the node 200 and situated respectively on the axis [200] and the row [200-220] (Figure 7.7). Of the three elastic waves associated with $P$, one has its amplitude vector $\boldsymbol{A}_1$ directed along $\boldsymbol{OP}$ for reasons of symmetry, and the other two are rigorously transverse. Since the vector $\boldsymbol{OP}$ is none other than $\boldsymbol{s}$, the three cosines of the angles $\alpha_i$ are equal to 1, 0, and 0. Thus at $P$ we are only concerned with the longitudinal wave. At $Q$, on the contrary, it is one of the two transverse vibrations which is nearly in the direction of $\boldsymbol{s} = \boldsymbol{OQ}$, the other two being normal or nearly so. From Eq. (*7.29*), the intensities at the points $P$ and $Q$ are therefore proportional to $1/v_{\text{long}}^2$ and $1/v_{\text{transv}}^2$, since $OP \cong OQ = s$.

In the case of sylvite the velocities of propagation of the transverse waves propagating along a fourfold axis is smaller than that of the longitudinal waves. The intensity at $Q$ is therefore larger than that at

**FIGURE 7.7.** *Anisotropy of temperature-diffuse scattering around a node.*

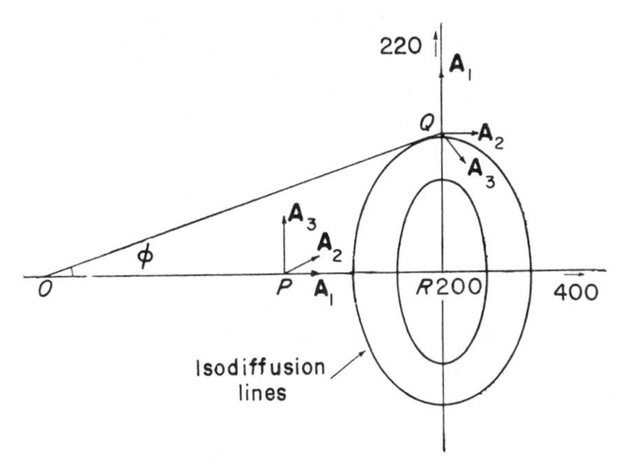

$P$, whence the elongated form of the scattering regions 200 normal to the axis [200]. Equation (*7.28*) shows that the variations of the scattered intensity around a node are due to complex causes. From one point to another of reciprocal space, the amplitudes and velocities of the elastic waves vary, and the relative importance of the three waves changes.

The scattering becomes weaker as the point under consideration moves away from the nodes, since $|\boldsymbol{k}|$ then increases. The scattering is not isotropic, for there exist regions of strong scattering along certain rows, such as ⟨100⟩ for sylvite. This means that along these axes the waves propagate with a particularly *low* velocity which, according to Eq. (*7.28*), increases the value of $I_2$ for a given distance to the node.

One could also say that the directions of the strong scattering streaks are those in which the crystal is most easily distorted. In a stratified crystal such as graphite, the vibrations which are perpendicular to the hexagonal atomic planes ($\pi$) are much more intense than those which are parallel to these planes, whatever the direction of propagation of the waves in the planes ($\pi$). In the region around the nodes corresponding to a reflection on the planes ($\pi$), the vector $\boldsymbol{s}$ is approximately parallel to the strong vibrations and the scattering is therefore strong. On the contrary, it is weak around the nodes for which the vector $\boldsymbol{s}$ is parallel to the planes ($\pi$).

In fibrous crystals the fibers are not easily distorted along the axes, while the bonds between fibers are relatively weak. Thermal vibrations have little effect on the periodicity along the axis of the fiber. In reciprocal space the domains of scattering have the shape of a disk which is perpendicular to the axis of the fiber.

The scattering intensity as given in Eq. (*7.28*) is proportional to the square of the modulus of $\boldsymbol{s}$. It therefore increases with the node indices, so long as the Debye factor of Eq. (*7.7*), which decreases as $s^2$, is not preponderant. For large values of $s$ the general background scattering increases in the unit cell and we must then take into account the higher order satellites $\pm 2\boldsymbol{k}, \pm 3\boldsymbol{k}, \cdots$ for each wave, as well as higher-order scattering (Section 7.1.3) in which several elastic waves are involved. The phenomena are then less clear and their interpretation is more difficult.

### 7.1.6. Determination of the Elastic Properties of Crystals from the Scattering Pattern

The measurement of the absolute value of the scattered intensity at a given point of reciprocal space, as described in Section 6.3, gives only the sum

$$\sum_i \frac{\cos^2 \alpha_i}{\nu_i^2},$$

where the three $\nu_i$'s and the three $\alpha_i$'s are unknown. Fortunately, as we have already indicated, there are particular positions of the vector $\mathbf{s}$ for which a single elastic wave is important. When $\mathbf{s}$ is directed along an axis of symmetry of the crystal, one of the waves is purely longitudinal, and the other two, which are purely transverse, have their vibrations normal to $\mathbf{s}$ (Figure 7.7). The measurement of the scattering power at $P$ can thus permit the determination of the velocity of the longitudinal wave (Curien [16]) propagating along the [100] axis of the crystal as a function of its wavelength, which can be determined from the geometry of the experiment:

$$\Lambda = \frac{1}{R_{200}P},$$

and from Eq. (7.30),

$$v_L^2 = \frac{f^2 s^2 \Lambda^2 kT}{mI_2}.$$

To use the exact Eq. (7.28), we calculate $\nu$ from the approximate value $v_L/\Lambda$, which gives $E_\nu$ of Eq. (7.13),, and we thus obtain a second approximation:

$$v_L^2 = \frac{f^2 s^2 \Lambda^2 E_\nu}{mI_2}.$$

It is also possible to determine the velocity of certain transverse waves [16]. This will be illustrated for the case of a cubic crystal. Let us situate the extremity of $\mathbf{s}$ on the row $R_{200}$-$R_{220}$. By symmetry, the three vibrations are parallel to the three axes $\langle 100 \rangle$. If $\varphi$ is the angle between $\mathbf{s}$ and [100] as in Figure 7.7,

$$I_2 = \frac{f^2 s^2 \Lambda^2 kT}{m}\left(\frac{\cos^2 \varphi}{v_T^2} + \frac{\sin^2 \varphi}{v_L^2}\right).$$

This equation gives $v_T$ if $v_L$ is known from the first measurement.

*It is therefore possible, in certain cases, to determine the absolute value of the velocity of propagation of an elastic wave as a function of its wavelength, and hence of its frequency,* $\nu = v/\Lambda$. It is found that the velocity depends on the wavelength and that it tends to a limit when $1/\Lambda$ tends to zero. This limit is the velocity of propagation of waves having very long wavelengths compared to the atomic dimensions, such as one can produce in crystals by mechanical or electrical devices (Figure 7.8). For these long wavelengths the crystal can be considered to be homogeneous and continuous, but anisotropic. The theory of elasticity gives the velo-

**FIGURE 7.8.** *Velocity and frequency of acoustic waves as functions of their wavelength* (*Curien* [*16*]). *Longitudinal waves along the fourfold axis in a crystal of α-iron. The dots and the circles correspond to two different experiments.*

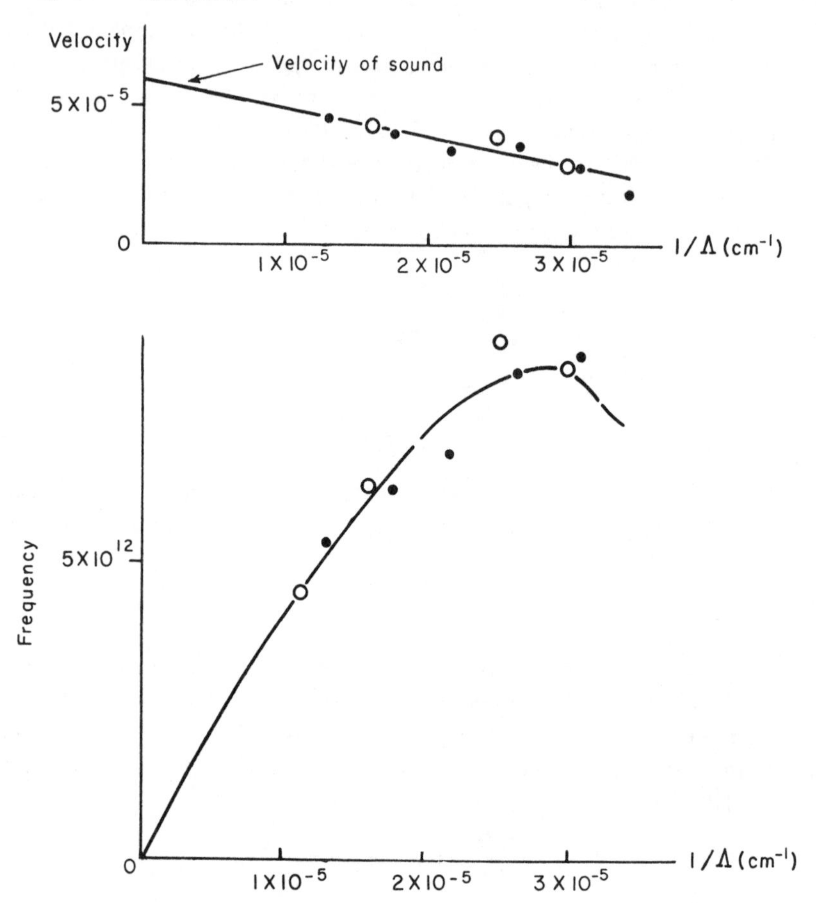


cities of propagation of the waves propagating in an arbitrary direction, in terms of the elasticity coefficients of Voigt. *It has been found that there is a satisfactory agreement between these values and those deduced from X-ray measurements* [19]. This agreement is quite remarkable since the absolute values are obtained from an equation which does not contain empirical coefficients.

This approach has rendered possible an extension of the classical methods for the investigation of the elasticity of solids where one determines the frequency of a wave of known wavelength. In the case of X-rays, one uses very high-frequency waves which need not be excited externally since they result from thermal agitation. The frequency of these waves is of the order of $10^{12}$, while mechanical vibrations have frequencies which do not normally exceed $10^8$. The very high frequencies are interesting because their wavelengths are so small that the atomic structure becomes important. The dynamic theory of lattices, as elaborated by Born [20], was applied by Curien [16] to the case of iron. Curien has shown how the curves of frequency as a function of wavelength—as obtained by X-ray scattering measurements—permit the calculation of the coefficients required in the Born theory. He can then calculate the elastic forces exerted on the iron atoms within the crystal. It is found that the forces are in general not central and that one must take into account not only the forces between first neighbors, but also between those between second and third neighbors. If the atom is displaced by $dr$ centimeters in the direction of a first neighbor, the elastic restoring force is $3.53 \times 10^4\ dr$ dynes. It is $1.16 \times 10^4\ dr$ dynes for second neighbors, and still $0.52 \times 10^4$ $dr$ for third neighbors. If the displacement $dr$ is normal to the vector between neighbors, in such a way that the distance between atoms remains constant, the elastic restoring force is weaker by a factor of about 5.

X-Ray measurements can also give the spectrum of the lattice frequencies [16], [21], or the number $dN_\nu$ of vibrations having a given frequency. This gives the total energy of vibration,

$$E = \sum dN_\nu E_\nu ,$$

where $E_\nu$ is given as a function of the temperature in Eq. (*7.13*), and the *specific heat* $c = dE/dT$. This value is obtained from a measured frequency distribution, and not from one based on more or less gross assumptions, as is the case for all the theories of the specific heat of solids [7]. These few examples indicate the role which X-rays can play in developing a theory of the field of forces in crystals, which is an essential basis for the theory of solids.

### 7.1.7. The Case of Molecular Crystals

If a crystal has $g$ atoms per unit cell, $g$ being larger than unity, the thermal agitation waves can be divided into two groups. For a given propagation vector, we first have the three *acoustic waves* discussed above for the case of the simple crystal, and also three $(g - 1)$ *optical waves.* At long wavelengths there exists a sharp distinction between the two groups: in the case of the acoustic waves, atoms of a given size are nearly in phase so the atoms of a unit cell move as a whole; in the case of the optical waves, these atoms move with respect to each other. Thus, in the case of a molecular crystal having more than one molecule per unit cell, vibrations within the molecules, or vibrations of one molecule with respect to the others, are classed as optical. While the frequency of the acoustical waves tends to zero as $\Lambda$ tends to infinity, that of the optical waves tends to a finite values which is of the order of those found in optics. It is these latter vibrations which produce the Raman effect.

Because of their high frequencies, the optical vibrations have little effect on X-ray scattering [Eq. (*7.28*)] and they can therefore be neglected as a first approximation. However, the scattering patterns for certain molecular crystals cannot be interpreted properly without them. As we have seen, the scattering due to the acoustic waves always increases in the vicinity of the nodes of the reciprocal lattice, but there are cases in which spots occur either between the nodes or around nodes of zero structure factor. One then assumes, as a second approximation, that this scattering is caused by a relative motion of rigid molecules. Vibrations within the molecules themselves give even higher frequencies, and thus lower amplitudes.

Ice crystals, for example, give intense narrow streaks along certain rows of the reciprocal lattice as in Figure 7.9 [22]. The diffraction patterns of organic crystals of benzene, anthracene, etc. have also been investigated

**FIGURE 7.9.** *Scattering pattern for an ice crystal showing diffuse streaks along certain axes of the reciprocal lattice (K. Lonsdale).*

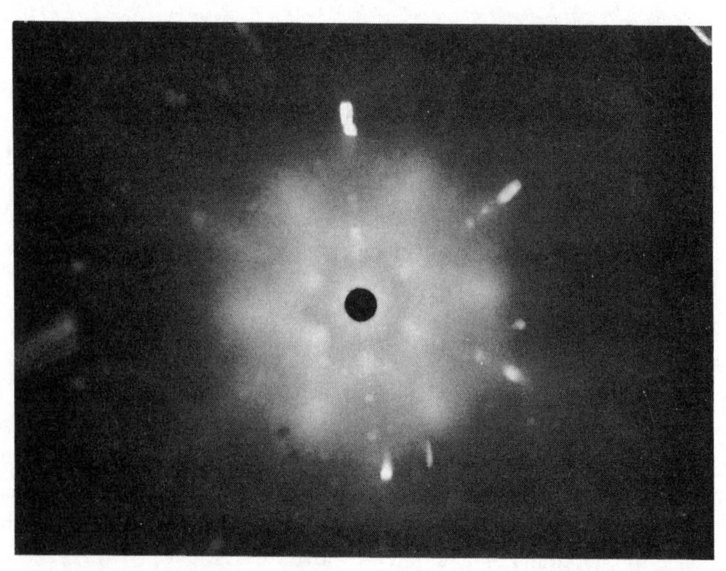

[23]. If the molecules vibrated independently, the scattered intensity would be proportional to the square of the modulus of the structure factor for the molecule which is the Fourier transform of the electronic density of the molecule and there are usually regions of intense scattering at the maxima of the transform. However, since the interactions between molecules are not negligible, the scattering is not exactly proportional to $|F|^2$.

There is no doubt that X-ray scattering from molecular crystals will become a useful method for the study of the relative motions of molecules, and thus of atomic bonds in organic crystals.

Temperature-diffuse scattering is therefore important, not only because of the information it can provide, but also because it is inevitable. It must be well understood because it must often be subtracted from the observed patterns. This problem is particularly important in the case of a crystalline powder, where one must calculate the average of the effect given by Eq. (*7.28*) for a single crystal oriented at random with respect to the incident beam. It is possible to perform this calculation with a fair accuracy in the case of a cubic crystal, and there exist formulas for this purpose which are easy to apply [24].

## 7.2. Planar Disorder

One of the most important cases of displacement disorder is planar disorder. We consider a system of lattice planes ($\pi$) which are left intact and parallel but irregularly displaced one with respect to the other. We have seen in Section 6.2.4 that the scattering is then limited *to the rows of the reciprocal lattice which are normal to the planes* ($\pi$). We shall now rediscover these results by applying the general Eq. (*6.5*):

$$I(\boldsymbol{s}) = \sum_m V(\boldsymbol{x}_m) y_m \exp(2\pi i \boldsymbol{s} \cdot \boldsymbol{x}_m) . \qquad (7.32)$$

We select the crystal axes in such a way that the ($\pi$) planes are the (001) planes, and we assume, for simplicity, that the axis $\boldsymbol{c}$ is normal to them, as in Figure 7.10. Under these conditions, the axis of the reciprocal lattice $\boldsymbol{c}^*$ is parallel to $\boldsymbol{c}$ and the plane (001) of the reciprocal lattice is the reciprocal lattice of the crystal plane (001). Since we have planar disorder, the displacement $\Delta\boldsymbol{x}_n$ depends only on the plane ($\pi$) where the unit cell $n$ is located, and therefore only on the parameter $n_c$. Thus

$$F_n = F \exp(-2\pi i \boldsymbol{s} \cdot \Delta\boldsymbol{x}_{n_c})$$

and, according to Eq. (*6.3*),

$$y_m = |F|^2 y_{m_c} = \overline{F_n F^*_{n+m}} = |F|^2 \overline{\exp[2\pi i \boldsymbol{s} \cdot (\Delta\boldsymbol{x}_{n_c+m_c} - \Delta\boldsymbol{x}_{n_c})]} . \qquad (7.33)$$

**FIGURE 7.10.** *Scattering calculation in the case of planar disorder.*

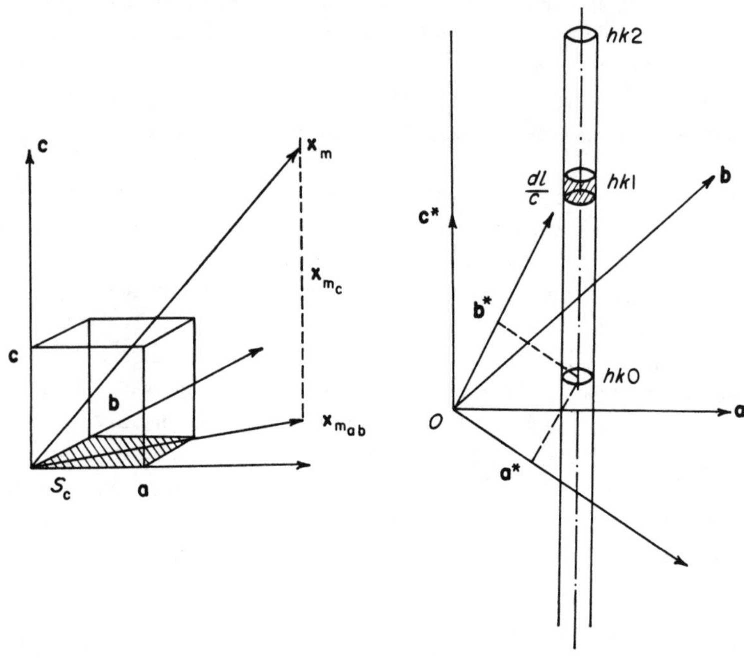


**FIGURE 7.11.** *The function $V(\boldsymbol{x}_m)$. Plan and elevation of the volume $VV(\boldsymbol{x}_m)$.*

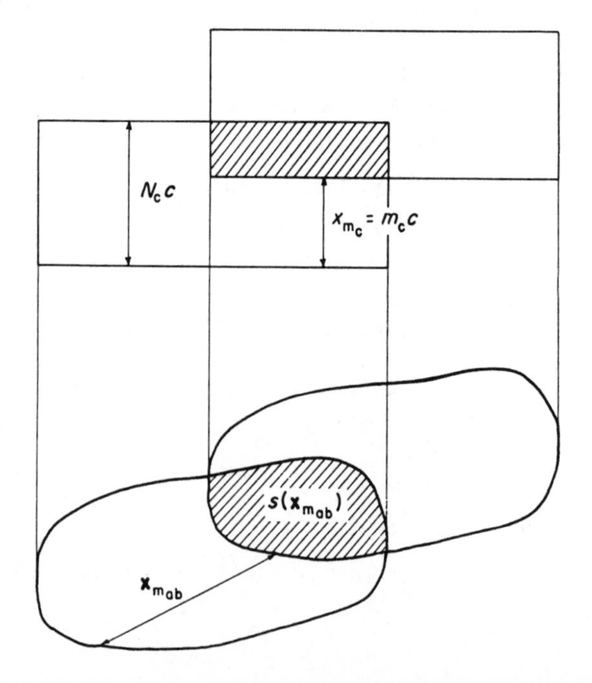

We also assume that the crystal is a cylinder whose elements are parallel to $\boldsymbol{c}$, so that it is formed of $N_c$ lattice planes (001) of the same shape. Let us write the vector $\boldsymbol{x}_m$ to show its component along $\boldsymbol{c}$:

$$\boldsymbol{x}_{m_a m_b m_c} = \boldsymbol{x}_{m_a m_b 0} + \boldsymbol{x}_{00 m_c} = \boldsymbol{x}_{m_{ab}} + \boldsymbol{x}_{m_c} .$$

We also decompose $\boldsymbol{s}$ as the sum of its projections on $\boldsymbol{c}^*$ and on the plane (001):

$$\boldsymbol{s} = \boldsymbol{s}_{a^*b^*} + \boldsymbol{s}_{c^*} .$$

Equation (7.32) involves first the function $V(\boldsymbol{x}_m)$, where $V$ is the volume of the object and $VV(\boldsymbol{x}_m)$ is the volume which is common to the crystal and its double displaced by $\boldsymbol{x}_m = \boldsymbol{x}_{m_{ab}} + \boldsymbol{x}_{m_c}$. This volume, which is shown in Figure 7.11, is equal to the product of the surface $S(\boldsymbol{x}_{m_{ab}})$, which is common to the contours of $(\pi)$ and of its double displaced by $\boldsymbol{x}_{m_{ab}}$, multiplied by the length $N_c c - |\boldsymbol{x}_{m_c}|$, or $(N_c - |m_c|)c$. Thus, if $S$ is the area of the section $(\pi)$, then $V = SN_c c$ and

$$V(\boldsymbol{x}_m) = \frac{S(\boldsymbol{x}_{m_{ab}})(N_c - |m_c|)c}{V} = \frac{S(\boldsymbol{x}_{m_{ab}})}{S}\left(1 - \frac{|m_c|}{N_c}\right). \qquad (7.34)$$

Equation (7.32) can therefore be rewritten as follows, using Eqs. (7.33) and (7.34)

$$I(\boldsymbol{s}) = \sum_{m_a}\sum_{m_b}\sum_{m_c}\{S(\boldsymbol{x}_{m_{ab}})/S\}\left(1 - \frac{|m_c|}{N_c}\right)|F|^2 y_{m_c} \exp[2\pi i\boldsymbol{s}\cdot(\boldsymbol{x}_{m_{ab}} + \boldsymbol{x}_{m_c})]$$

$$= [|F|^2 \sum_{m_a}\sum_{m_b}\{S(\boldsymbol{x}_{m_{ab}})/S\}\exp(2i\boldsymbol{s}_{a^*b^*}\cdot\boldsymbol{x}_{m_{ab}})]\sum_{m_c}\left(1 - \frac{|m_c|}{N_c}\right)y_{m_c}$$

$$\times \exp(2\pi i\boldsymbol{s}_{c^*}\cdot\boldsymbol{x}_{m_c}) . \qquad (7.35)$$

The expression

$$|F|^2 \sum_{m_a}\sum_{m_b}\{S(\boldsymbol{x}_{m_{ab}})/S\}\exp(2\pi i\boldsymbol{s}_*\cdot\boldsymbol{x}_{m_{ab}})$$

is the two-dimensional analogue of the term $I_1$ of Eq. (6.25), which gives the unit scattering power for a perfect three-dimensional crystal. It is different from zero only in the immediate neighborhood of the nodes of the reciprocal lattice of the plane lattice, if the object has a sufficiently large cross-section. We must introduce a form function for the plane $(\pi)$, namely $\sigma(\boldsymbol{x}_{ab})$, which is equal to unity within the crystal and zero outside. Corresponding to this function, which is defined in the plane $(\boldsymbol{a}, \boldsymbol{b})$, we have a Fourier transform defined in the plane $(\boldsymbol{a}^*, \boldsymbol{b}^*)$ of the reciprocal lattice:

$$\Sigma(\boldsymbol{s}_{a^*b^*}) = \int \sigma(\boldsymbol{x}_{ab})\exp(-2\pi i\boldsymbol{s}_{a^*b^*}\cdot\boldsymbol{x}_{ab})ds_x .$$

Around each node $hk0$, the intensity is defined by an expression which is analogous to Eq. (4.28). We shall call $S_c$ the surface of the unit cell of the plane lattice (001), so that $V_c = S_c c$, as in Figure 7.10. From Eq.

(*4.28*) the scattering power per unit cell for the plane lattice is given by

$$I(\mathbf{s}_{a^*b^*}) = \frac{|F|^2}{SS_c} |\Sigma(\mathbf{s}_{a^*b^*} - \mathbf{r}^*_{hkl})|^2 .$$

If the planes are sufficiently large, the reflection domain is limited to the nodes and, in any case, the integral of the scattering power over the domain related to the node [see Eq. (*4.28a*)] is

$$I_{hk} = \frac{F^2}{S_c} = \frac{F^2}{V_c} c . \tag{7.35a}$$

In reciprocal space the intensity diffracted by the imperfect crystal is therefore concentrated in columns whose sections are the plane reflection domain surrounding the nodes of the base plane. We shall call the integral of the scattering power over a section of this column the scattering power at a point of the row. This scattering power varies along the row according to the following law:

$$I(\mathbf{s}^*) = \frac{F^2}{V_c} c \, \Sigma \left(1 - \frac{|m_c|}{N_c}\right) y_{m_c} \exp(2\pi i \mathbf{s}_{c^*} \cdot \mathbf{x}_{m_c}) .$$

The axis of the reciprocal lattice $c^*$ is equal to $1/c$. Let us set $s_{c^*} = l/c$ and $x_{m_c} = m_c c$. Suppressing now the index $c$ to simplify the notation,

$$I(l) = \frac{|F|^2}{V_c} \Sigma_{-N}^{+N} \left(1 - \frac{|m|}{N}\right) y_m \exp(2\pi i l m) , \tag{7.36}$$

where $l$ is the ordinate along $\mathbf{c}^*$ expressed in units of the lattice parameter $1/c$ of the unit cell of the reciprocal lattice; $m$ is an integer lying between $-N$ and $+N$; $N$ is the total number of planes ($\pi$) in the object; $|F|^2 y_m$ is the average value[1] of the product $F_n F^*_{n+m}$ of the structure factors for two unit cells separated by the distance $mc$; and $I(l)$ is the scattering power along a row [001]. The quantity $I(l)\,dl$ is defined as the integral of the scattering power per unit cell of the crystal in the volume of reciprocal space which is a cylinder of height $dl/c$ and whose cross-section is the plane reflection domain surrounding the nodes $hk0$ of the base lattice (Figure 7.10).

Many authors have used the above equation as the starting point for the study of particular cases of planar disorder. It is rather complex because it takes into account both the size effect, represented here by the number of planes $N$, and the disorder. We shall dissociate these two effects, just as we have already done for the analogous three-dimensional Eq. (*7.32*) or (*6.5*).

If the function $I(l)$ has sharp peaks, as often occurs in crystal diffraction,

[1] Contrary to the notation used in Section 6.2, we have called this average $|F|^2 y_m$, and not $y_m$.

their width along the rows is determined by the thickness of the crystal $N$. For example, if there is no disorder ($\Delta \boldsymbol{x}_n = 0$), the stacking is completely regular, and all the coefficients $y_m$ are equal to unity. The quantity

$$\Sigma_{-N}^{+N} \left(1 - \frac{|m|}{N}\right) \exp(2\pi i l m) \tag{7.37}$$

is the interference function for a linear lattice of $N$ points. It has a series of peaks for integral values of $l$, and their profile is given by the function $(1/N)(\sin^2 \pi N l)/(\sin^2 \pi l)$, whose maximum value is $N$ and whose integral along the length of the row is equal to unity:

$$\int \frac{1}{N} \frac{\sin^2 \pi N l}{\sin^2 \pi l} dl = 1 .$$

From Eq. (*7.36*), the total intensity for the node is $I_{hkl} = F^2/V_c$, which is indeed the intensity for a node in a perfect crystal [Equation (*5.7*)]. In terms of $l$, the width of the peak is $1/N$.

If, on the contrary, the intensity varies slowly along a row, or if it has peaks whose width is large compared to that of the peaks of the regular lattice $1/N$, the coefficients $y_m$ decrease sufficiently rapidly with $m$ that we can neglect $|m|$ with respect to $N$, so long as $y_m$ is not zero. The unit scattering power then becomes simply

$$I(l) = \frac{|F|^2}{V_c} \Sigma\, y_m \exp(2\pi i l m) . \tag{7.38}$$

This equation is the exact one-dimensional analogue of the three-dimensional Eq. (*6.10*). We can use here the result of the discussion of Section 6.2 as applied to a single variable. If $y(x)$ is a continuous function of $x$ which has values $y_m$ for integral values $m$ of $x$, and which does not oscillate between two consecutive points $m$ and $m + 1$, we can replace the above summation by an integral for small values of $l$:

$$\int_{-\infty}^{+\infty} y(x) \exp(2\pi i l x)\, dx = Y(l) ,$$

where $Y(l)$ is the Fourier transform of $y(x)$. Equation (*7.38*) is equivalent to

$$I(l) = \frac{F^2}{V_c} \Sigma_p\, Y(l - p) , \tag{7.39}$$

where $p$ is any integer. If $Y(l)$ is different from zero only for $|l| \ll 1$, for each value of $l$ there is only one term of the above summation which is different from zero, and the intensity curve is built up from a succession of curves $Y(l)$ repeated around each node. The integral width of a peak is

$$\Delta(l) = \frac{\int I(l)\,dl}{I(0)} = \frac{y(0)}{\int y(x)dx} \,. \tag{7.40}$$

Inversely, we can find the coefficients $y_m$ by measuring the scattered intensity along the row. If $I(l)/F^2$ is a periodic function of $l$ whose periodicity is unity—which means that $y_m$ is independent of $l$—the inversion of Eq. (*7.38*) gives

$$y_m = \int_0^1 \frac{I(l)}{|F|^2} V_c \exp(-2\pi ilm)dl \,. \tag{7.41}$$

If $I(l)/|F|^2$ varies in the same manner but with different absolute values in each unit interval of $l$, we look for a function of $l$ which is such that $I(l)/[\varphi(l)|F|^2]$ is as nearly periodic as possible. Inverting this function then gives the coefficients $y_m$ (Section 6.2.3):

$$y_m = \varphi(l)\int_0^1 \frac{I(l)V_c}{\varphi(l)|F|^2} \exp(-2\pi ilm)dl \,.$$

It is only in the case where the peaks of the intensity function have widths due to the disorder and to the finite thickness of the object which are comparable—or in other words, when $\Delta l$ [Eq. (*7.40*)] and $1/N$ are comparable—that neither of the above solutions are acceptable. We then either perform the summation of Eq. (*7.36*) directly or apply the faltung theorem as we have done in the general Eq. (*6.5*). In the present case, we have the function

$$\begin{Bmatrix} 1 - \dfrac{|x|}{N} \\ 0 \end{Bmatrix} \quad \begin{matrix} \text{for} \quad |x| < N \\ \text{for} \quad |x| < N \end{matrix}$$

instead of $V(\boldsymbol{x})$. The transform of this function is $\sin^2 \pi Nl/\pi^2 Nl^2$ from the Equation of part 10 of Table A.1, Appendix A. Instead of being represented by the transform $Y(l)$ of $y(x)$, the intensity becomes [Eq. (*6.7*)]

$$I(l) = \frac{|F|^2}{V_c} \frac{\sin(\pi Nl)}{\pi^2 Nl^2} * Y(l) = \frac{|F|^2}{V_c} \int \frac{\sin^2(\pi Nu)}{\pi^2 Nu^2} Y(l-u)du \,.$$

It is possible to give a simple interpretation to Eq. (*7.36*) in the case where the disorder arises solely from displacements which are parallel to the axis $\boldsymbol{c}$. Returning to the definition of $y_m$ [Eq. (*7.33*)], we have

$$\Sigma_m \left(1 - \frac{|m|}{N}\right) y_m \exp(2\pi ilm) = \frac{1}{N} \Sigma_m \Sigma_{m'} \exp[2\pi i\boldsymbol{s}\cdot(\boldsymbol{x}_{m'} - \boldsymbol{x}_m)] \,,$$

where $x_m$ and $x_{m'}$ are the abscissas on the axis $\boldsymbol{c}$ of the planes $m$ and $m'$. The variation of the scattering power along the row corresponds to the diffraction pattern of $N$ points situated along a straight line at the positions of the successive planes (Figure 7.12). This is true whatever the arrangement of the $N$ points; they need not be in the neighborhood of the nodes of a perfect linear lattice. In this particular case, if these points form a regular linear lattice, the rows are occupied by periodic nodes. If, on the contrary, the points are distributed at random, we have "gaseous" scattering and the interference function is equal to unity. For the intermediate case where the distance between two successive points remains close to an average value $d_m$, we have a one-dimensional analogue of a

**FIGURE 7.12.** *Interference functions along rows of the reciprocal lattice normal to the plane* ($\pi$). (a) *Complete disorder. The interference function is equal to unity.* (b) *Slightly perturbed structure. The nodes have a width which increases with* $l$. (c) *"Liquid." The main maximum is near* $l = 1$.

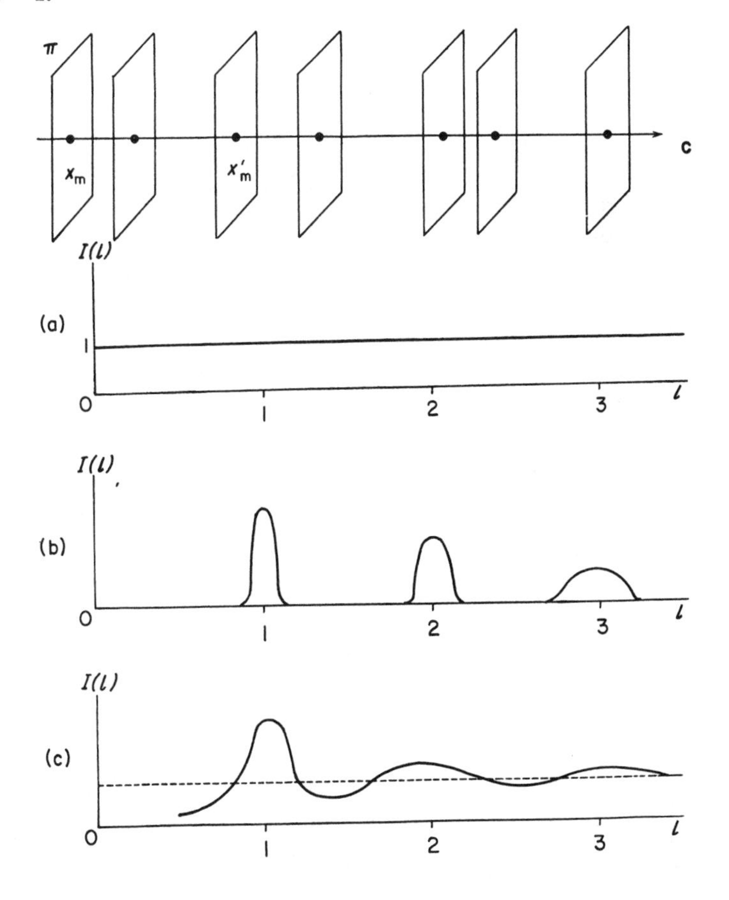

liquid (Figure 3.9). The intensity is weak for $l = 0$ and shows one or more broad maxima, the first of which is at a distance nearly equal to $1/d_m$.

We shall apply these results to some models of imperfect crystals which have been used in experimental work.

### 7.2.1. Periodic Variation in the Lattice Spacings

Let us consider a series of identical planes such that the distance be-

**FIGURE 7.13.** (a) *Parallel planes whose spacing varies as a sine function from* $a(1 - \varepsilon)$ *to* $a(1 + \varepsilon)$. *The intensity of the satellites as a function of the node index* $l$. (b) *Parallel planes, one-half of the spacings being* $a(1 - \varepsilon)$, *and the other half* $a(1 + \varepsilon)$. *Intensity of the satellites as a function of the node index* $l$.

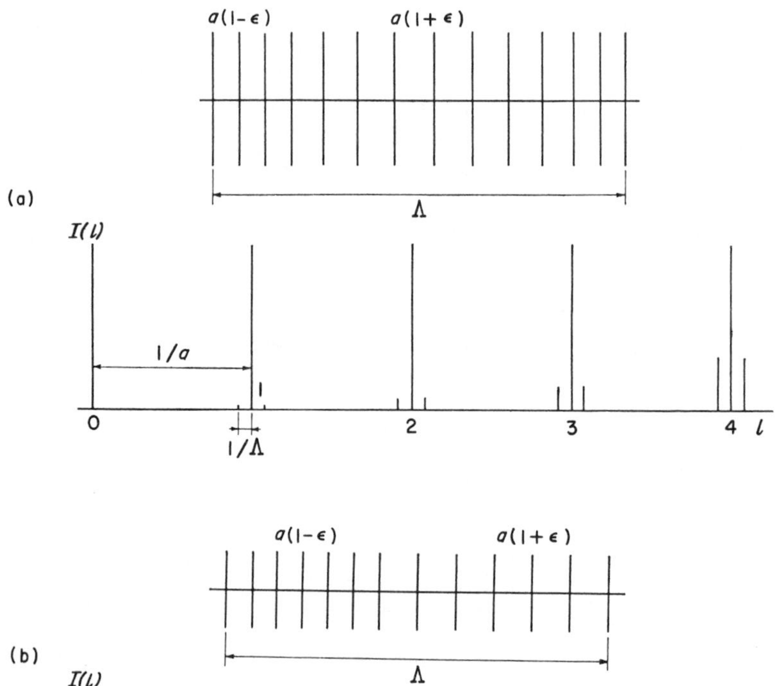


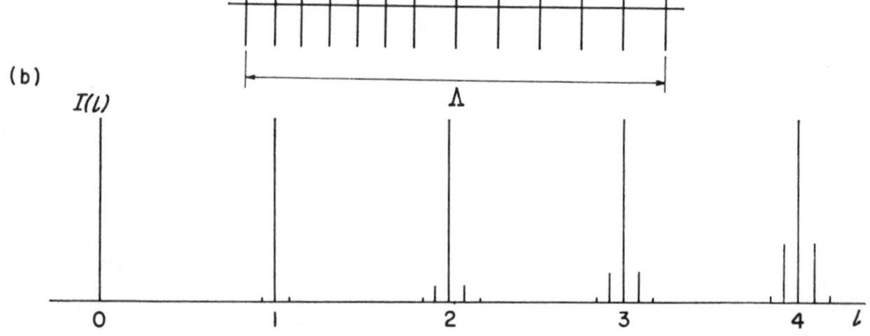

tween two neighboring planes is not a constant $a$, but varies sinusoidally from $a(1+\varepsilon)$ to $a(1-\varepsilon)$, with a period $\Lambda$ as in Figure 7.13a. These periodic compressions and decompressions are produced by a displacement $\Delta x_n$ normal to the planes:

$$\Delta x_n = \frac{\Lambda\varepsilon}{2\pi}\cos\left(\frac{2\pi na}{\Lambda}\right) \tag{7.42}$$

for small values of $\varepsilon$.

This is a particular type of disorder which was already studied in the case of thermal agitation in Section 7.1.3. We can consider here that we have a displacement wave parallel to a set of lattice planes. We have shown that, as a first approximation, each node in reciprocal space is accompanied by two symmetrical satellites at distances $\pm 1/\Lambda$ from the node in the direction normal to the displacement wave, or normal to the lattice planes, and thus along the corresponding row of the reciprocal lattice. Their intensity involves the scalar product of $\boldsymbol{s}$ by the amplitude of the displacement wave [Eq. (*7.24*)]; this is the product of the value of the amplitude $\Lambda\varepsilon/2\pi$ by $l/a$, where $l/a$ is the projection of $\boldsymbol{s}$ on the normal to the plane. The quantity $l$ is therefore the coordinate of the point of reciprocal space along the rows under consideration in units of the distance between two consecutive nodes.

According to Eqs. (*7.24*) and (*7.25*), the ratio of the intensity of the satellites to that of the normal node to which they are related is

$$\pi^2\frac{f^2}{V_c}(\boldsymbol{s}\cdot\boldsymbol{A})^2\Big/\frac{f^2}{V_c} = \pi^2\left(\frac{\Lambda\varepsilon}{2\pi}\,\frac{l}{a}\right)^2 = \frac{\Lambda^2\varepsilon^2 l^2}{4a^2}\,. \tag{7.43}$$

The intensity of the satellite therefore increases with the square of the coordinate $l$ of the node and *there are no satellites of appreciable intensity around the center.* This is one verification of the general principle stated at the beginning of this chapter.

Let us now consider a stack composed of successive sheets of $n/2$ planes with a spacing $a(1+\varepsilon)$, and an equal number with a spacing $a(1-\varepsilon)$, as in Figure 7.13b. The lattice spacing is then described by a square wave instead of a sine wave. According to the Fourier theorem, however, the square wave can be considered to be the sum of sine waves whose periods are $na=\Lambda, \Lambda/3, \Lambda/5,\cdots$ and whose amplitudes are respectively proportional to $\varepsilon^2, \varepsilon^2/9, \varepsilon^2/25,\cdots$. One can expect that the reciprocal space will contain on either side of each node and on the rows normal to the planes, pairs of satellites at distances $\pm 1/\Lambda, \pm 3/\Lambda, \pm 5/\Lambda,\cdots$, or at a fraction $1/n, 3/n, 5/n, \cdots$ of the distance between two nodes in the row. However, since their intensities decrease rapidly with order, one cannot expect to be able to observe more than the satellites of the first order and, experimentally, this case is very similar to the preceding one.

If we considered separately the two sets of sheets $a(1 + \varepsilon)$ and $a(1 - \varepsilon)$, they would correspond to reflection domains of width $2/na$, centered at $l/[a(1 + \varepsilon)]$ and $l/[a(1 - \varepsilon)]$, and thus spaced progressively further apart with increasing order of reflection, $l$. However, this would be true only if the waves diffracted by the groups of compressed and decompressed planes were independent. In fact, the successive sheets are coherent and interference makes the real diffraction pattern completely different.

This type of structure has been observed in the case of alloys which show side bands in their scattering patterns. The first example of this was the Cu Ni Fe alloy [25], but a number of other cases are now known [26]. The structure is produced when a homogeneous face-centered cubic phase separates below a certain temperature into two cubic phases of different composition, but with nearly equal lattice parameters. After quenching if the homogeneous solid solution is made to drift toward its state of equilibrium by successive annealings, several intermediate stages are observed. The first of these is characterized by powder patterns showing the diffraction lines for a single cubic phase, each line being flanked by weak and diffuse side bands.

In the case of the Cu Ni Fe alloy, the nodes of the cubic phase in the reciprocal space of the single crystal are accompanied by a pair of satellites situated symmetrically on the $\langle 100 \rangle$ rows, the distance between the satellites being the same for all the nodes. This is therefore an illustration of the theoretical calculation discussed in this paragraph. It is probable that the alloy becomes nonuniform as it drifts toward its state of equilibrium and that it becomes stratified, parallel to one of the (100) planes, due to a partial segregation of the atom into layers which are alternately enriched or partly depleted of their copper atoms. The scattering factors of the three types of atoms being quite similar, this segregation has no direct effect on the scattering pattern but the modulation in composition causes a modulation in the spacing of the (100) planes, and the main results of this section remain valid even if the modulation is not purely sinusoidal. The distance between satellites is a function of the period of the modulation, while the fluctuation in the lattice parameter determines only the intensity of the side bands, according to Eq. (*7.43*). Using this model, Hargreaves [27] gave the first quantitative interpretation of this phenomenon. The thickness of the lamellae depends on the heat treatment, and for the Cu Ni Fe alloy is of the order of 50–120 Å.

It is certain that the modulation is not strictly periodic because the side bands have an appreciable width. This could be explained either by irregularities in the modulation or by the small number of lamellae which scatter coherently. It has been shown that a single lamella with a larger than normal lattice parameter, flanked by a pair of parallel lamellae with a smaller than normal parameter, gives a diffraction pattern which agrees with the observations [28]. It is not yet possible to decide whether the perturbation is periodic or limited to a single zone. Arguments have been put forward in favor of both models [29].

For some alloys the scattering regions around the nodes of the reciprocal lattice are more complex than well-defined satellite points. The zones where the alloy is not homogeneous must then be correspondingly complex [30].

### 7.2.2. Stacking Faults

Lattice planes are in some cases stacked according to a definite law, except for stacking faults between two successive planes. We shall study various models which have been proposed to account for the observations.

As a first example we consider the planes (001), each one being spaced from the preceding one by a common vector $\boldsymbol{c}$. A fault at the $n$th plane involves an extra displacement $\Delta_n$, *parallel to the plane*, the modulus and direction of $\Delta_n$ being left unspecified. Beyond this fault, the stack resumes its original order as in Figure 7.14.

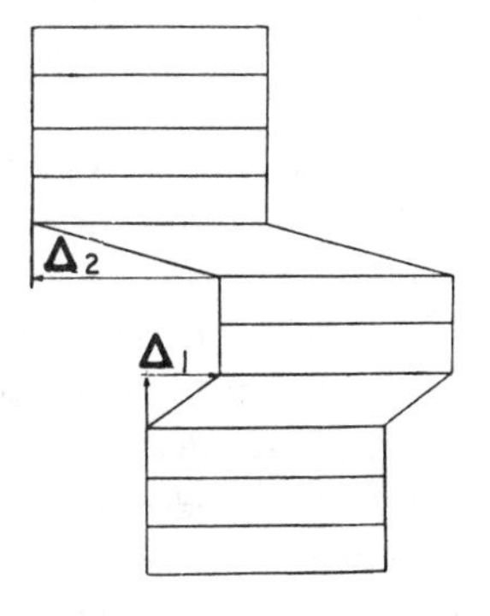


**FIGURE 7.14.** *Stacking faults for identical planes.*

These irregularities will necessarily give rise in reciprocal space to domains of scattering which are elongated along the [001] rows of the reciprocal lattice. We can calculate the angular distribution of the scattered radiation from the average $\overline{F_n F_{n+m}^*} = |F|^2 y_m$, according to Eq. (7.33). Then, if $\boldsymbol{s}$ is normal to the plane, i.e. for the axis [001], the displacement has no effect, since $\Delta_n$ is normal to $\boldsymbol{s}$. Thus $y_m = 1$ for this axis, whatever the value of $m$. For all the other rows, if there is no fault between the planes $n$ and $n+m$, then $\overline{F_n F_{n+m}^*}$ is equal to $|F|^2$, since $\Delta_n = \Delta_{n+m}$. If there is a single fault, and, a fortiori, if there are several, $\Delta_n - \Delta_{n+m}$ can have any value, and therefore the average $\overline{F_n F_{n+m}^*}$ for a sufficient number of such pairs of planes is zero. Thus $y_m$ is equal to the probability that there is no fault in a set of $m$ planes. If we call $\alpha$ the probability that there exists a fault at a given plane, then the probability that there is no fault is $(1-\alpha)$, and the probability that there is no fault on $m$ consecutive planes is $(1-\alpha)^m$. Thus $y_m = (1-\alpha)^{|m|}$ for all the rows, but $y_m = 1$ for the axis [001].

We can deduce from this the following two conclusions. (a) On the axis [001] the scattering is concentrated around the nodes of the regular lattice, and the width of the reflection domain depends only on the total number of planes $N$. On each side of the node the intensity decreases along the row according to the law $(\sin^2 \pi Nl)/(N\pi^2 l^2)$ and, for these particular conditions of diffraction, the crystal behaves like it were perfect. (b) On all the other rows, so long as $\alpha$ is not too small, so that $y_m = (1-\alpha)^{|m|}$ decreases sufficiently rapidly with $m$, the scattering power is given by Eq. (*7.38*),

$$I(l) = \frac{F^2}{V_c} \sum_{-\infty}^{+\infty} (1-\alpha)^{|m|} \exp(2\pi i l m)$$

or

$$I(l) = \frac{F^2}{V_c} [1 + 2 \sum_1^{\infty} (1-\alpha)^m \cos(2\pi l m)] \,. \qquad (7.44)$$

This example is important because many types of disorder give coefficients $y_m$ of the form $K^{|m|}$. The following results are therefore quite general.

The intensity $I(l)$ is periodic, and the period is unity; its maximum value corresponds to integral values of $l$ and thus *to the nodes of the average lattice*. It is *symmetrical* with respect to its nodes and the effect of the disorder therefore *to broaden the reflection domain symmetrically* without displacing its maximum. The effect is therefore analogous to a reduction in the thickness of a perfect crystal.

It is possible to perform the summation of Eq. (*7.44*) directly. As shown in Appendix B.2, we find that

$$I(l) = \frac{|F|^2}{V_c} \frac{\alpha(2-\alpha)}{1 + (1-\alpha)^2 - 2(1-\alpha)\cos(2\pi l)} \,. \qquad (7.45)$$

Figure 7.15 shows $I(l)V_c/|F|^2$ for values of $l$ ranging from $-1/2$ to $+1/2$ for various values of $\alpha$. When $\alpha$ is equal to unity, $I(l)$ is *constant* because the disorder is complete and the interference function is equal to unity along the row. When $\alpha$ decreases, a maximum appears at the lattice point, since the lattice is then regular between two faults. As the number of faults decreases, the peak rises and becomes narrower while the background decreases. Whatever the value of $\alpha$, the integral of the scattering power along a row between two consecutive nodes is constant. This results from the theorem of Section 2.3, and also from the following relation, which can be easily verified:

$$\int_{-1/2}^{+1/2} \frac{\alpha(2-\alpha)\,dl}{1 + (1-\alpha)^2 - 2(1-\alpha)\cos 2\pi l} = 1 \,.$$

**FIGURE 7.15.** *The scattered intensity along the rows of the reciprocal lattice for various values of the probability $\alpha$ of the existence of a fault at a given plane.*

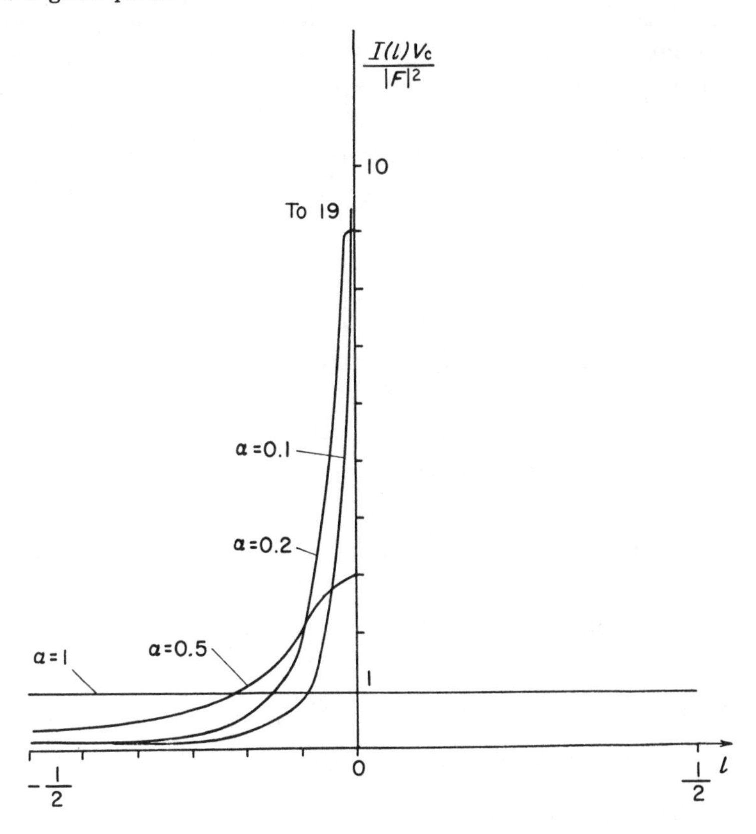


The maximum intensity is obtained by setting $l = 0$ in Eq. (*7.45*) and is $(2 - \alpha)/\alpha$. The integral width of the spot is therefore

$$\frac{\int I(l)dl}{I^{\max}} = \frac{\alpha}{2 - \alpha} \tag{7.46}$$

or approximately $\alpha/2$ when $\alpha$ is small. Thus, as a first approximation, the diffraction pattern for an irregular arrangement of an infinite number of planes—for the rows other than the axis [001]—is similar to that of a regular crystal whose thickness is such that it contains $2/\alpha$ planes. For such a crystal the intensity curve is

$$\frac{\sin^2(2\pi l/\alpha)}{(2/\alpha)\pi^2 l^2}.$$

This curve has the same width as in Eq. (7.45) and its profile is so little different that one cannot hope to distinguish between the two curves by experimental means. The essential difference is that the nodes $00l$ are broadened just like the others in the case of a small crystal, and that they remain very narrow for the infinite crystal with stacking faults.

When $\alpha$ becomes *very small*, the reflection domain becomes so narrow that Eq. (7.44) is no longer valid if the crystal is thin. The effect of the finite thickness of the crystal becomes appreciable whenever $2/\alpha$ is of the order of $N$. We must then perform the summation of Eq. (7.36):

$$I(l) = \frac{F^2}{V_c} \Sigma_{-N}^{+N}\left(1 - \frac{|m|}{N}\right)(1-\alpha)^{|m|} \exp(2\pi i l m)$$
$$= \frac{F^2}{V_c}\left[1 + 2\,\Sigma_1^N\left(1 - \frac{m}{N}\right)(1-\alpha)^m \cos(2\pi l m)\right]. \qquad (7.47)$$

This is possible with the methods described in Appendix B, but the resulting complex expression is not of general interest. As an approximation, we can assume that the true width is given by

$$\Delta l = \sqrt{\frac{\alpha^2}{4} + \frac{1}{N^2}}\,. \qquad (7.48)$$

When the size effect is unimportant, but when $\alpha$ is small, we can use the other method of calculation which was described in Sections 6.2.1 and 7.2. We introduce the continuous function

$$y(x) = (1-\alpha)^{|x|} = \exp[|x|\ln(1-\alpha)]$$

which, for integral values of $x$, is equal to $y_m$. Its Fourier transform is

$$\int \exp[|x|\ln(1-\alpha)]\exp(2\pi i l x)dx = \frac{-2\ln(1-\alpha)}{[\ln(1-\alpha)] + (2\pi l)^2}$$

and, according to Eq. (7.39),

$$I(l) = \frac{F^2}{V_c}\left\{\frac{-2\ln(1-\alpha)}{[\ln(1-\alpha)]^2 + (2\pi l)^2} + \frac{-2\ln(1-\alpha)}{[\ln(1-\alpha)]^2 + 2\pi(l-1)^2} + \cdots \right.$$
$$\left. + \frac{-2\ln(1-\alpha)}{[\ln(1-\alpha)]^2 + 2\pi(l+1)^2}\right\} + \cdots .$$

This equation is interesting only when all the terms, except the first one, are zero near the node $l = 0$. This requires that the peak be narrow, and $\alpha$ must therefore be small. We can then set $\ln(1-\alpha) = -\alpha$ and the scattering power along the row is

$$I(l) = \frac{F^2}{V_c}\,\frac{2\alpha}{\alpha^2 + 4\pi^2 l^2}\,, \qquad (7.49)$$

which is identical to Eq. (7.45) when $\alpha^2$ is negligible compared to $\alpha$, and when $\cos 2\pi l$ is replaced by $1 - 2\pi^2 l^2$. Equation (7.49) can thus be used near the center of narrow lines. This method of calculation has the advantage of being more generally applicable than the direct summation which was possible only because of the particular form of the functions $y_m$.

If single crystals are not available, then one can only obtain a powder pattern. It is easy to predict the general characteristics of such a pattern from the reciprocal lattice for the single crystal. The $00l$ lines are narrow if the crystals are very thick because they can be broadened only by the size effect, whatever the degree of order. The apparent size deduced from the integral width of the Debye-Scherrer lines through the Scherrer formula [Eq. (*5.4*)] then corresponds exactly to the thickness of the crystal, since the diameters normal to the plane (001) are all equal. All the other lines are broadened because of the disorder in the arrangement of the lattice planes. But although the reflection domains around all nodes are identical, lines having different indices can have different widths: a small segment $\Delta l/c$ of a row corresponds to a change $\Delta s$ of the modulus of $\mathbf{s}$, i.e. to a variation $\Delta(2\theta)$ of the diffraction angle which depends on the position of the segment on the row. From Figure 7.16,

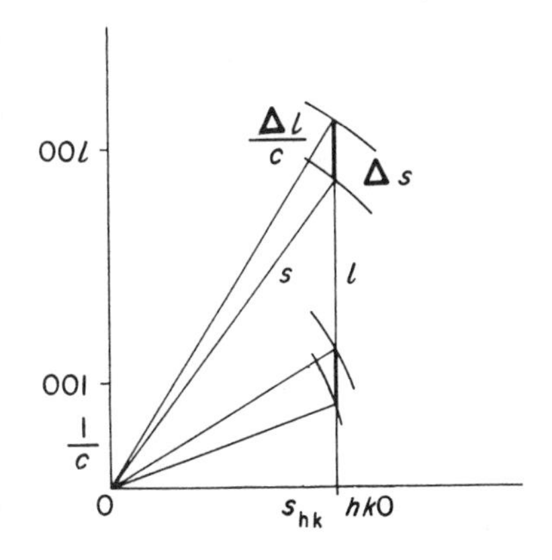


**FIGURE 7.16.** *Calculation of the powder pattern line widths for different indices, using the reciprocal lattice.*

$$s^2 = \frac{l^2}{c^2} + s_{hk}^2 ,$$

$$\Delta s = \frac{l}{c^2 s} \Delta l .$$

Since $s = (2 \sin \theta)/\lambda$, then

$$\Delta(2\theta) = \frac{\lambda^2}{c^2} l \frac{\Delta l}{\sin 2\theta} . \qquad (7.50)$$

Setting $\Delta l = \alpha/2$, this equation permits us to calculate the angular widths of all the $hkl$ lines other than $00l$. The broadening increases with the index $l$, and for a given value of $l$ it increases with decreasing $\theta$ or for decreasing $h$ and $k$.

According to Eq. (*7.50*), the width of the lines is zero when the index

**FIGURE 7.17.** (a) *The (hk0) line has a width which is of the second order with respect to the increment $\Delta l$ on the row.* (b) *Theoretical profile of the (hk0) powder pattern line for a crystal whose (00l) planes are stacked at random. The crystal is assumed to have a large area.* (c) *Profile for a crystal of area $L^2$ as calculated by Warren [31].*

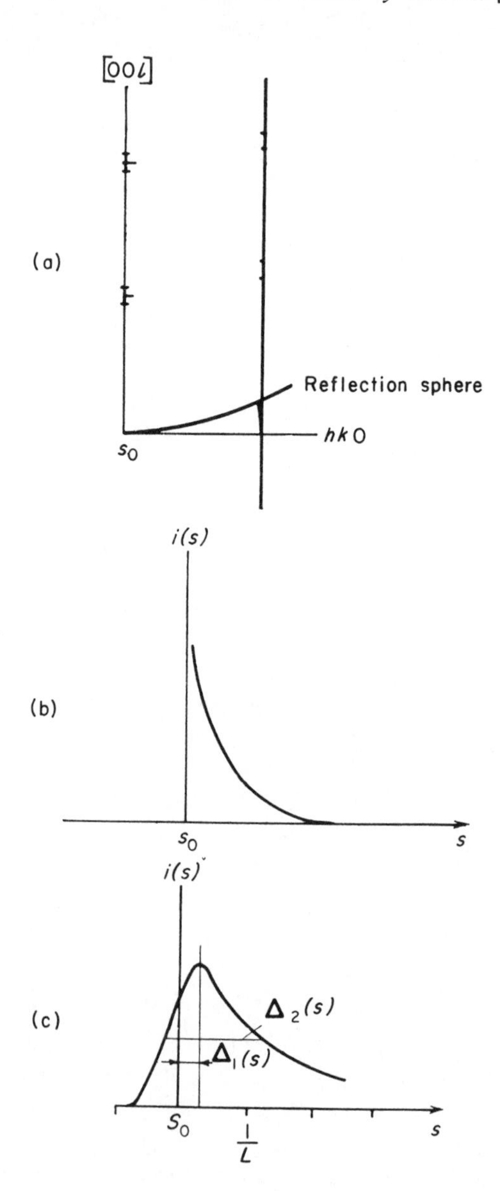

$l$ is zero. From Figure 7.17a, $\Delta(2\theta)$ is of the second order with respect to $\Delta l$. If the reflection domain is very extended the above differential is not valid. The line then has a measurable width and it is *asymmetric*, its center of gravity being displaced toward large angles with respect to the position of the lines of a perfect lattice. This displacement occurs only for the lines with $l = 0$. This shows that displacement and asymmetry in Debye-Scherrer lines do not necessarily indicate the presence of similar effects in the reciprocal lattice. One must therefore be careful not to ascribe the displacement of a line to a disorder in the interatomic distances.

To calculate the profile of a line, we must integrate the intensity distributed in reciprocal space between $s$ and $s + ds$ from the center. If $s_0$ is the distance to the node $hk0$, then

$$s^2 = s_0^2 + \frac{l^2}{c^2},$$

$$sds = \frac{ldl}{c^2},$$

$$dl = \frac{csds}{\sqrt{s^2 - s_0^2}}.$$

If the intensity is concentrated along the row itself with a distribution $I(l)$, the line profile becomes

$$i(s) \sim I(c\sqrt{s^2 - s_0^2})\,\frac{cs}{\sqrt{s^2 - s_0^2}}.$$

For example, if the planes are arranged at random, the scattering power is simply proportional to $F^2(s)$ and the profile is given by the function

$$\frac{F^2(s)s}{\sqrt{s^2 - s_0^2}}.$$

The line then has a sudden discontinuity at $s = s_0$ and a long tail toward large angles, as in Figure 7.17b.

This schematic calculation has no physical meaning, since it makes $i(s)$ infinite when $s$ tends toward $s_0$. To be more realistic, we must consider the distribution of the intensity around the row. This is given by $|\Sigma(\boldsymbol{s}_{a^*b^*})|^2$, where $\Sigma(\boldsymbol{s}_{a^*b^*})$ is the Fourier transform of the form function $\sigma(\boldsymbol{x}_{ab})$ for the contour of the planes (001). The integral which must be calculated is rather complex. It was shown by Warren [31] that, by substituting $\exp(-L^2r^2)$ for $\Sigma(\boldsymbol{s}_{a^*b^*})$, where $r$ is the distance from the point of reciprocal space to the row $hk$, and $L$ is the square root of the area of the crystal,[1] a line $hk0$ has the following characteristics when the

[1] The coefficient of the exponential is chosen in such a way that the integral width of the domain around the node $hk$ is $(1/L^2)$.

planes (001) are arranged at random (Figure 7.17c).

(a) The line is displaced away from the position for the perfect crystal toward large angles, and $\Delta_1 s = 0.32/L$.

(b) The apparent size calculated from the width of the line at half maximum (Section 5.1) is given by

$$\frac{1}{\Delta_2 s} = \frac{\lambda}{\Delta(2\theta)\cos\theta} = 0.57\,L\,.$$

The calculations of Warren have been improved to some extent by Wilson [32].

We have therefore shown that, if the plane spacing is constant and the planes are displaced laterally at random, the $00l$ lines of a powder pattern are narrow and symmetrical, and their width depends on the *thickness of the stack of planes*. We have also shown that the $hk0$ lines have sawtooth profiles with the abrupt side facing the center of the pattern. The shape of the $hk0$ lines depends on the *surface of the small crystal*. The lines $hkl$ are nonexistent.

The general characteristics of the patterns for certain types of graphite [31], [32] can be explained from the above considerations. Certain clays [33] also give sawtooth lines. This particular type of $hk0$ line, together with the narrow $00l$ lines, is characteristic of stacking faults in the (001) plane of the type studied in this section.

### 7.2.3. Stacking Faults in Close-packed Structures

The most compact arrangement of spherical atoms is obtained by stacking hexagonal lattice planes in such a way that each atom in a given layer is in contact with three spheres in the adjoining layer, so that it is situated above the center of the equilateral triangle formed by the centers of these three spheres. There are two ways in which the second layer can be placed on the first. The projection of a layer on the reference plane can take three different positions—designated by A, B and C in Figure 7.18. The only necessary condition for the stacking to be close-packed is that two consecutive layers be different. The two simplest arrangements which are periodic, and which can therefore give a crystal

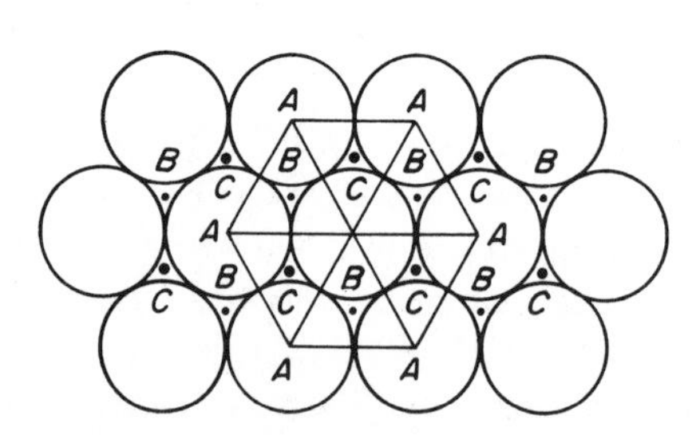


**FIGURE 7.18.** *Relative positions of the three layers* A, B, C *for a close-packed structure.*

lattice, are the *close-packed hexagonal* arrangement ABAB··· (the other arrangements BCBC··· and CACA··· are structurally identical), and the *face-centered cubic* arrangement ABCABC··· for which to the hexagonal planes are the (111) planes. The arrangement ACBACB··· is the twin of the latter and is symmetrical with respect to the reference plane. There are also other regular arrangements which are more complex and which give rise to unit cells extending over more than three layers—for example ABACAB··· or ABCBCAB···, which are found in certain forms of carborundum SiC.

The energy differences between these various arrangements are small. They would be zero if only neighboring layers showed an interaction energy. It is only the weak interactions between second neighbors which differentiate the hexagonal lattice from the cubic lattice, since the second layers are identical in one case (ABA) and different in the other (ABC). It is therefore possible for a compact lattice to contain faults, a layer A being substituted for a layer B or C. For example, in the hexagonal series

$$\mathrm{A\,B\,A\,B\,A\,B_{\perp}C\,B\,C\,B\,C\,B\,C\,B\,C\,B}\cdots,$$

there is one fault because there is one group of three consecutive layers which is of the cubic type (three different layers) while all the others are of the hexagonal type (two identical layers on either side of a different one). In the same way, the cubic arrangement

$$\mathrm{A\,B\,C\,A\,B\,C\,A\,B_{\perp}A\,C_{\perp}A\,B\,C\,A\,B\,C\,A\,B\,C\,A\,B\,C}$$

contains two hexagonal triplets, and therefore two faults. These types of imperfections have been observed in many crystals and it is therefore important to understand their effects on the diffraction pattern.

We select hexagonal axes in the reference plane, and set in the layer A the atom which has a scattering factor $f$ at a corner of the unit cell. The axis $c$ is equal to the distance between two neighboring layers; then $c=0.816a$. This is half the parameter $c$ of the hexagonal unit cell for the hexagonal arrangement, and one third of that for the cubic arrangement.

We know a priori that the diffracted intensity is concentrated on rows which are normal to the reference plane and which go through the nodes of the reciprocal lattice of the plane hexagonal lattice. This reciprocal lattice is built up on the vectors $\boldsymbol{a}^*$ and $\boldsymbol{b}^*$ normal to $\boldsymbol{a}$ and $\boldsymbol{b}$ and equal to $2/(a\sqrt{3})$. The problem is to calculate the scattering power at a point whose coordinate is $l/c$ on the row $hk$.

The position of the B atom with respect to those of layer A is given by the vector $(2\boldsymbol{a}/3) + (\boldsymbol{b}/3)$, and the structure factor for a unit cell of layer B is

$$F_B = f \exp\left[-2\pi i\left(\frac{2h}{3} + \frac{k}{3}\right)\right] = f \exp\left[2\pi i\left(\frac{h-k}{3}\right)\right]. \tag{7.51}$$

Similarly,

$$F_C = f \exp\left[-2\pi i\left(\frac{h}{3} + \frac{2k}{3}\right)\right] = F_B^* \tag{7.52}$$

for layer C. Therefore, for the node $hk$ such that $h - k =$ (multiplicity 3), the structure factor for any layer A, B, or C is $f$. In the case of these rows, whatever the number of faults, the structure factor is the same as if the lattice with axes $\boldsymbol{a}, \boldsymbol{b}, \boldsymbol{c}$ were regular. Along these rows, and in particular on the axis [001], we have a succession of nodes at intervals of $1/c$. These nodes can be broadened into a domain of reflection depending on the shape of the object, as if the object were a regular crystal. The anomalies arising from stacking faults appear only on the other rows.

We must now calculate the average value $y_m$ of $(\overline{F_n F_{n+m}^*})/f^2$ in the crystal. To define the arrangement of the layers, we only require the probability $P_m$ that two layers separated by a distance $mc$ be identical. We then have a probability $1 - P_m$ that the two layers are of the type AB or BA, AC or CA, BC or CB. It is not necessary to specify the nature of the pair because we shall assume that these six pairs are equally probable; since the notation A, B, C is purely arbitrary, the identification of the layers can be permuted. With this simplifying assumption, we can find the pattern for a mixture of a given crystal and of the other crystals obtained by the permutation of two layers. In the case of the hexagonal lattice, for example, we have the crystals ABAB···, ACAC···, etc. and these are identical. For the cubic lattice, we have the mixture ABCABC··· and ACBACB···. These are the two lattices which are symmetrical with respect to the reference plane, as we shall show later.

The $m$th neighbors of a plane A are planes A, B, C, which are found respectively in the proportions $P_m$, $(1 - P_m)/2$, and $(1 - P_m)/2$. On the other hand we have the the same total number of planes A, B and C. We therefore have

$$\begin{aligned} f^2 y_m = \frac{1}{3} F_A\left[P_m F_A^* + \frac{1-P_m}{2}(F_B^* + F_C^*)\right] \\ + \frac{1}{3} F_B\left[P_m F_B^* + \frac{1-P_m}{2}(F_A^* + F_C^*)\right] \\ + \frac{1}{3} F_C\left[P_m F_C^* + \frac{1-P_m}{2}(F_B^* + F_A^*)\right]. \end{aligned}$$

From the values which we found above in Eqs. (*7.51*) and (*7.52*) for the structure factors $F_A$, $F_B$, $F_C$,

$$y_m = P_m + (1 - P_m)\cos\left(2\pi\frac{h-k}{3}\right). \qquad (7.53)$$

When the layers are arranged at random, $P_m$ tends to 1/3 for large value of $m$ and we can set $P_m = (1/3) + Q_m$, where $Q_m$ tends to zero as the arrangement becomes random. Then the above equation can be rewritten as

$$y_m = \frac{1 + 2\cos[2\pi(h-k)/3]}{3} + Q_m\left[1 - \cos\left(2\pi\frac{h-k}{3}\right)\right]$$

and, according to Eq. (*7.36*), the scattering power on the rows, using the same notation, is given by

$$I(l) = \frac{f^2}{V_c}\left\{\left[\frac{1+2\cos[2\pi(h-k)/3]}{3}\right]\Sigma_{-N}^{+N}\left(1-\frac{|m|}{N}\right)\exp(2\pi ilm)\right.$$
$$\left. + \left[1-\cos\left(2\pi\frac{h-k}{3}\right)\right]\Sigma_{-N}^{+N}\left(1-\frac{|m|}{N}\right)Q_m\exp(2\pi ilm)\right\}. \qquad (7.54)$$

For rows having indices $hk$ such that $h - k =$ (multiplicity 3), then

$$\cos 2\pi\left(\frac{h-k}{3}\right) = 1,$$

and we have nodes with the periodicity $1/c$ of a perfectly regular crystal containing $N$ planes. For the other rows,

$$\cos 2\pi\left(\frac{h-k}{3}\right) = -\frac{1}{2}.$$

The first term in the summation of Eq. (*7.54*) is therefore zero. In the second term we neglect the size effect, as we did in the general Eq. (*7.38*). This is permissible if the crystal is not too small, and if it is sufficiently disordered. The scattering power along the row is given by

$$I(l) = \frac{3}{2}\frac{f^2}{V_c}\Sigma_{-\infty}^{+\infty}Q_m\exp(2\pi ilm). \qquad (7.55)$$

This equation shows that the variation of $I(l)/f^2$ is the same for all the rows such that $h - k \neq$ (multiplicity 3) and that it is periodic with a period $l = 1$, or $1/c$ in reciprocal space. The experimental measurement of I($l$) gives the values of $Q_m$, and therefore of $P_m$ by inversion:

$$Q_m = P_m - \frac{1}{3} = \int_0^1 \frac{V_c I(l)}{f^2}\exp(-2\pi ilm)dl, \qquad (7.56)$$

Since $P_0 = 1$ when $Q_0 = 2/3$, it is sufficient to find a relative value of $I(l)$ to determine all the coefficients $P_m$. *We must therefore necessarily find* $P_1 = 0$, since the arrangement is assumed to be close-packed.

Another way of using Eq. (*7.55*) is to calculate all the probabilities $P_m$ as functions of a small number of parameters. This is possible if we specify the nature of the faults and their respective probabilities. We shall give the results of some of the work which has been done in this direction [34].

(a) *We assume that the interaction is limited to neighboring layers.* Over a given layer we can set either of the other two different layers with an *equal probability*. Let us assume that A is the first layer. Then, if the $(m-1)$th is B or C with the probability $1-P_{m-1}$, we have one chance out of two that the $m$th will be A. Then

$$P_m = \frac{1-P_{m-1}}{2} . \tag{7.57}$$

Let us replace $P_m$ by $(1/3)+Q_m$. To solve this equation, we set $Q_m = q\rho^{|m|}$ and find that

$$P_m = 1/3 + 2/3(-1/2)^{|m|} .$$

Then, from Eq. (*7.55*), we have the scattering power:

$$I(l) = \frac{f^2}{V_c} \Sigma_{-\infty}^{+\infty} \left(-\frac{1}{2}\right)^{|m|} \exp(2\pi i l m) . \tag{7.58}$$

This equation is similar to Eq. (*7.44*). The above summation, according to Eq. (*7.45*), is given by

$$I(l) = \frac{f^2}{V_c} \frac{3/4}{(5/4)+\cos 2\pi l} . \tag{7.59}$$

The intensity has very large maxima at $l = p + (1/2)$, and minima at $l = p$, where $p$ is an integer. The ratio between the two is $3/(1/3) = 9$ (Figure 7.19).

**FIGURE 7.19.** *The function* $I(l)$ *giving the scattering power along the* $\langle 00l \rangle$ *rows for a close-packed structure in which the* A, B, *and* C *planes are arranged at random.*

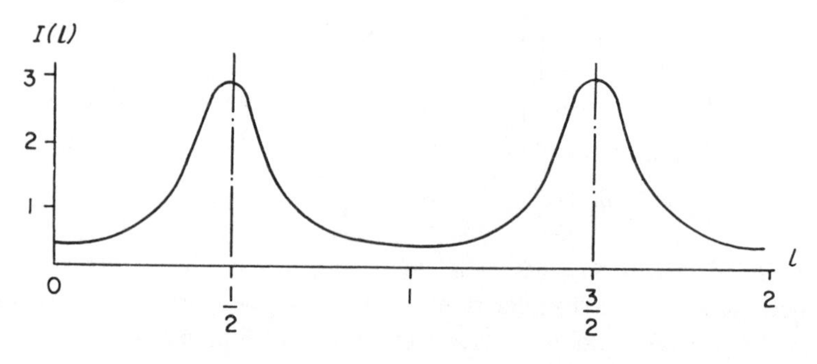

This is the greatest disorder which can exist in a close-packed arrangement. We can nevertheless see that the scattering already shows large variations along the rows.

(b) *The second neighbors are involved.* Two regular arrangements are possible, namely the hexagonal and the cubic arrangements. Let us consider the *close-packed hexagonal* arrangement, and let us assume that the probability of a fault is small (Wilson [36]). After the pair AB we should have A, but we assume that there is a probability $\alpha$ of finding C, with $\alpha$ being *small.* Under these assumptions, we find a recurrence relation between the probabilities which in this case will contain three terms, since the second neighbors are now involved. If the first layer is A, we calculate $P_m$ from the probability that the $m$th layer will be A. We have A in the $m$th layer if the $(m-2)$th layer is A and if there are no faults, since we have the series ABA or ACA. The probability for this is $(1-\alpha)P_{m-2}$. We can also have a fault (BCA or CBA) with A neither in the $(m-2)$th layer nor in the $(m-1)$th. The probability for this is $\alpha[1-P_{m-2}-P_{m-1}]$. Thus

$$P_m = (1-\alpha)P_{m-2} + \alpha(1-P_{m-2}-P_{m-1}) \qquad (7.60)$$

or

$$P_m + \alpha P_{m-1} + (2\alpha - 1)P_{m-2} = \alpha\,.$$

This equation can be solved by setting $P_m = (1/3) + q\rho^{|m|}$, if

$$\rho^2 + \alpha\rho - 1 + 2\alpha = 0\,. \qquad (7.61)$$

To satisfy the initial conditions $P_0 = 1$, $P_1 = 0$, we must use the two roots of the above equation and, neglecting the higher powers of $\alpha$, we finally obtain

$$P_m = \frac{1}{3} + \frac{1}{6}\left(1 - \frac{3}{2}\alpha\right)^{|m|} + \frac{1}{2}\left[-\left(1 - \frac{\alpha}{2}\right)\right]^{|m|}.$$

From Eq. (7.55), the scattering power is given by

$$I(l) = \frac{f^2}{V_c}\left\{\frac{1}{4}\Sigma_{-\infty}^{+\infty}\left(1-\frac{3}{2}\alpha\right)^{|m|}\exp(2\pi ilm) + \frac{3}{4}\Sigma_{-\infty}^{+\infty}\left[-\left(1-\frac{\alpha}{2}\right)\right]^{|m|}\exp(2\pi ilm)\right\}. \qquad (7.62)$$

This equation is easily interpreted by referring to the discussion of the analogous Eq. (7.58). Each sum corresponds to a maximum, the first one for integral values of $l$, and the second one for half integral values. Their intensities are in the ratio of one to three, and their integral widths are respectively $(3/4)\alpha$ and $(1/4)\alpha$; [see Eqs. (7.45) and (7.46), and Figure 7.20]. For $\alpha = 0$ we have a perfect close-packed hexagonal crystal with parameters $a$ and $2c$, whose reciprocal space has nodes separated by distances

**FIGURE 7.20.** *Reciprocal space corresponding to a hexagonal close-packed structure containing faults (Wilson [36]).*

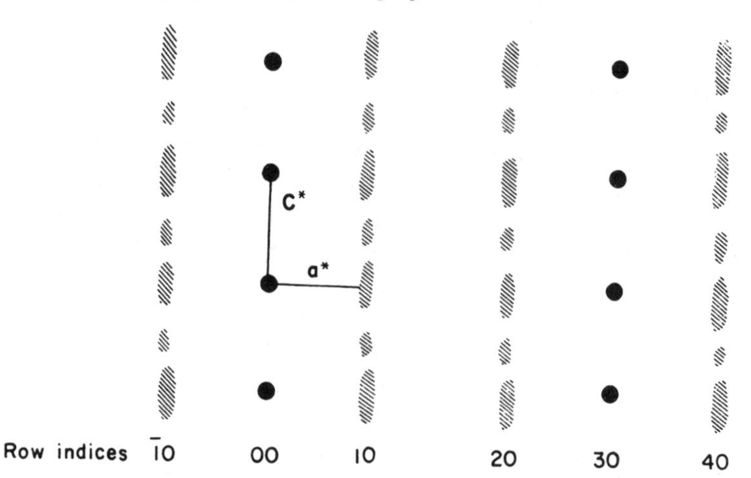


$c^*/2$ along the rows as shown in Fig. 7.20. It is easily shown that, along rows $hk$ such that $h - k =$ (multiplicity 3), nodes exist only for integral values of $l$ and that their intensity is 4; on the other rows there are alternate nodes of intensity 1 for $l$ integral and of intensity 3 for $l$ half integral.

The effect of stacking faults in the hexagonal lattice is therefore to broaden the nodes of the perfect lattice along the rows [001] without displacing them, as illustrated in Figure 7.20. The broadening, which is proportional to the disorder, is not the same for all the nodes: it is zero for $h - k = 3n$, it is $(3/4)\alpha$ for $h - k \neq 3n$ and integral values of $l$, and it is $(1/4)\alpha$ for $h - k \neq 3n$ and half integral values of $l$. The broadest nodes have an intensity which is three times smaller than the narrowest ones and they are therefore much more difficult to observe.

As shown in Section 7.2.2, it is possible to deduce the widths of the lines of a powder pattern from the reciprocal lattice of the single crystal. The lines $(00l)$ remain narrow if the crystal is sufficiently thick. The other lines have variable widths, partly because the broadening varies from one node to another, and partly because the effect of this broadening on the line depends on the position of the nodes in reciprocal space. Equations (7.50) can be used to apply the correction.

This type of imperfection has been observed experimentally in *cobalt* [35]. This crystal can be in equilibrium under either the hexagonal or the cubic form and there exist transition states where the hexagonal cobalt has faults where the structure has the cubic form. With the above method, the coefficient $\alpha$ was found to be equal to 0.1. This means that, on the average, there is one fault of the following type at every tenth plane:

$$\text{A B A B A B}_{\perp}\text{C B C B C}\,.$$

We have also assumed that the faults are distributed at random so they are completely independent of each other.

Another case which can be treated by analogous hypotheses is that of the cubic lattice containing a few rare faults, leading to a hexagonal arrangement such as A B C A $\text{B}_{\perp}$A C B A C. If $\beta$ is the probability of such a fault, then Eqs. (*7.60*) and (*7.61*) are still valid if we replace $\alpha$ by $1-\beta$. We also set $P_m = (1/3) + (q_1\rho_1{}^{|m|}) + (q_2\rho_2{}^{|m|})$, where $\rho_1$ and $\rho_2$ are the roots of the following equation, which is deduced from Eq. (*7.61*):

$$\rho^2 + (1-\beta)\,\rho + 1 - 2\beta = 0\,. \qquad (7.63)$$

These roots are the conjugates $\rho\exp(+i\Phi)$ and $\rho\exp(-i\Phi)$. Then the coefficients $q_1$ and $q_2$ must also be imaginary conjugates, so that $P_m$ is of the form $P_m = (1/3) + (q_r \cos m\varphi) - (q_i \sin|m|\varphi)$. If we take $\beta$ to be small, we find as a first approximation that

$$P_m = \frac{1}{3} + \frac{2}{3}(1-\beta)^{|m|}\cos\frac{2m\pi}{3}\,.$$

The intensity scattered on the rows $h-k \neq 3n$ is then

$$I(l) = \frac{f^2}{V_c}\left[\Sigma_{-\infty}^{+\infty}(1-\beta)^{|m|}\cos\frac{2m\pi}{3}\exp(2\pi i l m)\right]$$
$$= \frac{f^2}{2V_c}\left\{\Sigma_{-\infty}^{+\infty}(1-\beta)^{|m|}\exp\left[2\pi i\left(l+\frac{1}{3}\right)m\right]\right.$$
$$\left. + \Sigma_{-\infty}^{+\infty}(1-\beta)^{|m|}\exp\left[2\pi i\left(l-\frac{1}{3}\right)m\right]\right\}. \qquad (7.64)$$

There are two series of maxima of equal intensity situated at $l = -(1/3) +$ integer and $l = (1/3) +$ integer. Our hypotheses associate *the positions of the nodes of the face-centered cubic lattice* ABCABC and of its symmetric lattice ACBACB. The profile of the maxima is given by Eq. (*7.45*), where $\alpha$ is replaced by $\beta$ and their width is $\beta/2$, which is the average of the widths we have found for the nodes of the hexagonal lattice for the same degree of imperfection where $\alpha = \beta$.

This, however, is only an approximate solution since, if $\beta$ is different from zero, the argument $\Phi$ of the roots of Eq. (*7.63*) is very slightly different from $2\pi/3$, and $q_i$ in the expression for $P_m$ is not zero. The intensity then has a sine term of the form

$$\Sigma_{-\infty}^{+\infty}(1-\beta)^{|m|}\sin\frac{2\pi|m|}{3}\exp(2\pi i l m)\,, \qquad (7.65)$$

**FIGURE 7.21.** *Above*: *Function* $B = \sum_0^\infty (1-\beta)^m \sin 2\pi[l + (1/3)]m$. *Below*: *Function* $A = \sum_0^\infty (1-\beta)^m \cos 2\pi(l + (1/3)]m$. *The dashed curve* $A + KB$ *shows* $I(l)$ *near* $l = -1/3$.

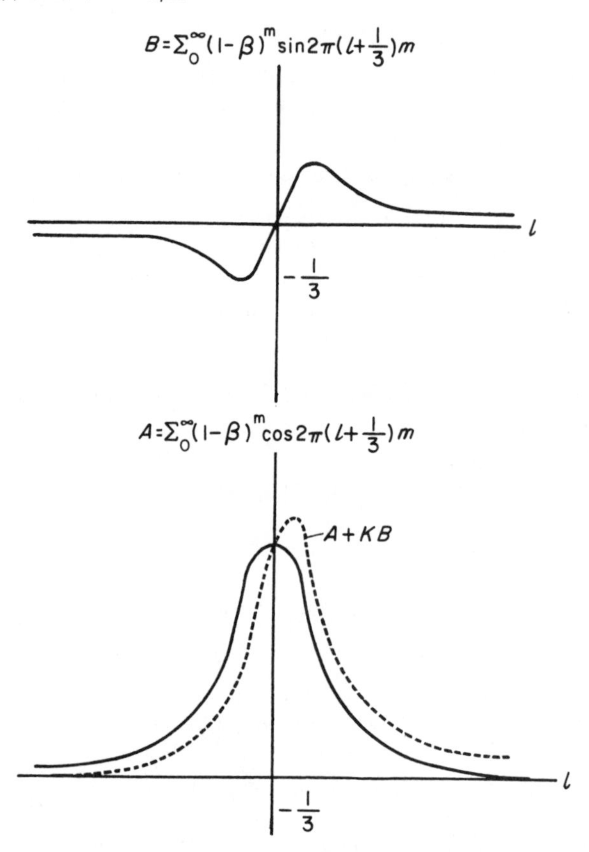


which can be written as

$$2\sum_0^\infty (1-\beta)^m \sin \frac{2\pi m}{3} \cos(2\pi l m)$$
$$= \sum_0^\infty (1-\beta)^m \sin\left[2\pi\left(l+\frac{1}{3}\right)m\right] - \sum_0^\infty (1-\beta)^m \sin\left[2\pi\left(l-\frac{1}{3}\right)m\right].$$

These summations can be calculated from Eqs. *B.3*, given in Appendix *B*. It will be seen from the shape of the curve in Figure 7.21 that, if we the add this sine term to the expression of Eq. *7.64*, *the maxima are displaced away from their position* $\pm 1/3$. Thus, in the case of the cubic lattice, the nodes of the perfect crystal are not only broadened but are also displaced by faults, contrary to what is seen in the case of the imperfect hexagonal lattice.

This is also observed in another type of stacking fault. We have so far considered only the so-called *growth* faults arising in the formation of the crystal by the stacking of successive planes. We can also consider *distortion faults*, starting with a perfect face-centered cubic crystal ABCABC$\cdots$ and assuming that the distortion has the effect of slipping one plane, thereby changing from one type to another while preserving the sequence for the other planes. This type of fault therefore never gives rise to twinning, as in growth faults.

Let $\alpha$ be the probability of a fault. Paterson [36] has shown that on the rows $h-k \neq$ (multiplicity 3) the nodes are broadened and displaced. The width of the maximum along the row is given by

$$\Delta l = \frac{1-\sqrt{1-3\alpha(1-\alpha)}}{1+\sqrt{1-3\alpha(1-\alpha)}} \tag{7.66}$$

and the position of the maximum (Figure 7.22) is

$$\begin{aligned} l &= \frac{1}{2} - \frac{1}{2\pi} \text{arc tan}\,[\sqrt{3}(1-2\alpha)] \qquad \text{for } h-k=3n+1\,, \\ l &= \frac{1}{2} + \frac{1}{2\pi} \text{arc tan}\,[\sqrt{3}(1-2\alpha)] \qquad \text{for } h-k=3n-1\,. \end{aligned} \tag{7.67}$$

If we set $\alpha = 1$ in the above equations, the widths of the maxima become zero and they are situated at $l = (1/2) \pm (1/6)$ for $h-k = 3n \pm 1$. These are the positions of the nodes for the perfect twin lattice which is symmetrical with respect to the reference plane. This result was to be expected, since if there is a fault at each plane the lattice ABCABC becomes ACBACB.

**FIGURE 7.22.** *Schematic diagram of the scattering on the rows* $\langle 00l \rangle$ *of a cubic lattice with distortion faults* (*Paterson* [36]). (a) *Cubic lattice* ABC$\cdots$, (b) *Cubic lattice with faults.* (c) *Twin cubic lattice ACB*$\cdots$.

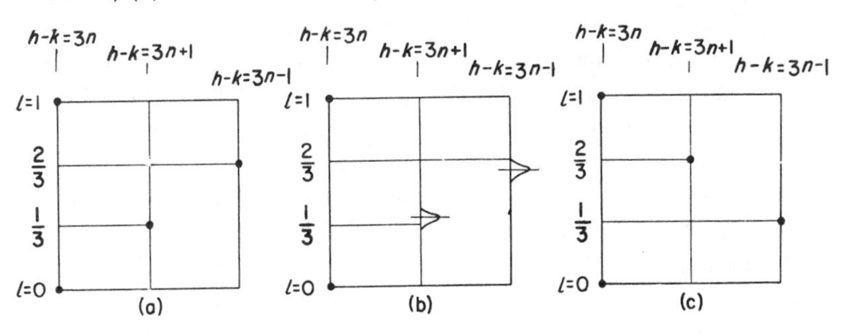

The process of passing from the reciprocal lattice of the crystal to the powder pattern is more complex than for the hexagonal lattice, since a line then corresponds to several nodes, and these do not all play the same role with respect to the hexagonal axes which we have adopted. Of the four planes, there is only one (111) which remains intact. Thus the line 111 corresponds to the node (00.1) and also to the nodes

$$\left(01.\frac{1}{3}\right), \quad \left(\bar{1}1.\frac{1}{3}\right), \quad \left(\bar{1}0.\frac{1}{3}\right).$$

The first one of these is normal, while the three others are broadened. The widths of the lines depend on their indices and their maxima are displaced; none of the lines remains narrow. Since the displacement can take place in either direction, it cannot be mistaken for a displacement due to a variation of the parameter. Figure 7.23 shows schematically the powder diagram for a face-centered cubic crystal in which the stacking faults come from the distortion of a single set of (111) planes (Paterson [36]).

**FIGURE 7.23.** *Positions and shapes of powder pattern lines for a cubic crystal with distortion faults on a (111) plane.*

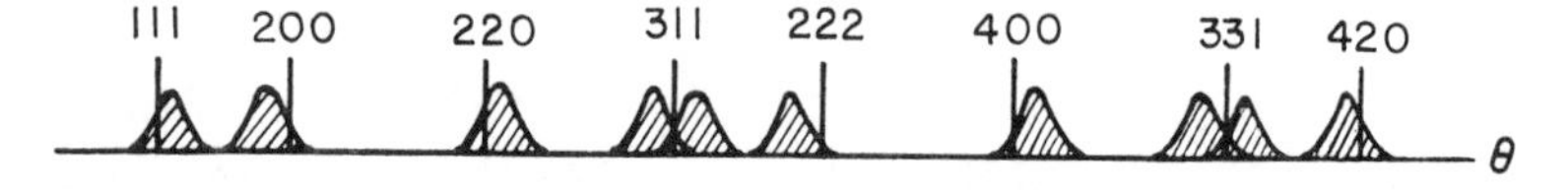


Many examples of imperfect cubic crystals have been observed, such as carborundum [37] and uranyl chloride [38], in which slip occurs parallel to the (111) planes and produces stacking faults. These faults can be either artificial or natural (mica [39]). Barrett [40] has observed diffuse streaks along the rows ⟨111⟩ in the case of single crystals of the cubic phase Cu-2% Si. It can thus be expected that stacking faults in the planes {111} can contribute to line broadening in the case of the pattern for a well work-hardened metallic powder. In fact, Warren and Warekois [41] have observed line displacement in the case of work-hardened $\alpha$ brass, which can be explained by the Paterson formulas. The probability of a fault appears to be of the order of $2.5 \times 10^{-2}$. Other measurements [42] on copper and aluminium have given results ten times smaller. The effects of stacking faults in body-centered lattices have also been investigated, both theoretically [43] and experimentally [44].

(c) Jagodzinski [45] has assumed an *interaction with the three first neighbors.* These form a set of the hexagonal or cubic types: ABA or ABC. The fourth layer can either follow the same pattern or not:

ABAB, ABAC, and ABCA, ABCB. He defines the nature of this fourth layer by introducing the two parameters $\alpha$ and $\beta$:

probability hexagonal → hexagonal: $1-\alpha$ ,
probability cubic → hexagonal: $1-\beta$ ,
probability hexagonal → cubic: $\alpha$ ,
probability cubic → cubic: $\beta$ .

The quantities $P_m$ can then be calculated, but the calculations are very complex. The intensity curve along the rows $h-k \neq$ (multiplicity 3) shows maxima (hexagonal) at *fixed* positions $l=0$, $l=(1/2)$, and maxima near the positions for the cubic lattice $l=(1/3)$, $l=(2/3)$, but the positions of these can *vary considerably*. It is even found that, for a certain value of $\alpha$ and $\beta$, the maxima corresponding to the regular arrangement ABAC occur near $l=(1/4)$ or $(3/4)$.

Jagodzinski has used these theoretical results to interpret the scattering from *wurtzite*. This is the hexagonal form of zinc sulphide, which can also exist under the cubic form (called *blende*). Since one form can change into another by changing the stacking of the Zn layers, he concluded that wurtzite must contain stacking faults and that cubic crystals could be formed using these faults as germs. This gives good spots, superposed over the background, at the positions of the cubic nodes.

Stacking faults have now been observed directly in many types of crystals by means of the electron microscope [46]. Despite this fact X-ray diffraction methods have maintained their usefulness both for detecting stacking faults and for determining their number in a given sample. The reason for this is that diffraction measurements are applicable to *any* sample—for example to a powder—while the electron microscope requires extremely thin samples and can therefore be used only in very special cases.

We have studied planar disorder but linear disorder has also been observed. In this case the periodicity is preserved in only one direction [47] and the scattering is then distributed along planes in reciprocal space (Section 6.2.4). This scattering is therefore much weaker and is difficult to observe. Until now it has not been possible to make sufficiently accurate measurements to establish quantitative models for such structures.

## 7.3. The Structure of Work-hardened Metals

We start with an aggregate of macroscopic crystals, like those found after the solidification of a molten metal or after its recrystallization. After a large plastic deformation, such as is obtained by rolling or filing, the pattern is that of a crystalline powder corresponding to the original lattice except that the intensity along a Debye-Scherrer ring is not uniform; this indicates that the crystals have orientations which are characteristic of the deformation (deformation texture).

Thus, as a first approximation, the patterns simply indicate that the original crystals have been broken into small fragments without any change in structure, and that they may have particular orientations. However, if the pattern for fine, well-annealed filings is carefully compared with that for the work-hardened metal, various differences are observed, the most evident of which is that *the lines for the work-hardened metal are broadened.* Sometimes very slight displacements of the centers of the lines are observed. Differences have also been found in the relative intensities of the lines and in the background. Using this fact, attempts have been made to determine the imperfections introduced into the lattice of the crystal blocks by the plastic deformation.

### 7.3.1. Original Experiments and Theories

In a very elementary way one can imagine two different schematic models to explain the broadening of the lines [48].

(a) *Fragmentation theory.* If the original crystal has been broken into very small fragments, the lines are broadened by the size effect. The calculations of Chapter 5, which were made on the hypothesis that the grains were well separated, can still be used, despite the fact that the elementary crystals touch each other. This is because the disorientation between two adjoining grains renders their scattering completely incoherent, so that they act as if they were separated. This does not apply to the 000 node, however, because the center of the reciprocal lattice is not surrounded by the reflection domain which surrounds all the other nodes; in this particular case intercrystalline interferences cancel completely the scattering at small angles. This question will be discussed in Chapter 10.

We can find the apparent size $L$ of a crystal from the width $\Delta(2\theta)$ of a line, using Eq. (*5.4*):

$$\Delta(2\theta) = \frac{\lambda}{L \cos\theta} . \tag{7.68}$$

The exact meaning of this average size of the elementary crystal was explained in Section 5.3.1. Since, in general, we have no reason to assume that the crystal fragments have very anisotropic shapes, all the lines should give approximately equal values of $L$, which is thus a characteristic of the structure of the work-hardened metal. It is assumed in this theory that the blocks do not contain any appreciable number of imperfections, the imperfections being limited to the regions between blocks.

(b) *Parameter fluctuation theory.* We assume that the crystal fragments are of a size ($>0.1\,\mu$) sufficient to preclude line broadening. We also assume that they have a regular lattice but that they are either compressed or stretched by the internal stresses. They therefore have parameters which are slightly different from the average value for the crystal in equilibrium, the distribution curve having a width $\Delta a$. For a given line the dependence of the diffraction angle on the value of the parameter is obtained by differentiating the Bragg law:

$$d\theta = \tan\theta \, \frac{d\, d_{hkl}}{d_{hkl}} .$$

In the case of a cubic lattice,

$$\frac{d\, d_{hkl}}{d_{hkl}} = \frac{da}{a} .$$

The profile of each line is thus similar to that of the distribution of the parameters, on a scale which changes from one line to another. The width of a line is related to $\Delta a$ as follows:

$$\Delta(2\theta) = 2\,\frac{\Delta a}{a}\,\tan\theta . \qquad (7.69)$$

The widths of the lines should therefore all yield the same value for $\Delta a/a$. According to this theory, the ratio $\Delta a/a$ characterizes a given work-hardened state.

Since the dependence of $\Delta(2\theta)$ on $\theta$ is different in Eqs. (*7.68*) and (*7.69*), it seems easy to decide which one is the most appropriate: one theory leads to a constant $L$, while the other leads to a constant $\Delta a/a$. Figure 7.24 shows that, in practice, the two functions are clearly distinguishable only for small values of $\theta$. In this region there are few lines and their widths are not much

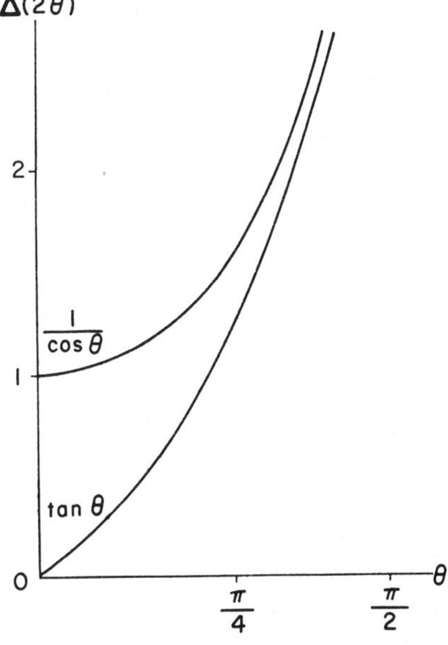


**FIGURE 7.24.** *The width of the Debye-Scherrer lines as a function of the Bragg angle* $\theta$. (a) *Broadening due to the size effect* $\Delta(2\theta) \propto (1/\cos\theta)$. (b) *Broadening due to a change in the parameter* $\Delta(2\theta) \propto \tan\theta$.

larger than the experimental widths for perfect crystals, so the measurements are very inaccurate.

There is also another difference between the two equations: for a given value of $\theta$, $\Delta(2\theta)$ is proportional to $\lambda$ in Eq. (*7.68*), while it is independent of $\lambda$ in Eq. (*7.69*). If we change the wavelength, the width of a line at a given angle $\theta$ should either vary or not, according to the theory which applies. Here again the experiments are difficult. To vary $\lambda$ ap preciably, we must compare Cr K$\alpha$ or Cu K$\alpha$ with Mo K$\alpha$. For such short wave-lengths, however, the back-reflection lines for which the effect should be the most marked are very weak because of the small value of the scattering factor. On the other hand, the relative spectral widths of the two radiations are different, so the corrections are very different in the two cases.

The question has long been a subject of controversy and no general agreement has been possible. The fragmentation theory has been used by Wood and his collaborators in many publications [49]. A theory of plastic deformation has been given by Bragg [50], starting from the values for the crystal sizes in the work-hardened metal as found by Wood. Yet it was found that the fragmentation theory was well verified in the case of work-hardened tungsten [51], for the line widths are in this case proportional to $\tan\theta$. Another argument in favor of the deformation theory is the following. If the various crystals are distorted elastically, the maximum of $\Delta a/a$ must correspond to the maximum distortion of the elastic region and this can be determined from the elastic limit and Young's modulus. In fact, metals having a low elastic limit, such as aluminum and lead, give narrow lines and the quantitative agreement is fair [52].

Finally, various authors assume that the two effects occur simultaneously, and they have interpreted the observed widths as the sum of a size effect characterized by a parameter $L$ and of a distortion effect characterized by a second parameter $\Delta a/a$ [53]. One can object that this simple addition is only a gross approximation, especially since it is a simple matter to arrive at a satisfactory agreement by using two arbitrary parameters to explain approximate measurements on a small number of lines. The two parameters do not therefore appear to have the physical significance which is attributed to them.

We can therefore conclude that the two models are probably much too simple. But if we only consider the line widths, it is always possible to construct a model which can explain them within the accuracy of the measurements. For this reason Warren and Averbach [54] have used the following approach. One measures the complete profile with the greatest possible accuracy, using the methods of Section 5.5, and then tries to determine objectively, without reference to any particular model, the information on the atomic structure which can be deduced directly from the experimental data.

### 7.3.2. The Line Profile

We shall start with Eq. (6.13) for the line profile of a distorted crystal. Within a constant factor, we have

$$i(s) = \int V(t)y(t)\exp(-2\pi ist)dt\ . \tag{7.70}$$

According to this equation, the line profile $i(s)$, drawn as a function of $s = (2\sin\theta)/\lambda$ using a conveniently chosen origin, is proportional to the Fourier transform of the product $V(t)y(t)$. The variable $t$ is a *length which is normal to the lattice planes* ($hkl$) on which the reflection occurs. If several planes have the same spacing, we must add their effects calculated separately. In the case of cubic crystals, this normal is an axis of the crystal which is chosen as the $\boldsymbol{c}$ axis. The product $VV(t)$ is the volume common to the crystal and to its double displaced by the distance $t$ (Section 2.4.2). Finally, $y(t) = \overline{F_n F^*_{n+m}}$ is the average over all the crystal of the product of the structure factors for two unit cells separated by the vector $\boldsymbol{t} = m\boldsymbol{c}$.

7.3.2.1. PURE DISTORTION EFFECT. Let us first neglect the size effect. As we have already seen, this is equivalent to setting $V(t)$ equal to unity for all values of $t$ for which $y(t)$ is different horn zero.

We consider a simple lattice containing a single atom of scattering factor $f$ per unit cell. If $\Delta\boldsymbol{x}_n$ is the displacement of the $n$th atom with respect to the *average regular lattice,*

$$F_n F^*_{n+m} = f^2 \exp[2\pi i \boldsymbol{s}\cdot(\Delta\boldsymbol{x}_{n+m} - \Delta\boldsymbol{x}_n)]\ .$$

But $\boldsymbol{s}$ is normal to the plane ($hkl$), since the extremity of $\boldsymbol{s}$ is situated inside the small domain of reflection surrounding the node ($hkl$). Therefore

$$\boldsymbol{s}\cdot(\Delta\boldsymbol{x}_{n+m} - \Delta\boldsymbol{x}_n) = sL_t\ ,$$

where $L_t$ is the difference be-

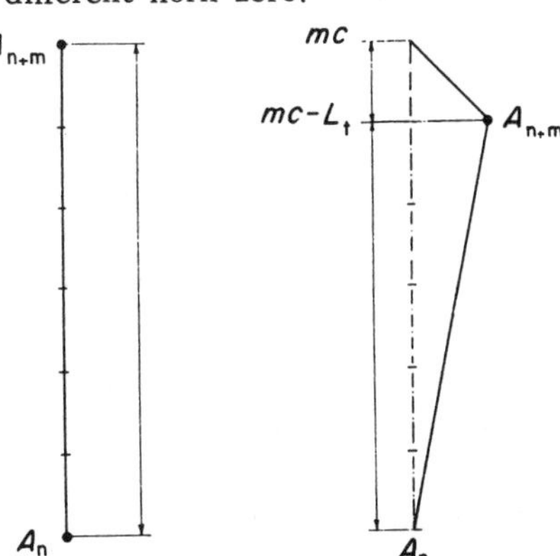


**FIGURE 7.25.** *Definition of $L_t$. The distance $A_nA_{n+m}$ in the unperturbed crystal (left) and in the perturbed crystal (right).*

tween the length of a column of $m$ unit cells parallel to $\mathbf{s}$ or to $\mathbf{r}^*_{hkl}$, and the length $mc = t$ which it would have in the undistorted crystal (Figure 7.25). We must therefore calculate the following average for all the columns of $m$ unit cells of average length $t$:

$$y(t) = f^2 \overline{\exp(2\pi i s L_t)} . \tag{7.71}$$

If the distortions are small and if $\mathbf{s}$ is not too large, which applies to a node with sufficiently small indices, we can replace the exponential function by the first three terms of the series and, since the average is equal to the sum of the averages,

$$y(t) = f^2(1 + 2\pi i s\overline{L_t} - 2\pi^2 s^2 \overline{L_t^2}) .$$

Referring the distorted crystal to the average lattice means that $\overline{L_t}$ is zero. This is equivalent to saying that we consider only the shape of the line and not its displacement with respect to the corresponding line for the undistorted crystal. Then

$$y(t) \cong f^2[1 - 2\pi^2 s^2 \overline{L_t^2}] \cong f^2 \exp(-2\pi^2 s^2 \overline{L_t^2}) .$$

Let us consider an example drawn from a paper by Warren [55], concerning the profile of the (111) line of *α brass* thoroughly work-hardened by filing. This profile is assumed to be corrected for experimental errors, as indicated in Section 5.5. We can deduce from it the curve *a* of Figure 7.26 by using a Fourier transformation (Appendix A.6.2). This transform is proportional to the function $y(t)$, according to Eq. (*7.70*). If we take the ordinate at the origin to be equal to unity, the function is

$$y_1(t) = \exp(-2\pi^2 s^2 \overline{L_t^2}) , \tag{7.72}$$

and we therefore have

$$\Delta L^2(t) = \overline{L_t^2} = \frac{-1}{2\pi^2 s^2} \ln y_1(t) . \tag{7.73}$$

Let us recall the exact meaning of $\Delta L(t)$. We consider the pairs of atoms $A_n$ and $A_{n+m}$ of the crystal separated by $m$ unit cells ($m = t/c$) in the direction normal to the (111) plane. Then $\Delta L(t)$ is the quadratic average of the fluctuations $L_t$ of the projection on [111] of the true distance $A_n A_{n+m}$ between the pairs of atoms. When $t$ tends to zero, $L_t/t$ tends toward the value of the local expansion $\varepsilon$, at the point $A_n$, in the direction [111]. The tangent to the curve of $\Delta L(t)$ at the origin (Figure 7.26) is therefore the quadratic average of the values of this expansion. It determines the curvature of $y_1(t)$ at the origin and its value is

$$\overline{\varepsilon^2} = \lim_{t=0} \frac{1 - y_1(t)}{2\pi^2 s^2 t^2} . \tag{7.73a}$$

**FIGURE 7.26.** (a) *Fourier transform of the 111 line profile of α brass.* (b) *Quadratic average of the fluctuations of $L_t$ as a function of t, as calculated from* (a) *by the approximate relation of Eq.* (7.72). (c) *The same quantity calculated from the distribution function* $\varphi(L_t)$.

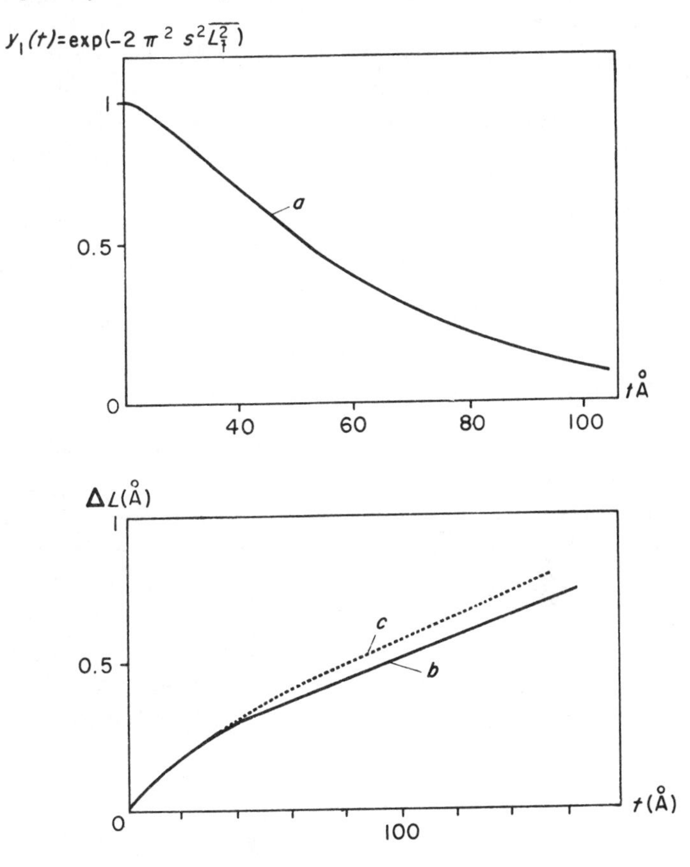


In the case of the model, according to which the structure is made up of regular blocks whose parameters vary slightly from one to the other, $\varepsilon = da/a$ and $L_t = \varepsilon t$ in each block, and thus $\Delta L^2 = \overline{\varepsilon^2} t^2$. The curve *b* of Figure 7.26 would then be a straight line. Let $f(\varepsilon)$ be a normalized function giving the frequency of the variation of the parameter $\varepsilon = da/a$ in the distorted crystal. Then the average of Eq. (7.71) is calculated by

$$y(t) = f^2 \int f(\varepsilon) \exp(2\pi i s \varepsilon t) d\varepsilon \propto \int f(\varepsilon') \exp(2\pi i \varepsilon' t) d\varepsilon' \, ,$$

where $\varepsilon' = s\varepsilon$. The function $f(s\varepsilon)$ is therefore the Fourier transform of

$y(t)$ within a multiplying factor. From Eq. (*7.70*), $i(s)$ is also proportional to the Fourier transform of $y(t)$, so the line profile gives the distribution curve of the parameters, and its width $\Delta s = s(\Delta a/a)$. This is equivalent to $\Delta(2\theta) = 2(\Delta a/a)\tan\theta$, which is the result found earlier in Eq. (*7.69*) in a very elementary manner.

It is important to note that the calculation in this chapter is much more general. When the atomic displacements are very heterogeneous, it is not possible to utilize the usual laws of crystal reflection, since the lattice planes are then distorted even at short range. Equations (*7.70*) and (*7.71*) provide a relation between the diffracted intensity and the averages of the fluctuations in the distances between pairs of individual atoms without reference to the positions of the neighbors.

In order to utilize Eq. (*7.71*) in the form of Eq. (*7.72*), we have assumed $sL_t$ to be small, but this approximation is not sufficient. For example, in his work on $\alpha$ brass, Warren calculated $\Delta L$ as a function of $t$ starting from the profile of the line (222), as he had already done starting from the line (111). The value of $\Delta L$ is the same in both cases, since we are dealing with the same lattice planes; the experimental values are nevertheless different even near the beginning of the curve.

Warren has suggested the following method for determining $\Delta L$. The method is rigorous, in principle. Let us consider the set of lines corresponding to the different orders of reflection on a given lattice plane and giving a set of curves $y_1(t, s)$. The quantity $L_t$ is the same for all these lines. Let us call $\varphi(L_t)$ the distribution function for $L_t$, such that $\varphi(L_t)dL_t$ represents the number of columns containing $m = t/c$ unit cells and whose effective lengths lie between $(mc + L_t)$ and $(mc + L_t + dL_t)$. We also set $\int \varphi(L_t)dL_t = 1$. We can calculate the average of Eq. (*7.71*) with the help of this function:

$$y_1(t, s) = \overline{\exp 2\pi i s L_t} = \int \varphi(L_t) \exp(2\pi i s L_t) dL_t . \qquad (7.74)$$

Thus, for a given values of $t_0$, the function $\varphi(L_{t_0})$ is the Fourier transform of $y_1(t_0, s)$. We therefore draw the curve of the function $y_1(t_0, s)$ *as a function of* $s$. Unfortunately, we can determine only a few points; we know that $y_1(t_0, 0) = 1$ and, except when the diffraction pattern is obtained with a very short wavelength, the number of observable lines $(nh, nk, nl)$ is small, and even some of them cannot be used. For example, the line (333) coincides with (511), and (600) with (442), so it is not possible to determine the profile of the (333) and (600) lines. It is, however, possible to use a work-hardened metal whose texture is such that the parasitic lines are of negligible intensity.

**FIGURE 7.27.** *The function $y_1(t_0, s)$ of Eq. (7.74) for various values of the parameter $t_0$.*

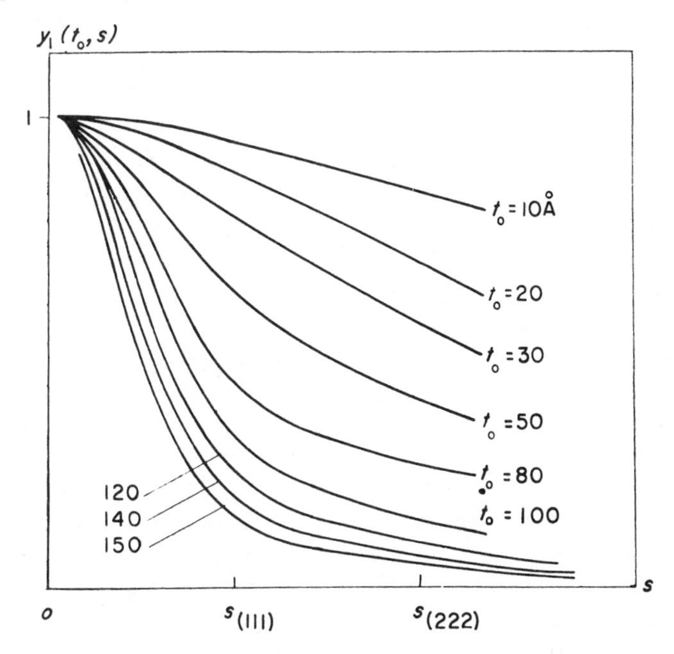


If it is possible to draw the curves $y_1(t_0, s)$, as in Figure 7.27, we find its transform

$$\varphi(L_{t_0}) = \int y(t_0, s) \exp(-2\pi i L_t s) ds ,$$

which gives the quadratic average

$$\Delta L^2(t) = \int \varphi(L_t) L_t^2 dL_t .$$

The curve $c$ of Figure 7.26 was obtained in this way; it is not very different from the curve $b$.

The curve of $\Delta L$ as a function of $t$ is the most complete information which can be extracted a priori from an analysis of the line profiles. Its tangent at the origin gives the quadratic average of the local expansions, and its curvature shows that the regions where the distortion is homogeneous are quite small—of the order of 30 Å in the case under consideration. Over larger distances the interatomic distances approach the normal value, since an expanded region is followed by a region of contraction, or conversely.

As is generally the case, X-rays give the statistical distribution of the interatomic distances. The mathematical analysis of the experimental results yields information which is accurate and independent of a priori hypotheses, but this is far from sufficient to determine completely the structure of a metal. For example, of all the possible displacements it is only those which are parallel to a given direction which affect the curve of $\Delta L(t)$.

7.3.2.2. THE COMBINED EFFECTS OF BLOCK SIZE AND OF DISTORTION. We have assumed above that the size of the coherent blocks was large enough not to affect the line profile. Under these conditions, the Fourier transform of the line profile should start with a horizontal tangent [Eq. (*7.72*)]. It will be recalled that the transform starts with a straight, oblique portion when the line is broadened exclusively by the size effect. It should therefore be possible to distinguish between these two types of broadening from the study of a single line profile.

If now both the size effect and the distortion come into play, the general Eq. (*7.70*) indicates that the experimental Fourier transform $A(t)$ is equal to the product $y(t)V(t)$. From Section 5.3.2, $V(t)$ is of the form $1-(t/\bar{M}_1)$ for small values of $t$, where $\bar{M}_1$ is an average diameter of the block measured in the direction perpendicular to the reflecting plane. Also, from Eq. (*7.73a*),

$$y(t) = 1 - 2\pi^2\overline{\varepsilon^2}s^2t^2 \,.$$

Thus

$$A(t) = 1 - \frac{t}{\bar{M}_1} - 2\pi^2 s^2\overline{\varepsilon^2}t^2 + \cdots \,. \tag{7.75}$$

The transform therefore has both a finite slope and a curvature at the origin; the slope gives the average size of the block, and the curvature gives the quadratic average of the expansion. The two effects can therefore be determined separately, at least in principle. In actual practice, this method is much too inaccurate to be of any use since the shape of the transform near the origin depends mostly on the tails of the line profile, where the experimental errors are the greatest. Bertaut [56] has shown indeed that the profile of a line which is broadened only by the size effect must be measured out to a distance equal to five or six times its width, in order to remove any false curvature in the transform at the origin.

Warren and Averbach [57] have devised a better method, involving the set of curves for the lines of increasing orders corresponding to a given lattice plane. The distortion effect decreases with $s$ and, for $s=0$, we should have $A(t, 0) = V(t)$. The procedure consists in determining $A(t, 0)$ by extrapolation, starting from the known curves $A(t, s)$. For small values of $s$, the curve $y(t, s)$ is of the form $\exp(-Ks^2)$ and

$$A(t, s) = V(t) \exp(-Ks^2)\,,$$

or

$$\ln A(t, s) = \ln V(t) - Ks^2\,. \tag{7.76}$$

We then draw the curve of $\ln A$ as a function of $s^2$ for different values of the parameter $t$. This curve terminates by a portion which is quite

**FIGURE 7.28.** *Top: the function $V(t)$ giving $\bar{M}_1 = 1{,}000$ Å for Cu-2% Si rolled to 50% of its original thickness* [57]. *Bottom:* $\ln A(t, s)$ *as a function of $s^2$ for various values of $t$.*

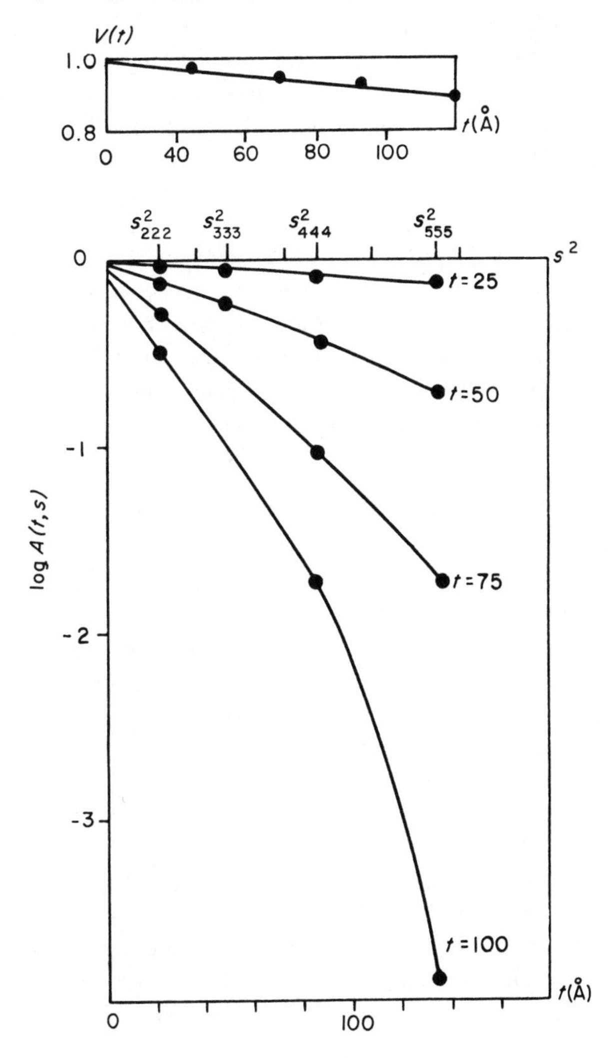

straight and which is easy to extrapolate down to $s = 0$, as in Figure 7.28. If there is a size effect, these curves do not start at the origin and the ordinate at $s^2 = 0$ gives $\ln V(t)$. We then draw a second curve of $V(t)$ as a function of $t$. This is nearly a straight line and the intersection of the tangent at the origin with the $t$ axis gives $\bar{M}_1$.

Warren has found values of $\bar{M}_1$ which are of the order of 200 Å in the case of certain work-hardened metals. This appears to indicate that the metal is broken up into independent blocks whose diameters are of the order of 200 Å. In fact, the only conclusion we can draw is that a line normal to the reflecting plane crosses faults such that, on the average, the diffraction is not coherent over more than 200 Å, but these faults need not be the boundaries surrounding well defined blocks.

These results illustrate the extent of the recent progress in the study of crystal imperfections. This progress is due both to the extension of the theory and to improvements in the experimental techniques, for mathematical calculations are meaningless unless the experimental data are sufficiently accurate. The situation is still unclear and the interpretation of the line widths of work-hardened metals by the above methods has led to difficulties because lines of different orders do not give the same results. It is now known that in some metals another type of distortion—namely stacking faults—may be relatively important [54]. We have discussed these in Section 7.2.3. It is in general possible to separate out their effects because stacking faults displace the centers of the lines, whereas other types of distortion do not displace but only broaden them.

### 7.3.3. Line Intensities for Work-hardened Metals

We have seen that the size effect broadens a line without diminishing its total intensity, and that there is no scattering outside the small domain of reflection surrounding the node. This also applies to the distortion effect when the coefficients $y(t)$ remain sufficiently large for appreciable values of $t$. When the curve of $y(t)$ is broad, its transform $i(s)$ is narrow; in other words, we must assume that there exists a correlation between the displacements of the atoms up to quite large distances (Section 6.2.3). If the displacements were random—even in unit cells which are not far from each other—we would then have general scattering, and therefore a reduced line intensity.

Some old measurements by Brindley and Spiers [58] on nickel filings had seemed to show that the line intensity decreases as the indices for the line increase, exactly as in the temperature effect. They therefore concluded that the disorder in a work-hardened metal is "frozen thermal agitation," or that the atoms of the work-hardened metal preserve the random positions which they have at a given instant under the influence of the thermal vibrations. According to Brindley, the ratio of the line intensity for the work-hardened metal and for a well-crystallized powder could be written as $\exp(-4\pi^2 s^2 \Delta x^2/3)$, like the Debye factor of Eq. (*7.7*), where $\Delta x$ is the quadratic average of the atomic displacements due to the work hardening. In the case of nickel filings, Brindley and Spiers found that $\Delta x = 0.08$ Å.

Recent measurements [59] of line intensities, on the contrary, have shown that the total intensity of a line does not decrease under the influence of work hardening. Since it is essential to extend the measurements out to a considerable distance from the center of the line, it is probable that Brindley's measurements were in error because the tails of the line profile had been neglected. In such a case, the error is largest for the lines at large angles which are the broadest. For the lines at small angles, it was found that the intensity even increased with work hardening. The distortion suppresses the extinction effects which can be important in the crystallized metal, but it is extremely difficult to investigate this phenomenon quantitatively, as was indicated in Section 4.10.

If the energy diffracted in the lines remains constant, work hardening should not produce scattering in regions which are far removed from the lines. Averbach and Warren [59] have found no appreciable increase in the background, which is due to thermal agitation and to Compton scattering, both in the work-hardened and in the crystallized metal.

## REFERENCES

1. P. DEBYE, *Ann. Phys.*, **43** (1914), 49.
2. I. WALLER, *Z. Phys.*, **17** (1923), 398; **51** (1928), 213.
3. R. W. JAMES and E. M. FIRTH, *Proc. Roy. Soc., A*, **117** (1927), 62.
4. C. W. HAWORTH, *Phil. Mag.*, **5** (1960), 1229.
5. D. R. CHIPMAN, *J. Appl. Phys.*, **31** (1960), 2012.
6. K. LONSDALE and H. J. MILLEDGE, *Acta. Cryst.*, **13** (1960), 499.
   K. LONSDALE, H. J. MILLEDGE and K. V. K. RAO, *Proc. Roy. Soc., A*, **255** (1960), 82.
7. C. KITTEL, *Introduction to Solid State Physics*, 2nd ed., Wiley, New York (1956).
8. R. W. JAMES, *The Optical Principles of X-Ray Diffraction*, Bell, London (1948). p. 238.
9. G. E. M. JAUNCEY and G. G. HARVEY, *Phys. Rev.*, **37** (1931), 1203.
10. J. LAVAL, *Bul. Soc. Franç. Minér. Cryst.*, **62** (1939), 137; **64** (1941), 1.
11. M. BORN and K. HUANG, *Dynamical Theory of Crystal Lattices*, Clarendon Press, Oxford (1954).
12. M. BORN, *Report on Progress in Physics*, **9** (1942–1943), 294.
13. L. BRILLOUIN, *Wave Propagation in Periodic Structures*, McGraw-Hill, New York (1946).

14. A. J. C. Wilson, *X-Ray Optics*, Methuen, London (1949), p. 104.
15. J. Laval, *J. Phys. Radium*, **15** (1954), 545 and 657.
16. H. Gurien, *Bul. Soc. Franç. Minér. Cryst.*, **75** (1952), 343.
G. N. Ramachandran and W. A. Wooster, *Acta Cryst.*, **4** (1954), 335, and 431.
17. M. Born and M. Göppert-Mayer, *Handdbuch der Phys.*, Vol. XXIV, Part II, Springer Verlag, Berlin (1933), p. 623.
18. H. A. Jahn, *Proc. Roy. Soc. London, A*, **179** (1942), 320; **180** (1942), 397.
19. S. C. Prasad and W. A. Wooster, *Acta. Cryst.*, **8** (1955), 506.
20. M. Born, *Proc. Roy. Soc. London, A*, **179** (1942), 69; **180** (1942), 397.
21. P. Olmer, *Acta. Cryst.*, **1** (1948), 57.
E. H. Jacobsen, *Phys. Rev.*, **97** (1955), 654.
M. Schwartz and L. Muldaver, *J. Appl. Phys.*, **29** (1958), 1661.
22. P. G. Owston, *Acta Cryst.*, **2** (1949), 222.
23. W. Hoppe, *Z. Elektro. Ber. Bunsengesel. Phys. Chem.* **63** (1959), 912.
J. L. Amoros, M. L. Canut and A. De Ache, *Z. Krist.*, **114** (1960), 39; *Bul. Soc. Franç. Minér. Crist.*, **84** (1961), 40.
24. B. E. Warren, *Acta Cryst.*, **6** (1953), 803.
D. R. Chipman and A. Paskin, *J. Appl. Phys.*, **30** (1959), 1992 and 1998.
F. H. Herbstein and B. L. Averbach, *Acta Cryst.*, **8** (1955), 843.
25. V. Daniel and H. Lipson, *Proc. Roy. Soc., A*, **181** (1943), 368; **182** (1944), 378.
E. Biedermann, *Acta Cryst.*, **13** (1960), 650.
26. A. Guinier, *Solid State Physics*, Vol. IX, Academic Press, New York (1958), p. 356.
27. M. E. Hargreaves, *Acta Cryst.*, **2** (1949), 259; **4** (1951), 301.
28. A. Guinier, *Acta Met.*, **3** (1955), 510.
29. M. Hillert, *Acta Met.*, **9** (1961), 525.
M. Hillert, M. Cohen and B. L. Averbach, *Acta Met.*, **9** (1961), 536.
I. J. Tiedema, J. Bouman and W. G. Burgers, *Acta Met.*, **5** (1957), 310.
J. Manenc, *Acta Met.*, **7** (1959), 124; *Rev. Met.*, **54** (1957), 867; *Acta Cryst.*, **10** (1957), 259.
30. A. Guinier, Solid State Physics, Vol. IX, Academic Press, New York (1958), p. 362.
31. B. E. Warren, *Phys. Rev.*, **59** (1941), 693.
32. A. J. C. Wilson, *Acta Cryst.*, **2** (1949), 245.
G. E. Bacon, *Acta Cryst.*, **7** (1954), 359.
33. S. B. Hendricks and E. Teller, *J. Phys. Chem.*, **10** (1942), 147.
G. W. Brindley and K. Robinson, *Miner. Mag.*, **28** (1948), 393.
G. W. Brindley and J. Mering, *Acta Cryst.*, **4** (1951), 441.
34. A. J. C. Wilson, *X-Ray Optics, Methuen,* London (1949), p. 68; *Proc. Roy. Soc., A*, **180** (1942), 277.
W. H. Zachariasen, *Phys. Rev.*, **71** (1947), 715.
J. Mering, *Acta Cryst.*, **2** (1949), 371.
B. E. Warren, *Progress in Metal Physics*, Vol. 8, Pergamon Press, London (1959), p. 168.
35. O. S. Edwards and H. Lipson, *Proc. Roy. Soc., A*, **180** (1942), 268.
36. M. S. Paterson, *J. Appl. Phys.*, **23** (1952), 805.
J. W. Christian, *Acta Cryst.*, **7** (1954), 415.
37. H. Jagodzinski, *Acta Cryst.*, **2** (1949), 298; **5** (1952), 518; **7** (1954), 17.

R. GEVERS, *Acta Cryst.*, **5** (1952), 518.
38. W. H. ZACHARIASEN, *Acta Cryst.*, **1** (1948), 277.
39. S. B. HENDRICKS, *Phys. Rev.*, **57** (1940), 448.
40. C. S. BARRETT, *Trans. A.I.M.E.*, **188** (1950), 123.
41. B. E. WARREN and E. P. WAREKOIS, *J. Appl. Phys.*, **24** (1953), 951.
42. G. B. GREENOUGH and E. M. SMITH, *Proc. Phys. Soc.*, *H*, **68** (1955), 51.
G. F. BOLLING, T. B. MASSALSKI and C. J. MCHARGREE, *Phil. Mag.*, **6** (1961), 491.
43. B. E. WARREN, *Progress in Metal Physics*, Vol. 8, Pergamon Press, London (1959), p. 187.
P. B. HIRSCH and H. M. OTTA, *Acta Cryst.*, **10** (1957), 447.
44. M. MCKEEHAN and B. E. WARREN, *J. Appl. Phys.*, **24** (1953), 52.
C. N. J. WAGNER, *Arc. Eisenhuttenw.*, **29** (1958), 489.
J. DESPUJOLS, *J. Appl. Phys.*, **29** (1958), 195.
45. H. JAGODZINSKI, *Acta Cryst.*, **2** (1949), 201 and 214.
46. W. BOLLMANN, *Phys. Rev.*, **103** (1956), 1588.
P. B. HIRSCH, R. W. HORNE and M. WHELAN, *Phil. Mag.*, **1** (1956), 677.
47. H. LAMBOT, *Rev. Met.*, **47** (1950), 709.
A. H. GEISLER and J. K. HILL, *Acta Cryst.*, **1** (1948), 238.
48. G. B. GREENOUGH, *Progress in Metal Physics*, Vol. 3, Pergamon Press, London (1952), p. 176.
49. W. A. WOOD and W. A. RACHINGER, *J. Inst. Met.*, **75** (1948), 571.
50. W. L. BRAGG, *Nature*, **149** (1942), 511.
51. A. R. STOKES, K. J. PASCOE and H. LIPSON, *Nature, London*, **151** (1943), 137.
C. S. SMITH and E. E. STICKLEY, *Phys. Rev.*, **64** (1943), 191.
52. H. D. MEGAW, H. LIPSON and A. R. STOKES, *Nature, London*, **154** (1944), 145.
53. U. DEHLINGER and A. KOCHENDÖRFER, *Z. Krist.*, **101** (1939), 134.
54. B. E. WARREN and B. L. AVERBACH, *Imperfections in Nearly Perfect Crystals*, (edited by W. SHOCKLEY), Wiley, New York (1953), p. 152.
B. E. WARREN, *Progress in Metal Physics*, Vol. 8, Pergamon Press, London (1959), p. 147.
C. R. HOUSKA and B. L. AVERBACH, *Acta Cryst.*, **11** (1958), 139.
C. N. J. WAGNER, *Acta Met.*, **5** (1957), 477.
55. B. E. WARREN and B. L. AVERBACH, *J. Appl. Phys.*, **21** (1950), 595.
56. E. F. BERTAUT, *Acta Cryst.*, **5** (1952), 117.
57. B. E. WARREN and B. L. AVERBACH, *J. Appl. Phys.*, **23** (1952), 497.
58. G. W. BRINDLEY and F. W. SPIERS, *Phil. Mag.*, **20** (1935), 882 and 893.
59. B. L. AVERBACH and B. E. WARREN, *J. Appl. Phys.*, **20** (1949), 1066.

CHAPTER

# 8

# MIXED CRYSTALS AND SUBSTITUTION DISORDER

It is well known that a mixed crystal can be obtained by crystallizing together two isomorphous salts like NaCl and NaBr. The resulting crystal then has physical properties which are completely analogous to those of a pure crystal. Similarly, certain single-phase alloys, called solid solutions, give diffraction patterns which are quite similar to those of pure substances having the same lattices, since the lines are narrow and in the same intensity ratios.

These observations indicate that the atoms of a mixed crystal are situated at the nodes of a regular lattice—or of several, as in the case of the NaCl-NaBr crystal.

Let us examine how the various atoms distribute themselves in a single lattice. The structure of most crystals can be more or less explained according to two extreme and relatively simple models, namely the *ordered* and the perfectly *disordered* solid solution.

In the disordered case, the diffraction pattern (lines or spots) is identical to that of a crystal having the same lattice and identical atoms in all the unit cells. Consider a simple crystal (one atom per unit cell) of a binary solid solution $AB$ whose atoms $A$ and $B$ are in the proportions $p_A$ and $p_B$, so that $p_A + p_B = 1$, and have scattering factors $f_A$ and $f_B$. Then the fictitious average type of atom has a scattering factor of $p_A f_A + p_B f_B$. It is possible, by means of energy considerations, to justify Végard's law relating the parameters of the solid solution to its composition and even, as a second approximation, to arrive at the deviations from this law [1]. The lack of periodicity arising from the irregular distribution of the $A$ and $B$ atoms at the nodes of the lattice has no effect on the diffraction lines, but rather produces a characteristic *scattering* which depends on the regularity of the atomic distribution. This is the case of short-range order. We shall examine this in detail later on.

In a perfectly ordered solid solution, the $A$ and $B$ atoms alternate at the nodes of the lattice according to a rigorous pattern. We then have a *superlattice*, and the diffraction patterns show new lines, which are called superlattice lines. Such superlattices are observed when the two constituent atoms occur in a simple ratio such as $AB$, $AB_2$, $AB_3$, $\cdots$

## 8.1. Long-range Order

### 8.1.1. Superlattice

Let us consider the relatively simple case of a one-dimensional lattice formed of atoms $A$ and $B$ arranged at regular intervals $u$ along a straight line. We shall assume that there are equal numbers of the two different types of atoms. If they are distributed at random as in Figure 8.1, then the diffraction pattern is the same as that of a lattice of periodicity $u$. If, on the contrary, the atoms $A$ and $B$ alternate regularly, $ABAB\cdots$, the period is $2u$. The diffraction pattern of a row such as $b$ in Figure 8.1 contains first the diffraction pattern of the row $a$, but new lines also

**FIGURE 8.1.** *Linear lattices containing equal numbers of atoms A and B.* (a) *Disordered lattice and schematic representation of its pattern.* (b) *Ordered lattice and its pattern.*

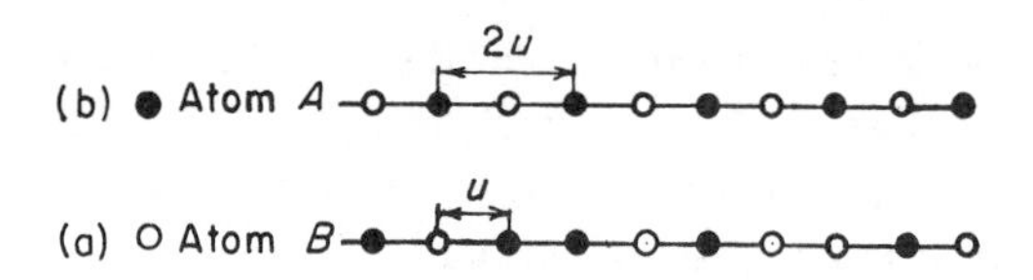


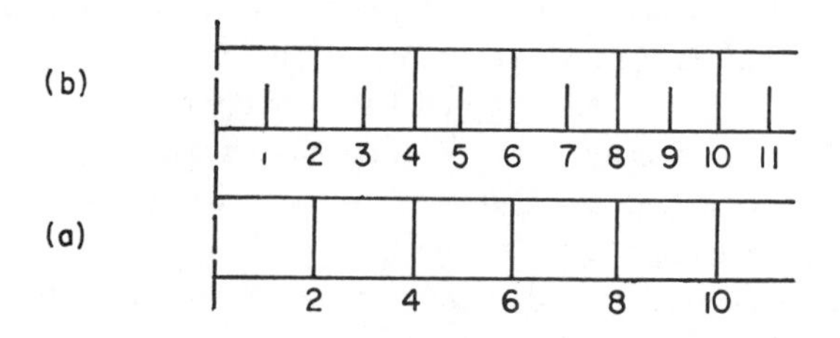

appear which correspond to the double period $2u$. This second set of lines is weaker because the wave scattered by $B$ is $\pi$ radians out of phase with that scattered by $A$, so that the structure factor is $f_A - f_B$ for the superlattice lines, and $f_A + f_B$ for the lines of the original lattice. The ratio of the intensities $[(f_A - f_B)/(f_A + f_B)]^2$ is less than unity and becomes very small when the scattering factors for $A$ and $B$ are nearly equal. In other words, the superlattice lines are of high intensity only if the atoms $A$ and $B$ are very different.

Similar characteristics are observed in the case of a three-dimensional lattice. Let us consider the classical example of the gold-copper alloy $AuCu_3$ [2]. The pattern for this alloy, quenched from a temperature above 425°C, is shown in Figure 8.2a. The pattern is that of a face-centered cubic lattice like that of gold or copper. If, instead of quenching the alloy, we allow it to cool gradually, so the solid-state reaction can take place, we obtain a pattern like that in Figure 8.2b. There are now lines of lower intensity as well as the lines of the pattern of Figure 8.2a. All the new lines correspond to lattice spacings which are twice as large as certain spacings of the first lattice. Lattice $b$ is a single cubic lattice with a unit cell identical to that of lattice $a$ and, since the density of the alloy is unchanged, the unit cell must contain four atoms, one atom of gold and three of copper.

The simplest explanation for these observations is the following. In the two cases, the atoms occupy the same positions in the unit cell. However, in lattice $a$, the atoms are distributed at random, so the probabilities of finding a gold or a copper atom at any one position are

**FIGURE 8.2.** *Schematic representation of diffraction patterns.* (a) *Disordered state.* (b) *Ordered state.*

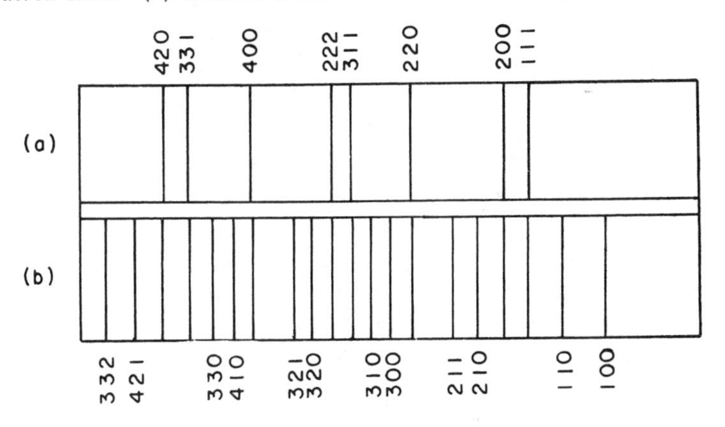

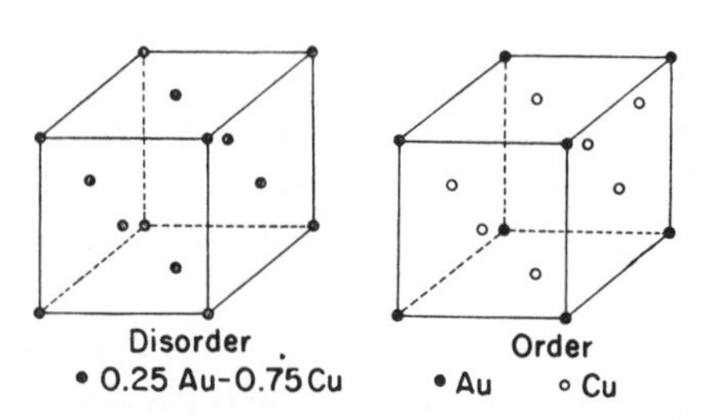


**FIGURE 8.3.** *Structure of the unit cell of $AuCu_3$ in the disordered* (a) *and ordered* (b) *states.*

respectively 0.25 and 0.75. But in lattice $b$ the corners are always occupied by a gold atom, and the centers of the faces are occupied by a copper atom. The solid solution is then said to be ordered (Figure 8.3).

In lattice $b$ the coordinates of the atoms are as follows:

| | | | |
|---|---|---|---|
| Gold | 0 | 0 | 0 |
| Copper | 1/2 | 1/2 | 0 |
| | 1/2 | 0 | 1/2 |
| | 0 | 1/2 | 1/2 |

Then, from Eq. (*4.29*), the structure factor is

$$F_{hkl} = f_{\mathrm{Au}} + f_{\mathrm{Cu}}\{\exp[-i\pi(h+k)] + \exp[-i\pi(k+l)] + \exp[-i\pi(l+h)]\} .$$

When $h$, $k$, $l$ have the same parity, $F_{hkl} = f_{\mathrm{Au}} + 3f_{\mathrm{Cu}}$ and the waves scattered by the four atoms of a given unit cell are in phase.

In the case of lattice $a$, in which the atoms are distributed at random, we also have

$$F_{hkl} = 4\left(\frac{1}{4}f_{\mathrm{Au}} + \frac{3}{4}f_{\mathrm{Cu}}\right) = f_{\mathrm{Au}} + 3f_{\mathrm{Cu}}$$

if $h$, $k$, $l$ have the same parity. The lines of pattern $a$ are found with the same intensity in pattern $b$.

If $h$, $k$, $l$ have different parities, $F_{hkl} = f_{\mathrm{Au}} - f_{\mathrm{Cu}}$ for the ordered lattice $b$. Since the four positions are equivalent in the disordered lattice, $F_{hkl} = 0$. The supplementary lines which appear in pattern $b$ are weak; since, for small angles, $f_{\mathrm{Au}}/f_{\mathrm{Cu}} = 2.8$, the ratio of the intensities of two sets of lines, which is proportional to the square of the ratio of their structure factors, is of the order of 10.

The superlattice lines are more easily observed when the scattering factors for the two types of atoms are quite different. In the case of atoms with neighboring atomic numbers, the intensity of the superlattice lines may be less than one-thousandth of that of the fundamental lines.

The method can be rendered more sensitive by using the anomaly in the atomic scattering factor which occurs at wavelengths slightly longer than the $K$ absorption edge of the element (Section 1.2.8). If the alloy under consideration is composed of two different elements having consecutive atomic numbers, the difference between their scattering factors is close to unity for small values of $(\sin\theta)/\lambda$, unless the wavelength is very close to one of the $K$ absorption edges. Since the anomaly occurs on only one of the scattering factors, the difference between these factors can be as large as three units, so the intensity of the superlattice lines is increased by a factor of 9. Jones and Sykes [3] have utilized this method to determine the structure of brass (CuZn) with Zn$K\alpha$ (1.44Å) radiation, which is very close to the absorption edge for Cu$K\alpha$ (1.37Å). The same method has also been used to observe the order in permalloy, which contains three atoms of nickel for each atom of iron [4].

### 8.1.2. Measurement of Long-range Order

A solution is considered to be ordered if superlattice lines appear on the diffraction pattern. Such a solution can nevertheless have a structure which is not exactly that of the ideal superlattice, either because the atomic distribution is somewhat different or because the atoms do not occur exactly in the correct proportions.

Let us consider again the binary solid solution $AB$ containing a fraction $p_A$ of $A$ atoms and $p_B$ of $B$ atoms, so that $p_A + p_B = 1$. If the solution is perfectly well ordered, there are two types of lattice points: atoms $A$ are on the lattice points of type 1 and atoms $B$ are on those of type 2. The fraction of lattice points of type 1 is therefore $p_A$ and those of type 2 is $p_B$. But if the alloy is not perfectly well ordered, the lattice points of type 1 contain only a fraction $n$ of atoms $A$; Table 8-1 shows the relative numbers of atoms $A$ and $B$ for both types of lattice points. If the order is perfect, $n$ is equal to unity, while if the disorder is perfect, $n$ is equal to $p_A$, since all nodes play the same role.

TABLE 8-1. The Relative Numbers of Atoms $A$ and $B$ for Lattice Points of Type 1 and Type 2

| Lattice Points | Number of Atoms of a Given Type in a Given Site | | Total Number of Sites |
|---|---|---|---|
| | Atoms of type $A$ | Atoms of type $B$ | |
| Site 1 | $np_A$ | $(1-n)p_A$ | $p_A$ |
| Site 2 | $(1-n)p_A$ | $p_B-(1-n)p_A$ | $p_B$ |
| Total number of atoms | $p_A$ | $p_B$ | |

The fraction $n$ therefore varies from unity to $p_A$ in going from perfect order to perfect disorder. The *long-range order parameter*[1] $S$ is defined so that, under these conditions, it varies from unity to zero:

[1] The concept of the long-range order parameter has been discussed and critized by Cowley [5]. It has been generalized for multicomponent alloys [6].

$$S = \frac{n - p_A}{1 - p_A} = \frac{n - p_A}{p_B} . \tag{8.1}$$

For the alloy $AB$, the parameter $S = 2n - 1$; for the alloy $AB_3$, the parameter $S = (4n - 1)/3$.

Table 8.1 shows the atomic distribution in terms of $n$, and thus in terms of $S$. We shall therefore be able to calculate the superlattice line intensities. The average scattering factor for the atoms situated at the lattice points of type 1 is given by

$$f_1 = \frac{np_A f_A + (1 - n)p_A f_B}{p_A} \tag{8.2}$$

and, for the atoms situated at the lattice points of type 2, by

$$f_2 = \frac{(1 - n)p_A f_A + [p_B - (1 - n)p_A]f_B}{p_B} \tag{8.3}$$

Then, if the structure factor for the superlattice lines is $f_1 - f_2$ as in the case of $AuCu_3$, from Eq. (*8.1*) and from $p_A + p_B = 1$ we have

$$F_{\text{super}} = f_1 - f_2 = (f_A - f_B)\left[n - (1 - n)\frac{p_A}{p_B}\right] = S(f_A - f_B) . \tag{8.4}$$

*The intensity of the superlattice lines is proportional to the square of the degree of long-range order.*

The parameter $S$ [7] can be measured by comparing the intensities of a normal line with that of a superlattice line since, from Eq. (*4.47*), this gives the ratio of the structure factors for the two lines or, in the case of $AuCu_3$,

$$S^2\left(\frac{f_{\text{Au}} - f_{\text{Cu}}}{0.25 f_{\text{Au}} + 0.75 f_{\text{Cu}}}\right)^2 .$$

Equation (*4.47*) involves the temperature factor

$$D = \exp\left(-\frac{16\pi^2 \sin^2\theta}{3\lambda^2}\,\Delta X^2\right),$$

given in Eq. (*7.7*). This factor can be determined in turn from the fact that, whatever the degree of order, the quadratic average of the atomic displacement $\Delta X^2$ remains constant, so that $\Delta X^2$ can be found by comparing two normal lines for which all the quantities in Eq. (*4.47*) are known, except $\Delta X^2$.

In the case of a series of alloys having the same composition, one of which is known to be perfectly ordered, $S$ can be found simply by comparing the ratios of the intensities for a given pair of lines, one of which is a normal line and the other a superlattice line:

$$S^2 = \left(\frac{i_{\text{super}}}{i_{\text{norm}}}\right) \times \left(\frac{i_{\text{norm}}}{i_{\text{super}}}\right)_{\text{for } S=1}$$

Keating and Warren [8] have investigated the order in $AuCu_3$ from the normal (200) line and the superlattice (210) line. The sample was a small plate of very fine filings pressed together, and a Geiger counter was used. The incident and reflected rays were kept symmetrical with respect to the surface of the plate; the absorption was therefore constant and did not appear in the ratios. A bent-quartz crystal monochromator was used. It was necessary to take into account the harmonics reflected by the monochromator, and especially the harmonic $\lambda/2$. For example, the normal (420) line for $\lambda/2$ coincided with the superlattice (210) line for $\lambda/2$. This harmonic was eliminated by performing two measurements with a Ross double filter (see Section 6.3.1.), which was exactly compensated for the $\lambda/2$ radiation.

### 8.1.3. Order in a Solid Solution whose Composition Differs from that of the Perfectly Ordered Solution

The structure cannot be perfectly ordered whenever the ratio of the numbers of atoms $A$ and $B$ is not exactly equal to that of the numbers of the two types of lattice points.

Bradley and Jay [9] have shown, in the case of a series of Fe Al alloys, that it is possible to determine the atomic distribution in the state which is closest to perfect order, as a function of the concentration. They assumed the alloy to be perfectly homogeneous but, according to another theory, there may exist two different phases in such cases, one of which is ordered, and the other disordered [10].

The alloys containing 25-50% of aluminum atoms were prepared in the form of ingots, reduced to a powder and passed through a very fine screen. The powder was then heated to 750°C and quenched in water. The Debye-Scherrer patterns were obtained with an iron anode because the presence of iron prevented the use of Cu K$\alpha$ radiation. The patterns showed the lines of a body-centered cubic crystal analogous to that of pure $\alpha$ iron, and also weak superlattice lines corresponding to a simple cubic unit cell having the same lattice constant. In the case of the Fe Al alloy there can be perfect order, the iron atom occupying the corners and the aluminum atom the center of the cube.

Let $f_1$ and $f_2$ be the average scattering factors of the atoms situated at the positions 1 and 2, the number 1 indicating a corner and 2 the center of the cube. The strong lines, which are those of the body-centered cubic lattice, have indices $hkl$ such that $h + k + l$ is an even number. The waves scattered by the atoms in positions 1 and 2 are in phase and the structure factor is given by

$$F_{\text{norm}} = f_1 + f_2 \,. \tag{8.5}$$

If the proportion of iron is $p$, there are, on the average, $2p$ iron atoms in the unit cell and $2(1-p)$ atoms of aluminum, so

$$F_{\text{norm}} = 2pf_{\text{Fe}} + 2(1-p)f_{\text{Al}} \,. \tag{8.6}$$

For the intense lines the structure factor can therefore be calculated from the composition of the alloy, whatever the arrangement of the atoms in the unit cell.

The superlattice lines are those for which $h+k+l$ is an odd number. The waves scattered by the atoms in positions 1 and 2 then have opposite phases and

$$F_{\text{super}} = f_1 - f_2 \,. \tag{8.7}$$

These lines depend on the degree of order, and $F_{\text{super}}$ varies from zero for perfect disorder to $f_{\text{Fe}} - f_{\text{Al}}$ for the perfectly ordered Fe Al alloy.

It is possible to determine a particular value of $F_{\text{super}}$ from the intensity of a superlattice line. However, since an absolute measurement would be very difficult, it is preferable to compare two lines such that $h+k+l$ is even for one and odd for the other. The ratio of their structure factors is then proportional to the square root of the ratio of their intensities, the constant of proportionality for two given lines being the same for the complete set of alloys. Bradley and Jay chose the $(100)\alpha$ line and the neighboring $(110)\beta$ line due to the K$\beta$ radiation, so the superlattice line could be compared to a normal line of comparable intensity, the K$\beta$ radiation being weak with respect to K$\alpha$:

$$\frac{F_{(100)\alpha}}{F_{(110)\beta}} = C\sqrt{\frac{I_{(100)\alpha}}{I_{(110)\beta}}} \,. \tag{8.8}$$

Bradley and Jay were able to determine the quantity $C$ by applying this relation to an alloy containing 50% of Al which was known to be perfectly ordered from a previous experiment and for which, therefore,

$$F_{(100)\alpha} = f_{\text{Fe}} - f_{\text{Al}} \,,$$

and

$$F_{(110)\beta} = f'_{\text{Fe}} + f'_{\text{Al}} \,,$$

where the numbers $f_{\text{Fe}}$, $f_{\text{Al}}$, $f'_{\text{Fe}}$, $f'_{\text{Al}}$ are respectively the structure factors[1] for the value of $(\sin\theta)/\lambda$, corresponding to the lines (100) and (110).

We can then deduce the value of $F_{(100)\alpha}$ for any alloy from the measurement of $I_{(100)\alpha}/I_{(110)\beta}$ and from the value of $F_{(110)\beta}$ using Eqs. (8.8) and

[1] Since the Fe K$\alpha$ and Fe K$\beta$ lines are close to the K absorption edge of iron, a correction is required in the case of iron.

(8.6). Two equations are therefore available to determine the scattering factors of the "average" atoms in sites 1 and 2. One equation, according to Eq. (8.7), is

$$f_1 - f_2 = F_{(100)\infty}, \tag{8.9}$$

and the other, from Eq. (8.6), for the same scattering angle, is

$$f_1 + f_2 = 2pf_{Fe} + 2(1-p)f_{Al}. \tag{8.10}$$

TABLE 8-2. Values of $f_1$ and $f_2$ Calculated from Equations (8.9) and (8.10) for Various Alloys

| Composition $p$(% Al) | $f_1 + f_2$ | $f_1 - f_2$ | $f_1$ | $f_2$ | $p$ | |
|---|---|---|---|---|---|---|
| | | | | | Calculated | Chemical analysis |
| 50 | 27.5 | 9.1 | 18.3 | 9.2 | 50 | 50 |
| 45 | 28.4 | 8.3 | 18.3 | 10.0 | 45.5 | 45 |
| 40 | 29.3 | 7.6 | 18.4 | 10.8 | 41 | 40 |
| 35 | 30.2 | 6.7 | 18.4 | 11.7 | 36 | 35 |
| 30 | 31.1 | 5.65 | 18.4 | 12.7 | 30.5 | 30 |
| 25 | 32.0 | 4.65 | 18.3 | 13.7 | 25.5 | 25 |

Table 8-2 shows the value of $f_1$ and $f_2$ calculated from these two equations for various alloys. In the case of the (100) reflection, $f_{Fe}=18.3$ and $f_{Al}=9.2$.

It can be seen that $f_1$ is a constant and is equal to $f_{Fe}$, so site 1 is occupied exclusively by iron atoms. In site 2 there are the $2(1-p)$ atoms of aluminum in the unit cell and the $(2p-1)$ remaining iron atoms. We must therefore have

$$f_2 = (2p-1)f_{Fe} + 2(1-p)f_{Al}. \tag{8.11}$$

The sixth column of Table 8.2 shows the value of $p$ calculated from Eq. (8.11) and from the value of $f_2$. The agreement between the experimental and theoretical values of $p$ is quite satisfactory.

The presence of superlattice lines in the diffraction patterns of these alloys therefore indicates that the two atoms in the unit cell are not identical, on the average, and the quantitative study of the diffracted intensity shows that one position is occupied by an iron atom, and the other by the remaining iron atoms and by all the aluminum atoms. The order becomes more and more perfect as the composition of the alloys approaches Fe Al.

## 8.2. Short-range Order in Disordered Solid Solutions

In a mixed crystal in which there is no long-range order, a lattice translation can superpose two atoms of different types. This lack of periodicity we shall call *pure substitution disorder,* since the crystal would become perfectly ordered if all the atoms were identical. We shall assume that the thermal agitation is the same as in a crystal having the same lattice but whose nodes are occupied by an "average" type of atom. To arrive at the diffraction pattern which is characteristic of substitution disorder, we must subtract from the observed pattern the thermal diffuse scattering which is due to the perfect "average" crystal.

As in the case of displacement disorder, the lack of periodicity produces only second-order effects which are difficult to detect. We shall now calculate these effects and show what information the patterns can provide on the real structure of mixed crystals.

### 8.2.1. Scattering Due to Substitution Disorder. Comparison with Displacement Disorder

The general equations of Section 6.2 are in this case particularly simple to use. We consider a simple lattice whose nodes are occupied by atoms $A, B, \cdots$ in the relative proportions $c_A, c_B, \cdots$, the scattering factors being $f_A, f_B, \cdots$. The scattering factor of the atom occupying the $n$th node is $f_n$, which is equal to one of the above factors. The average structure factor for the unit cell is therefore

$$\bar{F} = c_A f_B + c_A f_B + \cdots . \tag{8.12}$$

In general the scattering produced by this kind of disorder is quite distinct from the diffraction spots or lines, so Eqs. (*6.25*) and (*6.26*) can be used.

Therefore, (a) the diffraction pattern is that of a crystal of the same shape and lattice as the object, but whose nodes are occupied by the fictitious average atom of scattering factor $\bar{F}$ [Eq. (*8.12*)].

(b) The scattering distributed in the unit cell of the reciprocal lattice is

$$I_2 = \sum \Phi_m \exp(2\pi i \boldsymbol{s}\cdot\boldsymbol{x}_m) ,$$

where

$$\Phi_m = \overline{\varphi_n \varphi_{n+m}} = (f_n - \bar{F})(f_{n+m} - \bar{F}) . \tag{8 13}$$

In this case the quantities $\varphi_n$ and $\bar{F}$ are real and, from Eq. (*6.27*),

$$I_2 = \Phi_0 + 2\sum_1^\infty \Phi_m \cos(2\pi \boldsymbol{s}\cdot\boldsymbol{x}_m) , \tag{8.14}$$

where

$$\Phi_0 = \overline{f_n^2} - (\bar{F})^2 .$$

It is important to notice that, contrary to displacement disorder, the quantities $\bar{F}$ and $\Phi_m$ depend on $\boldsymbol{s}$ only through the scattering factors of the individual atoms, so that, aside from a slow decrease with $s$, they are constant in reciprocal space. We are therefore led to the following conclusions.

(a) The diffraction spots of the average crystal have relative intensities, which are analogous to those of a perfect crystal. The irregularity in the atomic distribution does not cause a weakening of the lines at large angles.

(b) The scattering is *periodic*, the periods being those of the reciprocal lattice. The scattering repeats itself from unit cell to unit cell, except for a general decrease in intensity with $s$. It will be recalled that, on the contrary, displacement disorder gives rise to a scattering the intensity of which increases, starting from the center and that the scattering near the center, i.e. at small scattering angles, is zero. In the case of substitution disorder we have the same scattering at the center as at the other nodes, and the intensity is even maximum at the center because the atomic scattering factors are maximum at $s = 0$. This general rule is very useful for distinguishing between the two types of disorder. We shall discuss it in more detail in Section 8.3.

Let us illustrate this rule by means of an example [11]. We consider a regular lattice in which the atomic scattering factors, instead of being constant, vary as a sine wave from $f(1 + \eta)$ to $f(1 - \eta)$. We studied a similar type of modulation for the case of displacement disorder in Section 7.2.1. Then

$$f_n = f(1 + \eta \cos 2\pi \boldsymbol{k} \cdot \boldsymbol{x}_n) , \tag{8.15}$$

where $\boldsymbol{k}$, the propagation vector for a wavelength $\Lambda$ has a length $1/\Lambda$ and is normal to the wave fronts over which $f_n$ is a constant.

The diffracted amplitude is then

$$\sum f_n \exp(-2\pi i \boldsymbol{s} \cdot \boldsymbol{x}_n) = \sum f \exp(-2\pi i \boldsymbol{s} \cdot \boldsymbol{x}_n) + \frac{f\eta}{2} \exp[-2\pi i (\boldsymbol{s} - \boldsymbol{k}) \cdot \boldsymbol{x}_n] + \frac{f\eta}{2} \exp[-2\pi i (\boldsymbol{s} + \boldsymbol{k}) \cdot \boldsymbol{x}_n] ,$$

where the first term gives the lattice composed of atoms having the average scattering factor $f$. The other two give *satellites at distances* $\pm \boldsymbol{k}$ of each node, with intensities which are $\eta^2/4$ times that of the principal node. *The satellites are therefore situated at the same positions as those which we found for a sinusoidal displacement disorder having the same propagation vector.* In the present case, however, the ratio of the intensity of the satellites to that of the principal node is *a constant for all the nodes.*

### 8.2.2. Scattering and Short-range Order

In the simplest case there is no short-range order, i.e. there is no relation between the nature of neighboring atoms. We have shown in Section 6.2.2 that the coefficients $\Phi_m$ of Eq. (*8.14*) were in this case all zero, except $\Phi_0$. The scattering power per atom is then given by

$$I_2 = \overline{f_n^2} - \overline{F^2} = (c_A f_A^2 + c_B f_B^2 + \cdots) - (c_A f_A + c_B f_B + \cdots)^2 , \qquad (8.16)$$

as shown previously [Eq. (*6.31*)]. For the solid binary solution *AB*,

$$I_2 = c_A c_B (f_A - f_B)^2 = c_A(1 - c_A)(f_A - f_B)^2 . \qquad (8.17)$$

Equation (*8.17*) is the *Laue formula*. The scattering is a constant throughout reciprocal space, if we neglect the variation of $f_A - f_B$, and this is *characteristic of perfect disorder*. The scattering is greatest for solutions having equal numbers of atoms *A* and *B*, and it increases as the difference between the atomic numbers of the two atoms increases. For example, for a disordered solid solution of AuCu, the scattering power at the center is 625 electrons per atom, which is that of a gas of atomic number 25. This scattering is weak and is of the order of magnitude of Compton scattering at large angles, or of thermal-diffuse scattering near the nodes of the reciprocal lattice. In practice this is too small to be observed, unless stray scattering is completely eliminated. On a Debye-Scherrer pattern taken under ordinary conditions, the background between the lines is not appreciably different for a disordered solid solution from what it is for a pure crystal. There have been, in fact, few experimental confirmations of the Laue formula because perfectly disordered solid solutions are rare, as could be expected from thermodynamic considerations [12].

The integral of the scattering power over a unit cell of the reciprocal lattice is

$$c_A c_B (f_A - f_B)^2 \frac{1}{V_c} ,$$

since $1/V_c$ is the volume $V^*$ of the unit cell of the reciprocal lattice, $V_c$ being that of the unit cell of the crystal lattice. The total scattering power for the node is

$$\frac{1}{V_c} (c_A f_A + c_B f_B)^2 ,$$

from Eq. (*4.39*), and the sum of these two terms is

$$\frac{1}{V_c} (c_A f_A^2 + c_B f_B^2) .$$

The terms $f_A^2/V_c$ and $f_B^2/V_c$ are the intensities diffracted by the perfect lattices $A$ and $B$ at the position of a node, and the above expression is the average value, each term being multiplied by a weighing factor $c_A$ and $c_B$. This is one consequence of the theorem discussed in Section 2.3, according to which the integral of the scattering power in reciprocal space is a constant, regardless of the atomic arrangement.

When the arrangement is modified, the intensity of the diffraction at the nodes remains constant, and the total scattered intensity must therefore also be a constant. The scattered intensity, however, is no more uniform within the unit cell; it changes from point to point according to the *degree of order* of the solid solution.

Let us define this concept of order. We have already seen that there exist ordered solid solutions (Section 8.1) which are characterized by superlattice lines, the atoms forming a regular pattern in a unit cell which is a multiple of that of the disordered lattice. Consider for the moment that such a long-range order does not exist, but let us assume that, for any two atoms which are distant by at most a few interatomic distances, the nature of one atom is correlated to that of the other. This is called *short-range order*. In the general case this order is defined by means of a parameter related to the lattice vectors.

In the case of a binary solid solution $AB$ in which the proportions of the atoms $A$ and $B$ are $c_A$ and $c_B$, we consider the pairs of atoms situated at the extremities of the vectors $\boldsymbol{x}_m$ with origin on an atom $B$. Let $n_{BA}$ be the proportion of the $BA$ pairs. Then the order parameter is

$$\alpha_m = 1 - \frac{n_{BA}}{c_A} . \tag{8.18}$$

We could also define it in terms of $n_{AB}$:

$$\alpha_m = 1 - \frac{n_{AB}}{c_B} . \tag{8.19}$$

This can be shown as follows. If we call $N_{AA}$, $N_{AB}$, $N_{BB}$, $N_{BA}$ the proportions of the various types of pairs on the vectors $\boldsymbol{x}_m$, then

$$n_{AB} = \frac{N_{AB}}{N_{AA} + N_{AB}} ,$$

$$n_{BA} = \frac{N_{BA}}{N_{BB} + N_{BA}} .$$

However,

$$c_A = N_{AB} + N_{AA} ,$$

$$c_B = N_{BB} + N_{BA} .$$

Since $\boldsymbol{x}_m$ and $\boldsymbol{x}_{-m}$ play identical roles in the lattice, $N_{AB} = N_{BA}$ and also $n_{AB}c_A = n_{BA}c_B$, so the above two expressions for $\alpha_m$ are equal.

In the state of perfect disorder, $\alpha_m = 0$ since $n_{BA} = c_A$, and $\alpha_m$ is equal to unity if all the pairs are identical couples $AA$ or $BB$. It is now easy to calculate $\Phi_m$ from the order parameter $\alpha_m$ [Eq. (*8.13*)]:

$$\Phi_m = (f_n - \bar{F})(f_{n+m} - \bar{F}) .$$

We have

$$f_n - \bar{F} = \begin{cases} f_A - c_A f_A - c_B f_B = c_B(f_A - f_B) & \text{if } n \text{ is an } A \text{ atom,} \\ f_B - c_A f_A - c_B f_B = c_A(f_B - f_A) & \text{if } n \text{ is a } B \text{ atom.} \end{cases}$$

Table 8.3 shows the frequency of occurrence of the various types of pairs, as well as the value of

$$\frac{(f_n - \bar{F})(f_{n+m} - \bar{F})}{(f_A - f_B)^2} .$$

We therefore have

$$\Phi_m = c_A c_B (f_A - f_B)^2 [c_B(1 - n_{AB}) - (c_A n_{AB} + c_B n_{BA}) + c_A(1 - n_{BA})] .$$

TABLE 8-3. Frequency of Occurrence of the Various Types of Pairs

| Type of Pair | Frequency | $\frac{(f_n - \bar{F})(f_{n+m} - \bar{F})}{(f_A - f_B)^2}$ |
|---|---|---|
| $AA$ | $c_A(1 - n_{AB})$ | $c_B^2$ |
| $AB$ | $c_A n_{AB}$ | $- c_A c_B$ |
| $BA$ | $c_B n_{BA}$ | $- c_A c_B$ |
| $BB$ | $c_B(1 - n_{BA})$ | $c_A^2$ |

If we replace $n_{AB}$ and $n_{BA}$ by their values in terms of $\alpha_m$ from Eqs. (*8.18*) and (*8.19*), remembering that $c_A + c_B = 1$, we find that

$$\Phi_m = c_A c_B (f_A - f_B)^2 \alpha_m \tag{8.20}$$

and the scattering power is therefore, from Eq. (*8.14*),

$$I_2(\mathbf{s}) = c_A c_B (f_A - f_B)^2 \sum \alpha_m \exp(2\pi i \mathbf{s} \cdot \mathbf{x}_m) ,$$

or, since $\alpha_m = \alpha_{-m}$ and $\alpha_0 = 1$ ,

$$I_2(\mathbf{s}) = c_A c_B (f_A - f_B)^2 (1 + 2 \textstyle\sum_1^\infty \alpha_m \cos 2\pi \mathbf{s} \cdot \mathbf{x}_m) . \tag{8.21}$$

The intensity is therefore the sum of the constant intensity predicted by the Laue formula and of a series of terms which oscillate periodically with the period of the reciprocal lattice, as shown in Figure 8.4.

Since the knowledge of the order parameter $\alpha_m$ determines completely the scattering pattern, then, inversely, all the order parameters can be determined from the scattering pattern [13]. For, once the scattering power has been mapped throughout a unit cell of the reciprocal lattice, inversion of the summation in Eq. (*8.21*) gives

**FIGURE 8.4.** *Schematic representation of the intensity scattered by a solid solution.* (a) *Disordered solution.* (b) *Solution with a certain degree of short-range order. The pairs of vertical lines represent the normal diffraction nodes.*

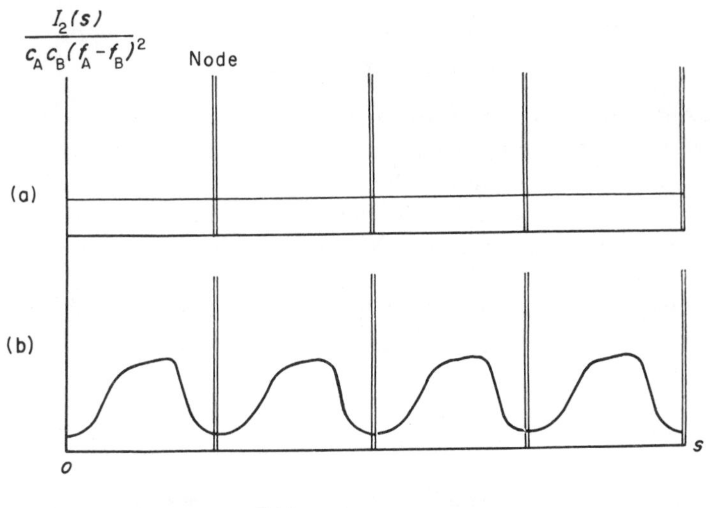


$$\alpha_m = \int_{\text{unit cell}} \frac{I_2(\boldsymbol{s})}{c_A c_B (f_A - f_B)^2} \exp(-2\pi i \boldsymbol{s} \cdot \boldsymbol{x}_m)\, dv_s \,. \tag{8.22}$$

It is important to note that the symmetry of the lattice is such that it is only necessary to explore a small fraction of the unit cell. Also, it is sufficient to determine relative values of the scattered intensity, since $\alpha_0$ is known to be equal to unity.

Another way of using Eq. (*8.21*) is to calculate the order parameters $\alpha_m$ in terms of a small number of variables which are characteristic of the nature of the atoms and of the lattice of the mixed crystal. This complex calculation can be performed starting from the values of the interaction energy between the atoms, and assuming that the solid solution is in thermodynamic equilibrium at a given temperature. The calculation is necessarily approximate. For example, one can take into account only the interactions between first and second neighbors. This was done for the $AuCu_3$ alloy by Fournet [14].

### 8.2.3. The Scattered Intensity in the Powder Pattern of a Mixed Crystal

The scattered intensity is easily deduced in this case from the Debye formula [Eq. (*2.54*)]. In a crystal an atom has $c_1$ neighbors at a distance $r_1$, with $c_2$ neighbors at a distance $r_2$, etc. If the sample is composed of a set of crystals of identical structure and random orientations, the scattering power per atom as a function of $s = (2\sin\theta)/\lambda$ is given by

$$I(s) = \sum_i c_i \overline{f_n f_{n+m_i}} \frac{\sin 2\pi s r_i}{2\pi s r_i} . \tag{8.23}$$

From Eqs. (*6.18*) and (*8.20*),

$$\overline{f_n f_{n+m_i}} = (\bar{F})^2 + \Phi_{m_i} = (\bar{F})^2 + c_A c_B (f_A - f_B)^2 \alpha_i .$$

We assume that all the nodes of a layer $i$ are at the extremities of vectors $\boldsymbol{x}_{m_i}$ which, by symmetry, play the same role in the lattice. For each layer we therefore have only *one* of the previously defined order parameters $\alpha_i$. For example, in the face = centered cubic lattice, the first layer contains the twelve nodes 110, the second contains the six nodes 200, etc.

Equation (*8.23*) can be divided into two parts, the first of which is the powder pattern for the average lattice and gives the normal diffraction lines. The second corresponds to the continuous background:

$$I(s) = c_A c_B (f_A - f_B)^2 \sum c_i \alpha_i \frac{\sin 2\pi s r_i}{2\pi s r_i} . \tag{8.24}$$

This formula is very useful for predicting a powder pattern when the order coefficients $\alpha_i$ are known, for it would be much more complicated to calculate the reciprocal lattice and to integrate over all the possible positions of the sample. The formula takes into account possible dissymmetries of broad lines with respect to the positions of superlattice lines for an ordered lattice. However, it is mathematically impossible to invert this equation in order to find the coefficients $\alpha_i$ from a powder pattern. Only by assuming that the order coefficients are all zero except those for the first layer, or for the first few, is it possible to determine coefficients $\alpha_1, \alpha_2, \cdots$ by trial and error by adjusting the calculated curve to fit the observations.

### 8.2.4. The Linear Equiatomic Lattice

We shall only discuss one very schematic example to illustrate the manner in which the scattering pattern depends on the degree of order. We consider a linear lattice ($x_m = ma$) of atoms $A$ and $B$ which occur in equal numbers [15], [16], and assume that interactions exist only between close neighbors. Let us call $p$ the probability that a pair of neighbors is different. For complete disorder, $p = 1/2$ and, if the order is perfect, we have $ABABAB\cdots$, and $p = 1$. In this particular linear lattice, all the order parameters can be expressed as functions of a single parameter $p$. *This is not possible for three-dimensional lattices.* According to Eq.

(8.18), the degree of order for the first neighbors is $\alpha_1 = 1 - 2p$. For the $m$th neighbors,

$$(n_{AB})_m = (n_{AB})_{m-1}(1 - p) + [1 - (n_{AB})_{m-1}]p \ .$$

These recurrence relations yield

$$\alpha_m = (1 - 2p)^m = \alpha_1^m \ ,$$

and, from Eqs. (8.21) and (7.45), the scattering power is

$$I_2(s) = \frac{(f_A - f_B)^2}{4} [1 + 2 \textstyle\sum_1^\infty (1 - 2p)^m \cos(2\pi sma)]$$

$$= \frac{(f_A - f_B)^2}{4} \frac{1 - \alpha_1^2}{1 - 2\alpha_1 \cos(2\pi sa) + \alpha_1^2} \ . \tag{8.25}$$

This function is illustrated in Figure 7.15 of the preceding chapter. The intensity is a constant for $p = 1/2$, corresponding to perfect disorder, and as the order increases the intensity exhibits a higher and higher maximum at $s = (2k+1)/2a$. These are the positions of the superlattice nodes corresponding to perfect order, since the ordered linear lattice $ABABAB\cdots$ has "normal" nodes of intensity $(f_A + f_B)^2/4$ for $s = 0$, $1/a$, $2/a, \cdots$ and superlattice nodes of intensity $(f_A - f_B)^2/4$ for $s = 1/2a, 3/2a, \cdots$.

As we have already seen in analogous cases, the above equation is not valid if the order is close to perfect ($\alpha_1 \cong -1$), since the peak is then too sharp and its shape is affected by the size of the crystal (Section 2.4.2).

It is possible to perform an exact calculation [16] by applying the general Eq. (6.4). If $N$ is the total number of atoms and $\alpha_1 = -1 + (z/N)$, the profile of the superlattice line is given by

$$I_2(s) = \frac{(f_A - f_B)^2}{4} \left[1 + \frac{2N}{z}\left(1 - \frac{1 - \exp(-z)}{z}\right)\right],$$

as a first approximation.

*Disorder therefore has the effect of broadening the superlattice peaks symmetrically*, and the broadening is already considerable, even if the disorder is weak. Thus, for $p = 0.9$, corresponding to $\alpha_1 = -0.8$ or to one fault in ten, the peak width is equivalent to that obtained with a crystal of 15 atoms. As the line becomes broader, its maximum intensity decreases because its integral intensity remains constant. This shows up experimentally in the following manner. So long as the theoretical profile is narrow enough that the observed width can be ascribed to geometrical factors in the experiment, its intensity remains constant but, when the experimental width is negligible compared to the theoretical width of the line, the maximum intensity decreases as the degree of order decreases. Thus, if the experimental width is $1/500a$, the peak is thirty times smaller for $\alpha_1 = -0.8$ than for perfect order.

If $p$ is smaller than 1/2, atoms of a given type have a tendency to group together. This phenomenon is called *segregation*. In Eq. (*8.25*), $\alpha_1$ becomes positive and the broad maximum occurs at integral value of $s$, while the minimum occurs at half-integral values. We then have a series of small fragments $AAAA\cdots$ and $BBBBB\cdots$, so the small crystals give a broad diffraction maximum at the normal positions of the nodes. These maxima become stronger as segregation increases.

This simple example leads to a conclusion which is valid for three-dimensional solid solutions: whenever there is a tendency for a certain type of order, *there appear scattering maxima which are centered on the positions of the superlattice nodes of the ordered lattice*. These maxima are broad and weak when there is only a small degree of order. Whenever there are scattering regions surrounding the normal nodes, and especially around the center of the reciprocal lattice, there is a tendency to segregation in a solid solution.

### 8.2.5. Short-range Order in the $AuCu_3$ Alloy

We have already discussed $AuCu_3$ in our study of long-range order. It is one of the best known alloys because the large difference between the scattering factors of the constituent atoms renders the scattering phenomena relatively easy to observe. There are many alloys of metals having neighboring atomic numbers, such as CuZn and NiFe, for which X-rays give no useful results. We have already described the ordered and disordered lattices for $AuCu_3$ in Section 8.1.1, but there also exist intermediate states where the superlattice lines are broad and the normal lines narrow. Jones and Sykes [17] first interpreted these observations in terms of an "antiphase domain" theory, according to which the alloy is formed of four different types of domains, in which the gold atom occupies one of the four possible positions in the unit cell of the disordered face-centered cubic lattice. These domains are all ordered and the lattice is unchanged, except for the fact that the gold atoms do not occupy the same sites in the various domains.

These domains have no effect on the normal lines, but they should broaden the superlattice lines because of their small size. As a first approximation, the width of these lines was calculated as if the domains were incoherent. But this assumption is too simple, since the antiphase domains diffract coherently for the normal lines but not quite incoherently for the superlattice lines. If we attempt to deduce the dimensions of these domains from the line widths by the Scherrer formula, we find results which differ systematically, according to the indices of the line considered. Moreover, for states with a small degree of order we observe

**FIGURE 8.5.** *Intensity in a powder pattern for a partially ordered $AuCu_3$ alloy.*

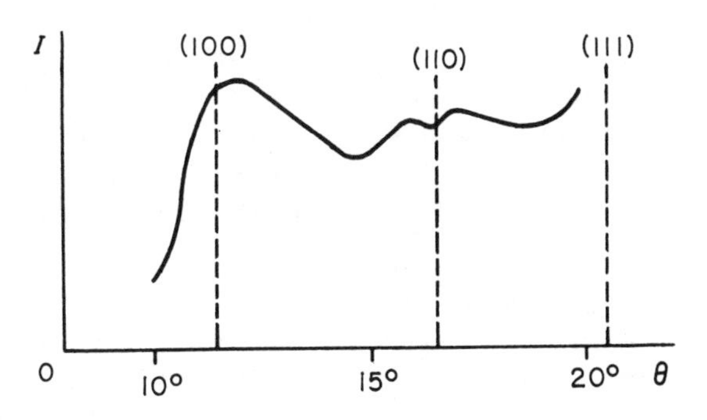


**FIGURE 8.6.** *Intensity in a powder pattern for $AuCu_3$ in equilibrium above the critical temperature, at 550°C.*

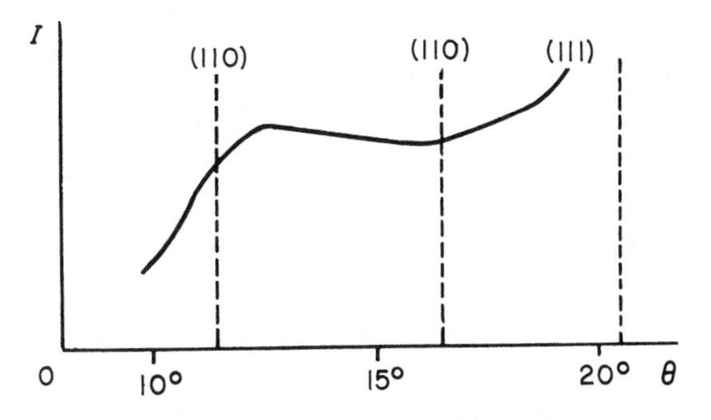


that the center of the 100 line is displaced and that the 110 line becomes a doublet, as in Figure 8.5. These facts cannot be explained by a simple size effect.

Another observation is that the powder diagrams for disordered alloys at temperatures higher than the critical temperature [18] give a large ring, similar to that given by liquids and situated approximately at the position of the 100 line. This is illustrated in Figure 8.6.

An exact study of the atomic distribution requires a detailed mapping of the scattering from a single crystal (Section 6.3). This has been done for $AuCu_3$ alloys under two different conditions: in equilibrium above the critical temperature of 395°C, and in nonequilibrium. The disordered state is preserved by annealing at room temperature and it then drifts under the influence of progressive annealings toward the stable ordered state.

(a) *Disordered Alloy in Equilibrium.* The scattering of X-rays has demonstrated the existence of short-range order in the absence of any long-range order, as was predicted theoretically by Bethe [19]. One finds scattering domains in the vicinity of the superlattice nodes, for example the 100, 210, 110, etc. nodes in the first unit cell. The domains centered on these nodes are shaped like thin lenses situated in the (100) planes. Figure 8.7 shows curves of equal intensity in the (001) plane. Cowley [13] has determined the distribution of the scattering power within the unit cell and applied Eq. (*8.22*) to calculate the coefficients $\alpha_m$. For reasons of symmetry, the six coefficients $\alpha_{100}$, $\alpha_{\bar{1}00}$, $\alpha_{010}$, $\cdots$ and the twelve coefficients $\alpha_{110}$, $\alpha_{\bar{1}10}$, $\alpha_{101}$, $\cdots$ are all equal. Table 8-4 shows the results Cowley obtained, as well as the coefficient $\alpha$ deduced from the structure of Au Cu$_3$ (Section 8.1.1) and from Eq. (*8.18*) for the perfectly ordered state. It will be observed that the $\alpha$ tend to zero rather rapidly,

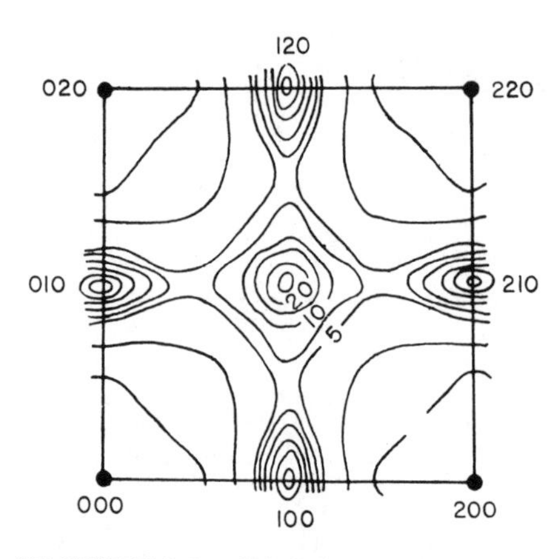


**FIGURE 8.7.** *Distribution of the scattering power of a single Au Cu crystal in equilibrium at 450°C; lines of equal scattered intensity in the plane $l = 0$ of the reciprocal lattice.*

TABLE 8-4. Short-range Order Coefficients Measured for the Au Cu$_3$ Alloy in Equilibrium at 405°C

| Order of the Shell $m$ | Number of Atoms on the Shell $c_m$ | Coordinates of $x_m$ | $\alpha_m$ for Perfect Order | $\alpha_m$ Measured at 405°C |
|---|---|---|---|---|
| 1 | 12 | 110 | −0.333 | −0.152 |
| 2 | 6 | 200 | 1.000 | 0.186 |
| 3 | 24 | 211 | −0.333 | 0.009 |
| 4 | 12 | 220 | 1.000 | 0.095 |
| 5 | 24 | 310 | −0.333 | −0.053 |
| 6 | 8 | 222 | 1.000 | 0.025 |
| 7 | 48 | 321 | −0.333 | −0.016 |
| 8 | 6 | 400 | 1.000 | 0.048 |
| 9 | {12 | 330 | −0.333 | −0.026 |
|  | {24 | 411 | −0.333 | −0.011 |
| 10 | 24 | 420 | 1.000 | 0.026 |

that the order has practically disappeared at a distance equal to three times that between close neighbors, and that even for close neighbors $\alpha$ has only one-half the theoretical value corresponding to the perfectly ordered state. It is observed that the degree of order decreases as the temperature increases.

These coefficients have been found theoretically by Fournet [14], starting from energy considerations deduced from the study of long-range order in alloys above the critical point. This is a good illustration of the importance of such X-ray studies, which can serve to confirm theories otherwise impossible to verify experimentally.

(b) *Partially Ordered Alloys.* The transformation of a disordered alloy into a stable ordered state by tempering can be extremely complex. Order is established within small antiphase regions, and experiment has led to the unexpected result that these regions can be arranged regularly to form a new superlattice with a large unit cell. In the case of the Au Cu alloy, for example, Johannson and Linde [20] found a phase called Au Cu II, which is stable below the critical temperature and which is formed of a column of 10 unit cells divided into two antiphase regions of 5 unit cells each.

Ordered regions are also observed in the case of $AuCu_3$ if the ratio between the numbers of Au and Cu atoms is slightly different from 3. The most clear-cut results are obtained from electron diffraction patterns for thin samples of alloy prepared by evaporation (Perio and Tournarie [21]). The reciprocal lattice of the crystal is shown schematically in Figure 8.8. It is found that two successive regions are separated by a plane of faults parallel to a (100) plane and that the relative positions

**FIGURE 8.8.** *Reciprocal space of an* $AuCu_3$ *crystal with regular antiphase domains.*

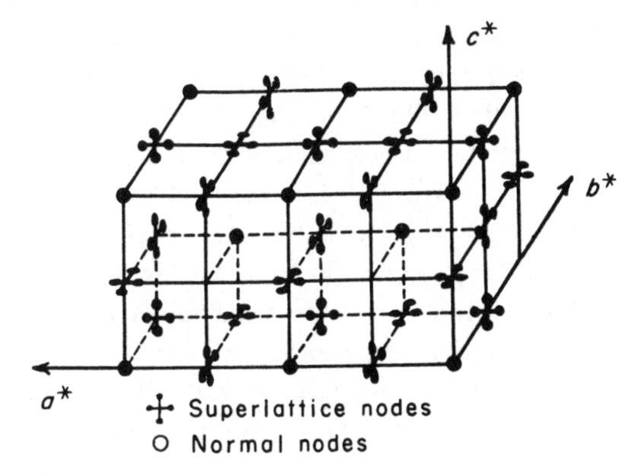

**FIGURE 8.9.** *Structure of two antiphase domains separated by a (001) plane.*

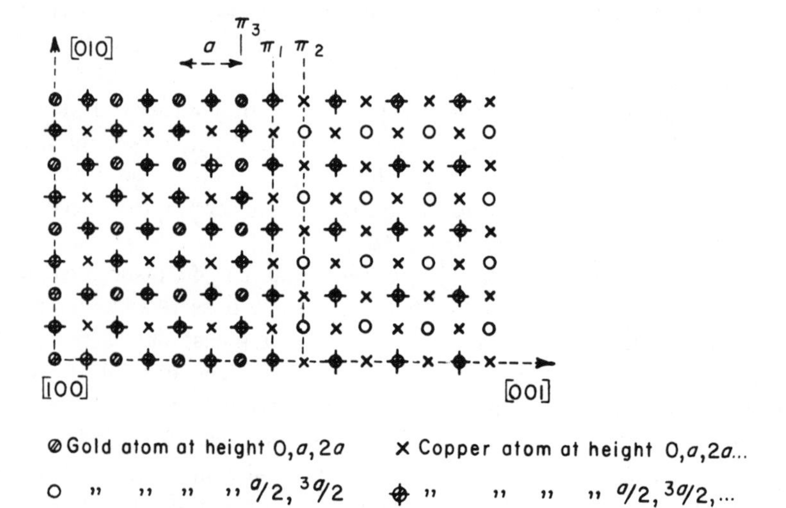


of the two regions is given by the vector $(a/2, a/2, 0)$, as in Figure 8.9. The two regions therefore meet on a plane of copper atoms ($\pi_1$). On the adjacent planes ($\pi_2$) and ($\pi_3$), it is only the second neighbors which occupy incorrect positions. The regular spacing of such regions of constant thickness gives rise to the satellite nodes situated near the nodes of the ordered phase along the ⟨100⟩ rows. The distance between satellites shows that the thickness of the regions is about five (100) planes. Such stratified regions have been observed directly by electron microscopy [22], but their existence had been suspected from the evidence provided by X-ray scattering in thick samples [18], [23]. It had been found that the reciprocal lattice has satellites situated like those obtained by electron diffraction, except that they are diffuse, indicating that the arrangement is not perfectly periodic. The formation of ordered antiphase regions is probably more difficult in a thick sample than in a thin one. The reciprocal lattice shown in Figure 8.9 is related to the powder patterns for the alloy of Figure 8.6. It is seen that the lines due to the satellites are displaced toward large angles, but not the lines due to the 100 node itself, and that the 110 lines must be split symmetrically about the position of the 110 line for the simple ordered phase.

Order-disorder transformations appear to be quite generally accompanied by periodic antiphase regions. The theoretical explanation for this is as yet unknown.

### 8.2.6. Segregated Solid Solutions

When an alloy of aluminum and silver, containing for example 20% of silver, is rendered homogeneous by heating and then quenched to room temperature, the solution becomes supersaturated and the atoms of silver tend to segregate. The scattering pattern then exhibits, around each node of the matrix, small scattering domains which have closely similar shapes [24]. As a first approximation it may be assumed that they are all identical, and it is then sufficient to study the one surrounding the center of the reciprocal lattice. (Small-angle scattering will be studied in Chapter 10.)

Since a scattering domain is spherically symmetrical around its node, it shows up undistorted at the center of a powder pattern, so it is unnecessary to use a single crystal. Figure 8.10 shows a section of such a domain along a meridian plane; it has the shape of a ring whose diameter is of the order of 5-10% of the unit cell of the reciprocal lattice. The intensity seems to become zero near the center.

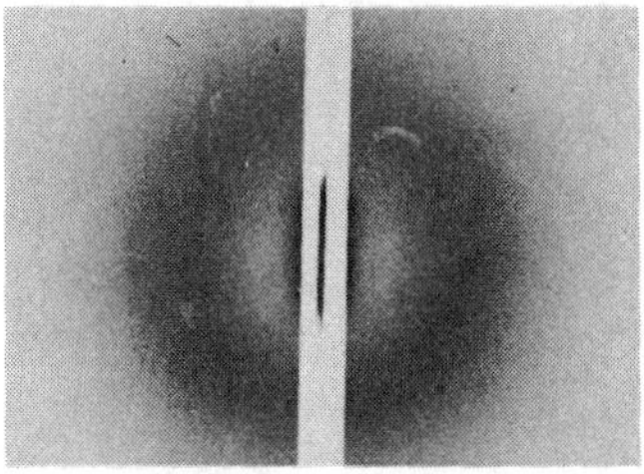

**FIGURE 8.10.** *Small-angle scattering pattern for an Al Ag (20% Ag) alloy after aging. A scattering region of similar shape surrounds each node of the reciprocal lattice (Walker and Guinier [24]).*

The disorder may be interpreted as due solely to substitution if it is assumed that the rings are all identical. It is certain that the distortions are very small, and this is to be expected since the diameters of the Al and Ag atoms differ by only one percent (2.86Å and 2.89Å, respectively).

The pattern may be interpreted by calculating the coefficients $\alpha_m$, using Eq. (*8.22*). The volume of integration chosen is a unit cell whose center is the center of the reciprocal lattice. Since the scattered intensity depends only on $s = |\boldsymbol{s}|$, a calculation analogous to that of Section 2.5 gives

$$\alpha_m = \int_{\text{unit cell}} \frac{I_2(\boldsymbol{s})}{c_{\text{Ag}}c_{\text{Al}}(f_{\text{Ag}} - f^{\text{Al}})^2} \exp(-2\pi i \boldsymbol{s} \cdot \boldsymbol{x}_m)\, dv_s\,,$$

$$\alpha(x) = \frac{1}{c_{\text{Ag}}c_{\text{Al}}(f_{\text{Ag}} - f_{\text{Al}})^2} \int_0^\infty I_2(s) \frac{\sin(2\pi s x)}{2\pi s x} 4\pi s^2\, ds\,. \tag{8.26}$$

The coefficient $\alpha_m$ depends only on the modulus of the vector $\boldsymbol{x}_m$. Since $\alpha(0) = 1$, it is sufficient to determine relative values of $I_2(s)$:

$$\alpha(x) = \frac{1}{2\pi x} \frac{\int I_2(s) \sin(2\pi xs) s\, ds}{\int I_2(s) s^2\, ds} . \qquad (8.27)$$

Setting $n_{AgAg}(x)$ as the probability that a pair of atoms separated by a distance $x$ is formed of two Ag atoms, then, from Eq. (*8.18*),

$$\alpha(x) = 1 - \frac{1 - n_{AgAg}(x)}{c_{Al}} = \frac{n_{AgAg} - c_{Ag}}{c_{Al}} .$$

Figure 8.11 shows $n_{AgAg}(x)$ as observed on an Al Ag alloy after quenching. The fact that $n_{AgAg}$ is larger than the average value $c_{Ag}$ over a distance of about 15Å demonstrates the existence of segregated groups of silver atoms. For larger values of $x$, $n_{AgAg}$ becomes smaller than $c_{Ag}$, indicating that the groups are surrounded by spherical shells depleted of part of their silver atoms, as could be expected. It may be shown that the absence of scattering at the center of a ring comes from the fact that the surplus of atoms in the central region of a group is just equal to the number of atoms missing in the shell.

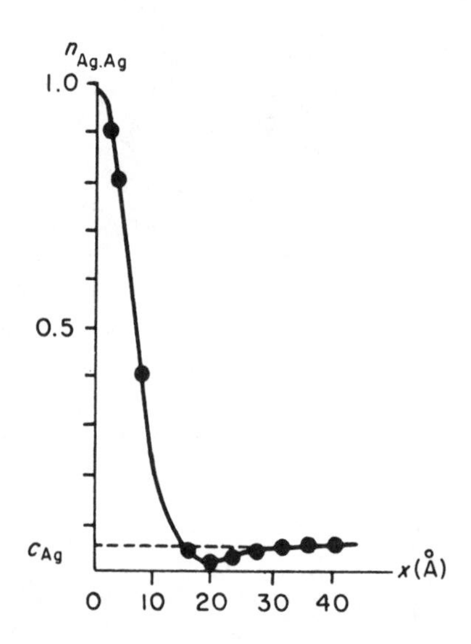


**FIGURE 8.11.** *Probability of finding a silver atom at a distance x from a given silver atom. This curve was deduced from the pattern of Figure 8.10, using Eq.* (*8.18*).

Much work has been done on the structure of this alloy after various types of heat treatment. It was found that the scattering domains vary slightly with the node indices and that they are not quite spherically symmetrical except near the center [25]. This indicates that there exists a slight displacement disorder associated with the segregation.

Doubt has been expressed about the above interpretation of the diffraction patterns in terms of a complex zone because there is also another possible explanation [26]. Let us assume that the alloy contains regions where the silver is highly concentrated and which are distributed in a matrix of uniform density. With this model there are only two densities involved, that of the groups of silver atoms and that of the matrix. If it is also assumed that the groups of silver atoms are distributed like the molecules in a liquid and not at random, the ring which is observed is the result of the interference between groups. For groups of given dimensions the analysis of the scattering curve gives the distribution function $P(x)$ for the groups in the matrix (Section 2.5). It is impossible to decide which model is correct unless the number of groups is known, and this requires absolute measurements of the scattering intensity. Certain experimental results [27], [28] appear to favor the second model.

## 8.3. Combination of Displacement Disorder and of Substitution Disorder

The general equations of Section 6.2.2 again apply to this case, but it is often difficult to interpret in terms of a plausible model the disorder coefficients $y_m$ or $\Phi_m$, which can alone be deduced experimentally. Now that we have analyzed separately the general characteristics of displacement disorder and of substitution disorder, we shall show by means of a few examples the effects of their combination.

### 8.3.1. Displacement Disorder Independent of Substitution Disorder

In this case the crystal is defined as follows.

(a) The different unit cells are displaced irregularly with respect to an *average geometrical lattice*.

(b) The content of the unit cells varies from one to the other, fluctuating about an *average unit cell*. There is, however, no relation between the displacement of a given unit cell and the fluctuation in its composition. From Eq. (*6.2*) the general expression for the scattering power per unit cell is

$$I(\boldsymbol{s}) = \sum_m \overline{F_n F^*_{n+m} \exp [2\pi i \boldsymbol{s} \cdot (\boldsymbol{x}_m + \Delta \boldsymbol{x}_{n+m} - \Delta \boldsymbol{x}_n)]}$$

and, as we have already noted for an analogous calculation (Section 2.7), the fact that the two types of disorder are independent permits us to state that the average of the product is equal to the product of the averages of the factors. *This would not be true if the two types of disorder were not really independent*. Therefore,

$$I(\boldsymbol{s}) = \sum_m \overline{(F_n F^*_{n+m})}\ \overline{\exp [2\pi i \boldsymbol{s} \cdot (\boldsymbol{x}_m + \Delta \boldsymbol{x}_{n+m} - \Delta \boldsymbol{x}_n)]}$$

and, from Eqs. (*6.23*) and (*6.24*),

$$\begin{aligned} \overline{F_n F^*_{n+m}} &= |\bar{F}|^2 + \Phi_m , \\ I(\boldsymbol{s}) &= |\bar{F}|^2 \sum_m \overline{\exp [2\pi i \boldsymbol{s} \cdot (\boldsymbol{x}_m + \Delta \boldsymbol{x}_{n+m} - \Delta \boldsymbol{x}_n)]} \\ &\quad + \sum_m \Phi_m \overline{\exp [2\pi i \boldsymbol{s} \cdot (\boldsymbol{x}_m + \Delta \boldsymbol{x}_{n+m} - \Delta \boldsymbol{x}_n)]} . \end{aligned} \qquad (8.28)$$

If the coefficients $\boldsymbol{\Phi}_m$ are small with respect to $|\bar{F}|^2$, we can, as a first approximation, consider $\Delta\boldsymbol{x}$ to be negligible compared to $\boldsymbol{x}_m$, and then

$$I(\boldsymbol{s}) = |\bar{F}|^2 \sum_m \overline{\exp[2\pi i\boldsymbol{s}\cdot(\boldsymbol{x}_m + \Delta\boldsymbol{x}_{n+m} - \Delta\boldsymbol{x}_n)]} + \sum_m \boldsymbol{\Phi}_m \exp(2\pi i\boldsymbol{s}\cdot\boldsymbol{x}_m) \,. \qquad (8.29)$$

This result can be interpreted as follows. The pattern is the *superposition* of (a) the pattern for a crystal in which there is pure displacement disorder, the unit cell being the *average* unit cell of the real crystal, and (b) the pattern for a crystal in which there is pure substitution disorder, the real unit cells occupying the normal positions for the average perfect lattice. This result is rigorous if there is no correlation between the content of two unit cells and their mutual positions. This is because all the coefficients $\boldsymbol{\Phi}_m$ of Eq. (*8.28*) are zero except $\boldsymbol{\Phi}_0$, which is equal to $\overline{|F|^2} - |\bar{F}|^2$ [Eq. (*6.30*)]. We therefore add to the pattern of the fictitious crystal whose unit cells are all replaced by a single average unit cell a continuous background $\overline{|F|^2} - |\bar{F}|^2$, which varies slowly in reciprocal space.

In the case of the mixed crystal $AB$ perturbed by thermal agitation, the pattern is the superposition of: (a) the pattern due to thermal agitation, as in Section 7.1, where all the unit cells are assumed to contain a single average atom of structure factor $c_A f_A + c_B f_B$; and (b) the pattern of a mixed crystal, as in Section 8.2. It is therefore possible to correct for thermal agitation before calculating the order parameters by means of Eq. (*8.22*). One may assume that a mixed crystal produces the same type of thermal-diffuse scattering, whatever the degree of order [13]. At any given point of reciprocal space, except at the nodes of the superlattice, the scattering power due to the substitution disorder is the difference between the scattering powers for the solid solution under consideration and for the same solution when it is perfectly ordered and at the same temperature [29]. The thermal scattering can be measured with the ordered alloy or calculated theoretically, both for single crystals (Section 7.1.5) and for powders [30]. Walker and Keating have shown that the scattering due to the local order itself is affected by the thermal agitation; its intensity is reduced as the scattering angle increases [31].

The assumption that the effects of displacement and substitution disorder are additive is particularly simple, but it is based on a hypothesis which is, in most cases, rather unlikely. In general, the different atoms forming the mixed crystal differ both in scattering power and in the extent of their fields of force. Their real positions in the crystal are thus related to their scattering factors. The mixed crystal Al Ag, in which the two atoms have very different scattering powers but nearly the same size, is rather exceptional; we have seen that it nevertheless shows some signs of lattice distortion (Section 8.2.6).

### 8.3.2. Displacement Disorder Correlated with Substitution Disorder

This is the case of a solid solution in which the two types of atoms have different sizes. In the case of close neighbors, for example, a pair of small atoms is closer than a pair of different atoms or a pair of large atoms. In a solid metallic solution which is known to be close-packed from its macroscopic density, it is highly probable that a displacement disorder accompanies substitution disorder, and the two are then strongly correlated. We shall illustrate the effect of such a correlation on the scattering patterns by means of a few simple examples.

8.3.2.1. LINEAR LATTICE WITH SIMULTANEOUS MODULATION OF THE LATTICE PARAMETER AND OF THE STRUCTURE FACTOR. We have already seen that a periodic modulation of the lattice parameter (see Section 7.2.2) and of the scattering factor (see Section 8.2.1) both give rise to symmetrical pairs of satellites situated at distances $\pm \boldsymbol{k}$ from the nodes of the average lattice, where $\boldsymbol{k}$ is the propagation vector and $|\boldsymbol{k}| = 1/\Lambda$. We have also seen that the ratio of the intensities of the satellites to that of their node were respectively $\Lambda^2\varepsilon^2 s^2/4$ and $\eta^2/4$, when the lattice parameter oscillates between $a(1+\varepsilon)$ and $a(1-\varepsilon)$, and when the structure factor oscillates between $f(1+\eta)$ and $f(1-\eta)$.

Let us now consider a perturbed lattice composed of planes containing a certain proportion of atoms having a larger scattering factor and also a larger size, the proportion being a sine function of position. The planes which scatter the most are then separated by the largest spacing. In such a case the structure factor $f_n$ and the coordinates of the successive planes can be written as follows, from Eqs. (*8.15*) and (*7.42*):

$$f_n = f\left(1 + \eta \sin \frac{2\pi na}{\Lambda}\right),$$

$$x_n = na + \Delta x_n = na - \frac{\Lambda \varepsilon}{2\pi} \cos 2\pi \frac{na}{\Lambda}. \tag{8.30}$$

It is essential to have a sine term in one expression and a cosine term in the other so that the maximum of $f_n$ corresponds to the maximum spacing, as in Figure 8.12a. The intensity is calculated by performing the summations

$$\textstyle\sum_n f_n \exp(-2\pi i s x_n),$$

on the assumption that $\varepsilon$ and $\eta$ are small [32].

This result can be attained by using the properties of the Fourier

**FIGURE 8.12.** (a) *Lattice with simultaneous variation of scattering factor and of reticular spacing.* (b) *Position and intensity of the satellites near the normal nodes.*

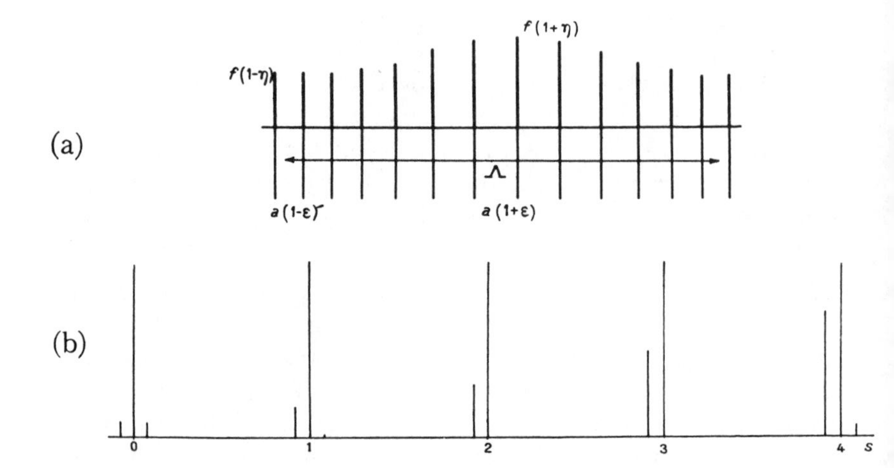


transforms and of the faltung product. The diffracted amplitude is the transform of the function representing the distribution of matter, or

$$F(x) = f\left(1 + \eta \sin\frac{2\pi x}{\Lambda}\right) \Sigma_n \,\delta(x - x_n)\,, \tag{8.31}$$

where $\delta(x - x_n)$ is the Dirac delta function. Let us first calculate the transform of $\Sigma\, \delta(x - x_n)$:

$$\begin{aligned}\text{transf}\ \Sigma\,\delta(x - x_n) &= \int \Sigma\,\delta\left(x - na + \frac{\Lambda\varepsilon}{2\pi}\cos\frac{2\pi na}{\Lambda}\right)\exp(2\pi isx)\,dx \\ &= \Sigma \exp\left[2\pi is\left(na - \frac{\Lambda\varepsilon}{2\pi}\cos\frac{2\pi na}{\Lambda}\right)\right] \\ &= \int \Sigma\,\delta(x - na)\exp\left(-i\Lambda\varepsilon s\cos\frac{2\pi x}{\Lambda}\right)\exp(2\pi isx)\,dx\,.\end{aligned}$$

It is therefore the transform of the product $\Sigma\,\delta(x - na)$, by

$$\exp\left(i\Lambda\varepsilon s\cos\frac{2\pi x}{\Lambda}\right) \cong 1 - i\frac{\Lambda\varepsilon s}{2}\exp\left(\frac{2\pi ix}{\Lambda}\right) - \frac{i\Lambda\varepsilon s}{2}\exp\left(-\frac{2\pi ix}{\Lambda}\right).$$

The respective transforms of the two factors (Appendix A) are

$$\begin{aligned}g_1(s) &= \frac{1}{a}\Sigma\,\delta\left(s - \frac{a}{n}\right), \\ g_2(s) &= \delta(s) - \frac{i\Lambda\varepsilon s}{2}\delta\left(s + \frac{1}{\Lambda}\right) - \frac{i\Lambda\varepsilon s}{2}\delta\left(s - \frac{1}{\Lambda}\right),\end{aligned}$$

so the transform of $\sum \delta(x-x_n)$ is the faltung

$$g_3(s) = g_1(s) * g_2(s)$$
$$= \frac{1}{a}\sum \delta\left(s-\frac{n}{a}\right) - \frac{i\Lambda\varepsilon s}{2a}\sum \delta\left(s-\frac{n}{a}+\frac{1}{\Lambda}\right) - \frac{i\Lambda\varepsilon s}{2a}\sum \delta\left(s-\frac{n}{a}-\frac{1}{\Lambda}\right). \tag{8.32}$$

The function $g_3(s)$ in reciprocal space contains the normal nodes of the lattice at $s = n/a$, and next to each node there is a pair of satellites at the distance $\pm 1/\Lambda$, of amplitude $\Lambda\varepsilon s/2$ times that of the normal node. This is the result which we found directly in Section 7.2.2.

We now require the transform of

$$f\left(1+\eta\sin\frac{2\pi x}{\Lambda}\right),$$

or

$$f - \frac{\eta f i}{2}\exp\left(\frac{2\pi i x}{\Lambda}\right) + \frac{\eta f i}{2}\exp\left(\frac{-2\pi i x}{\Lambda}\right)$$

This is the sum of three functions,

$$g_4(s) = f\delta(s) - \frac{\eta f i}{2}\delta\left(s+\frac{1}{\Lambda}\right) + \frac{\eta f i}{2}\delta\left(s-\frac{1}{\Lambda}\right). \tag{8.33}$$

The required amplitude is the transform of $F(x)$ [Eq. (*8.31*)], and therefore the faltung of $g_4(s)$ [Eq. (*8.33*)] by $g_3(s)$ [Eq. (*8.32*)]. This is obtained by taking the sum of the functions $g_4(s)$ centered on each one of the peaks of $g_3(s)$, each one being multiplied by its weighing factor. Neglecting the product of $\varepsilon$ and $\eta$, which are both small, we find that each node is accompanied by a pair of satellites situated at $\pm 1/\Lambda$, or *at the same positions as though either the plane spacings or the structure factors oscillated at a wavelength* $\Lambda$, *the other quantity remaining constant. The amplitudes of these satellites are different*, however, as illustrated in Figure 8.12b. The amplitudes are as follows:

$$-\frac{i\Lambda\varepsilon s}{2}\frac{f}{a} - i\frac{\eta}{2}\frac{f}{a} \quad \text{for the satellite } s = \frac{n}{a}-\frac{1}{\Lambda}$$
$$-\frac{i\Lambda\varepsilon s}{2}\frac{f}{a} + i\frac{\eta}{2}\frac{f}{a} \quad \text{for the satellite } s = \frac{n}{a}+\frac{1}{\Lambda}$$
$$\frac{f}{a} \quad \text{for the normal node} \quad s = \frac{n}{a}$$

The ratio of the intensities of the satellites to that of the normal node are thus respectively

$$\left(\frac{\Lambda\varepsilon s+\eta}{2}\right)^2 \quad \text{for the satellite } s = \frac{n}{a}-\frac{1}{\Lambda}$$
$$\left(\frac{\Lambda\varepsilon s-\eta}{2}\right)^2 \quad \text{for the satellite } s = \frac{n}{a}+\frac{1}{\Lambda} \tag{8.34}$$

*This asymmetry of the scattering patterns with respect to the nodes of the average lattice is characteristic of correlation between position and substitution disorder.* This can be interpreted physically as follows: pairs of the atoms with the larger scattering factor have more effect than pairs of the atoms with the smaller scattering factor. If the atoms which scatter more strongly are also larger, as in the preceding example ($\varepsilon$ and $\eta > 0$), they contribute to the maximum at small angles, which is thus the most intense. If, on the contrary, the smaller atoms have the larger scattering factors ($\varepsilon > 0$, $\eta < 0$), the more intense satellite is the one corresponding to the larger angle. There is no such asymmetry for the satellites which are close to the center of the reciprocal lattice, for then $s = 0$ in Eq. (*8.34*). We have already seen that displacement disorder has no effect in this region. *The scattering region near the center is similar to that obtained by neglecting the displacement disorder.* This is the second general conclusion which we can draw from this example.

### 8.3.3. Solid Solution with Atoms of Different Sizes

It is to be expected that a nonuniform distribution of the atoms will produce a lattice distortion, but such a distortion has not yet been calculated as a function of the properties of the atoms in the solution. On the other hand, it is generally difficult to obtain a complete and accurate description of the lattice disorder from the experimental data. We shall state the results of a few attempts which have been made at solving this important problem.

Warren, Averbach, and Roberts [33] have used certain gross simplifying assumptions for a binary solution $AB$. In the real lattice the vector $\boldsymbol{x}'_m$ separating two atoms is equal to $\boldsymbol{x}_m(1+\varepsilon)$, where $\boldsymbol{x}_m$ is the vector corresponding to the average perfect lattice, $\varepsilon$ having only the three different values $\varepsilon_{AA}$, $\varepsilon_{AB}$, and $\varepsilon_{BB}$, according to the nature of the atoms at its extremities. Since $\boldsymbol{x}_m$ is the average value of the vector for all pairs of indices $m$,

$$N_A\varepsilon_{AA} + N_{AB}\varepsilon_{AB} + N_B\varepsilon_{BB} = 0\,, \tag{8.35}$$

Then $N_A$, $N_{AB}$, $N_B$ are the proportions of the pairs of different types $AA$, $AB$ or $BA$, $BB$. Although this hypothesis is reasonable for close neighbors, it is of doubtful value for others, since the nature of the intermediate atoms necessarily comes into play. We also neglect changes in the directions of the vectors $\boldsymbol{x}_m$. The above authors take into account only the first neighbors and assume that there is no correlation between other pairs of atoms.

To calculate the scattering power,

$$I(\boldsymbol{s}) = \frac{1}{N} \sum_m \sum_n f_n f_{n+m} \cos 2\pi \boldsymbol{s} \cdot \boldsymbol{x}'_m ,$$

we require the average value of

$$f_n f_{n+m} \cos 2\pi \boldsymbol{s} \cdot \boldsymbol{x}'_m$$

for a given value of $m$. For a given type of pair such as $AA$, for example,

$$f_n f_{n+m} \cos 2\pi \boldsymbol{s} \cdot \boldsymbol{x}'_m = f_A^2 \cos 2\pi \boldsymbol{s} \cdot \boldsymbol{x}_m - f_A^2 2\pi \varepsilon_{AA} \boldsymbol{s} \cdot \boldsymbol{x}_m \sin 2\pi \boldsymbol{s} \cdot \boldsymbol{x}_m .$$

Thus, for the set of the $AA$, $AB$ and $BA$, $BB$ pairs, the average is:

$$\begin{aligned}\overline{f_n f_{n+m} \cos 2\pi \boldsymbol{s} \cdot \boldsymbol{x}'_m} &= \cos (2\pi \boldsymbol{s} \cdot \boldsymbol{x}_m)(N_A f_A^2 + N_B f_B^2 + N_{AB} f_A f_B) \\ &\quad -2\pi \boldsymbol{s} \cdot \boldsymbol{x}_m \sin (2\pi \boldsymbol{s} \cdot \boldsymbol{x}_m)(N_A f_A^2 \varepsilon_{AA} + N_B f_B^2 \varepsilon_{BB} + N_{AB} f_A f_B \varepsilon_{AB}) .\end{aligned}$$

This expression is calculated as a function of the order parameter $\alpha_m$ with the help of Eqs. (8.18), (8.19), and (8.35). Then we set

$$\beta_m = \frac{f_A}{f_A - f_B} \varepsilon_{AA} \left( \frac{c_A}{c_B} + \alpha_m \right) + \frac{f_B}{f_B - f_A} \varepsilon_{BB} \left( \frac{c_B}{c_A} + \alpha_m \right) . \tag{8.36}$$

Finally,

$$\begin{aligned}\overline{f_n f_{n+m} \cos (2\pi \boldsymbol{s} \cdot \boldsymbol{x}'_m)} &= (c_A f_A + c_B f_B)^2 \cos (2\pi \boldsymbol{s} \cdot \boldsymbol{x}_m) \\ &\quad + c_A c_B (f_A - f_B)^2 [\alpha_m \cos (2\pi \boldsymbol{s} \cdot \boldsymbol{x}_m) - \beta_m 2\pi \boldsymbol{s} \cdot \boldsymbol{x}_m \sin (2\pi \boldsymbol{s} \cdot \boldsymbol{x}_m)] .\end{aligned} \tag{8.37}$$

In this expression the first term is the contribution to the normal nodes, the second is due to substitution disorder [Eq. (8.21)], and the third is a corrective term arising from displacement disorder. We now perform the summation over all values of $m$ and find that the unit scattering power is given by

$$I(\boldsymbol{s}) = c_A c_B (f_A - f_B)^2 [\textstyle\sum_{-\infty}^{+\infty} \alpha_m \cos (2\pi \boldsymbol{s} \cdot \boldsymbol{x}_m) - \sum_{-\infty}^{+\infty} 2\pi \beta_m \boldsymbol{s} \cdot \boldsymbol{x}_m \sin (2\pi \boldsymbol{s} \cdot \boldsymbol{x}_m)]. \tag{8.38}$$

The second term is zero at the center and increases with $s$. Moreover, at the superlattice nodes, $\boldsymbol{s} \cdot \boldsymbol{x}_m = (2k + 1)/2$, so this term is zero and changes sign at these nodes. But this is just where the first term is a maximum in the state of partial order. Displacement disorder therefore displaces the maximum away from its normal position and renders the scattering regions asymmetric with respect to the superlattice node. And for pure substitution disorder, the term $\sum \alpha_m \cos 2\pi \boldsymbol{s} \cdot \boldsymbol{x}_m$ is periodic, since $\alpha_m$ is independent of $\boldsymbol{s}$. On the other hand, displacement disorder gives rise to scattering which varies from one homologous point to another in neighboring unit cells, since the coefficient of $\sin 2\pi \boldsymbol{s} \cdot \boldsymbol{x}_m$ is a function of $\boldsymbol{s}$. The effect of displacement disorder is zero at the center and increases with $s$, as already seen in Section 8.3.2.

In the case of powder patterns the scattered intensity due to substitution disorder is given by Eq. (*8.24*):

$$I(s) = c_A c_B (f_A - f_B)^2 \sum c_i \alpha_i \frac{\sin(2\pi s r_i)}{2\pi s r_i} .$$

If we now introduce displacement disorder, the correction term is proportional to the derivative of $\sin(2\pi s r_i)/2\pi s r_i$ with respect to $r_i$, just as in Eq. (*8.38*) the second term is proportional to the derivative of the first with respect to $\boldsymbol{x}$. The complete calculation [33] shows that we again have the same coefficient $\beta_m$:

$$I(s) = c_A c_B (f_A - f_B)^2 \left\{ \sum \alpha_i c_i \frac{\sin(2\pi s r_i)}{2\pi s r_i} - \sum \beta_i c_i \left[ \frac{\sin(2\pi s r_i)}{2\pi s r_i} - \cos(2\pi s r_i) \right] \right\} . \tag{8.39}$$

Warren and his collaborators [33] have illustrated the use of these equations in the case of scattering by a single crystal of $AuCu_3$ along the [100] axis and by a powder of $Ni_3Au_2$. This is illustrated in Figure 8.13. For the first neighbors, neglecting the higher-order terms, it is possible to account for the general shape of the experimental curve by choosing an empirical value for $\beta$, but the agreement is not quantitative [34].

Much work has been done during the past several years on the detailed investigation of the structure of distortions in solid solutions. The general theories as described in this book, or as expressed under a different form [35], have been applied to particular cases, such as the elastic distortions which are assumed to exist in the vicinity of every substituted atom [36], [37]. In analyzing the scattering around a node, Borie [38] showed the existence of the asymmetric term which is characteristic of

**FIGURE 8.13.** *Scattering intensity in a powder diagram for a $Ni_3Au_2$ alloy quenched from 850°C, Co Kα radiation [34]. The solid curve is experimental while the dashed curve was calculated from Eq. (8.39) with $\beta = -0.005$ and $\alpha = 0.05$.*

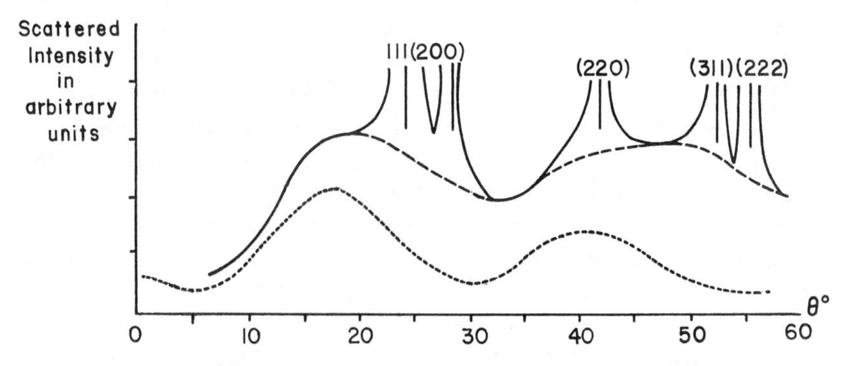

the phenomenon, as we have already indicated. The theoretical equations have been applied to a set of solid solutions, one of the objectives being to determine simultaneously both the degree of order and the degree of distortion resulting from the size effect. The experimental methods are well described in the literature [39], [40], [41].

The next section illustrates by means of a simple example the relations which exist between diffuse scattering and the two types of disorder.

Another approach to the problem is to measure the decrease in intensity of the diffraction lines. The Debye factor [Eq. (*7.7*)] for a solid solution comprises a term which is a function of the temperature as well as a term which depends on the displacement disorder and which arises from the size effect of the substituted atoms. If the intensities of a set of lines are measured at two different temperatures, it is possible to separate the two terms and to evaluate an average distance between a substituted atom and the corresponding node of the average perfect lattice [42], [43], [44].

8.3.3.1. LINEAR EQUIATOMIC ALLOY WITH ATOMS A AND B OF DIFFERENT SIZES. Let us consider a linear close-packed lattice containing equal numbers of atoms *A* and *B*. If the atoms are all of the same size, we have the linear lattice of Section 8.2.4. If they are of different sizes, we have displacement disorder. The scattering pattern for this highly simplified lattice can be calculated rigorously and will lead to results which are of interest for real alloys.

**FIGURE 8.14.** *Linear lattice containing equal numbers of A and B atoms.* (a) *Regular lattice with atomic distortions.* (b) *Close-packed arrangement.*

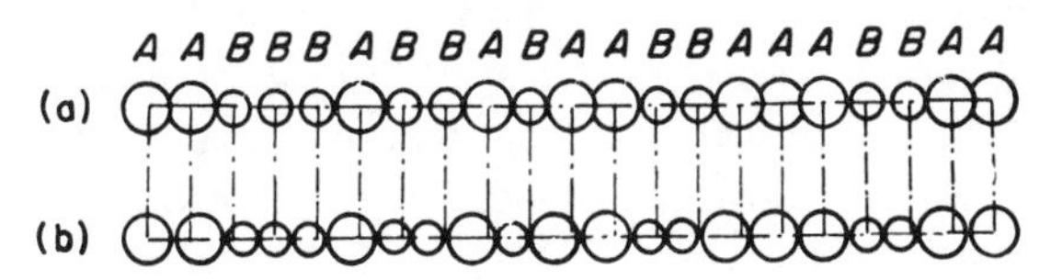


The linear lattice is defined by its average parameter $a = r_A + r_B$, where $r_A$ and $r_B$ are the radii of the two atoms and $a$ is the parameter which follows the Végard law. Such a linear lattice is illustrated in Figure 8.14. As in Section 8.2.4, the order is defined by the probability $p$ that a pair of neighbors is different. The position disorder depends on $\delta = (r_A - r_B)/(r_A + r_B)$. Since the scattering power must be calculated from Eq. (*6.10*), we must find $y_m = \overline{F_n F^*_{n+m}}$ as a function of $m$. The calculation is simple in the two extreme cases corresponding to complete disorder or to near-perfect order with sufficiently narrow superlattice lines.

For intermediate cases the numerical calculations can be performed with a fair approximation.

The calculations can be found in the literature [45]; we shall limit ourselves to the results.

*Perfect Disorder* ($p = 1/2$). The intensity is expressed as a function of $l$, which is the distance along the axis of reciprocal space (one-dimensional, like the object), the distance between two neighboring nodes of the average lattice being chosen as unity; an analogous notation was used in Section 7.2.1. Then

$$I(l) = \frac{\left\{\left(\frac{f_A + f_B}{2}\right)^2 4\sin^2(\pi l\delta)\cos^2(\pi l) + (f_A - f_B)^2\cos^2(\pi l\delta)\sin^2(\pi l) - \frac{f_A^2 - f_B^2}{2}\sin(2\pi l\delta)\sin(2\pi l)\right\}}{1 - 2[\cos(2\pi l\delta)][\cos(2\pi l)] + \cos^2(2\pi l\delta)} . \tag{8.40}$$

Let us discuss the meaning of the three terms in the numerator. If the scattering factors $f_A$ and $f_B$ of the two atoms are equal, we are left only with the first term. In this case we have only position disorder. (We may note in passing that this term was neglected in Section 8.3.3 because, according to Eq. (*8.39*), scattering disappears completely if $f_A = f_B$.) This first term gives a scattering curve which is similar to that discussed in the following chapter (Section 9.1). It involves a series of symmetrical maxima for integral values of $l$, these maxima being centered on the nodes of the average lattice: corresponding to the node of index $l_0$, we have a spot whose integral width $\pi^2 l_0^2 \delta^2$ increases rapidly with $l_0$, so that its maximum intensity decreases rapidly. This is characteristic of a crystal in which long-range order disappears gradually.

The second term corresponds to a weakly modulated continuous background. It is the term which, in the absence of position disorder ($\delta = 0$), tends to the uniform scattering of the Laue Eq. (*8.17*), $(f_A - f_B)^2/4$, for the linear disordered lattice.

As to the third term in the above numerator, it is negligible except for values of $l_0$ which are close to integers, but it becomes zero and changes sign at the nodes. When added to the first term, it produces an asymmetry and also a displacement of the maximum which is given approximately by

$$\Delta l = -\pi^2 l_0^3 \delta^3 \frac{f_A^2 - f_B^2}{(f_A + f_B)^2} .$$

The direction of this displacement depends on the sign of the product $\delta(f_A - f_B)$. The line is therefore displaced toward small angles if the

**FIGURE 8.15.** *Scattering power for an equiatomic, linear perfectly disordered alloy AB where,*

$$\frac{f_A}{f_B} = 2.72\,, \qquad \frac{r_A - r_B}{r_A + r_B} = 0.06\,.$$

*The dashed curve shows the Laue scattering which is observed when there is no displacement disorder. Curve b is similar to curve a, but drawn on a reduced scale.*

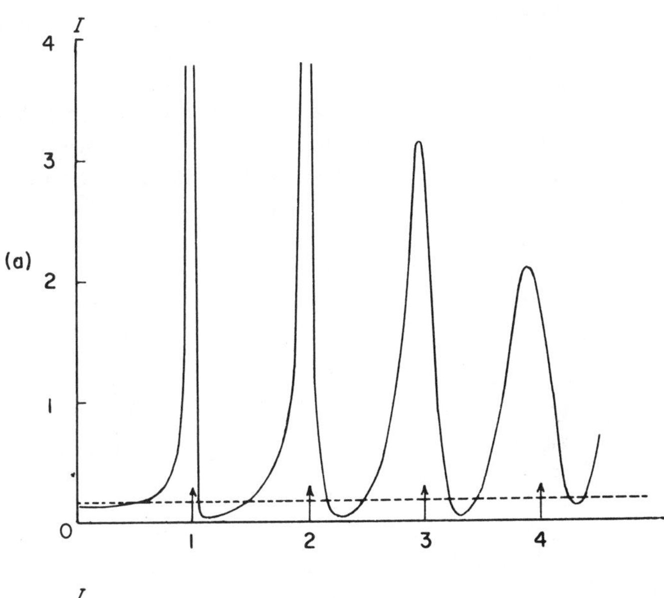


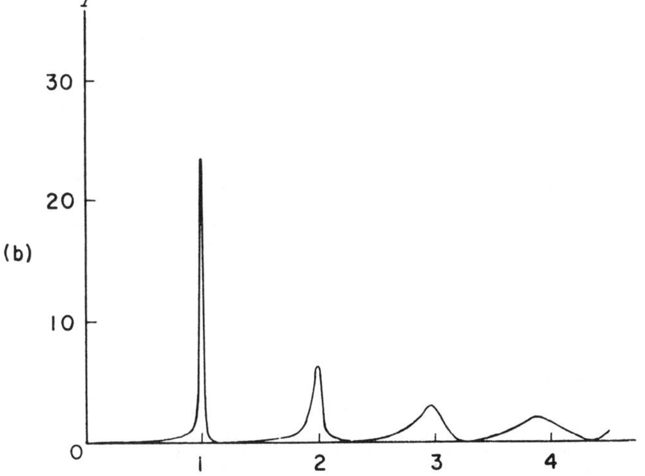

atoms which scatter the most are also the largest. This is the rule which we have already given in Section 8.3.2.

Figure 8.15 shows the total curve calculated by assuming that the atom $A$ is gold and that $B$ is copper: $f_A/f_B = 79/29$ and $\delta = (r_A - r_B)/(r_A + r_B) = 0.06$. Aside from the lines which gradually become broader and weaker, we also have a *continuous background which increases from one node to the next*. This would vary in the opposite direction if the atoms having the larger scattering factor were smaller. At the center, scattering is approximately given by the Laue formula, which neglects displacement disorder.

**FIGURE 8.16.** *Scattering power for a partially ordered ($p = 0.75$) linear alloy AB, calculated as in the preceding figure. The dashed curve shows the scattering for the same degree of order, but without displacement disorder. Curve b is curve a drawn on a reduced scale. The units for I are the same as in the preceding figure.*

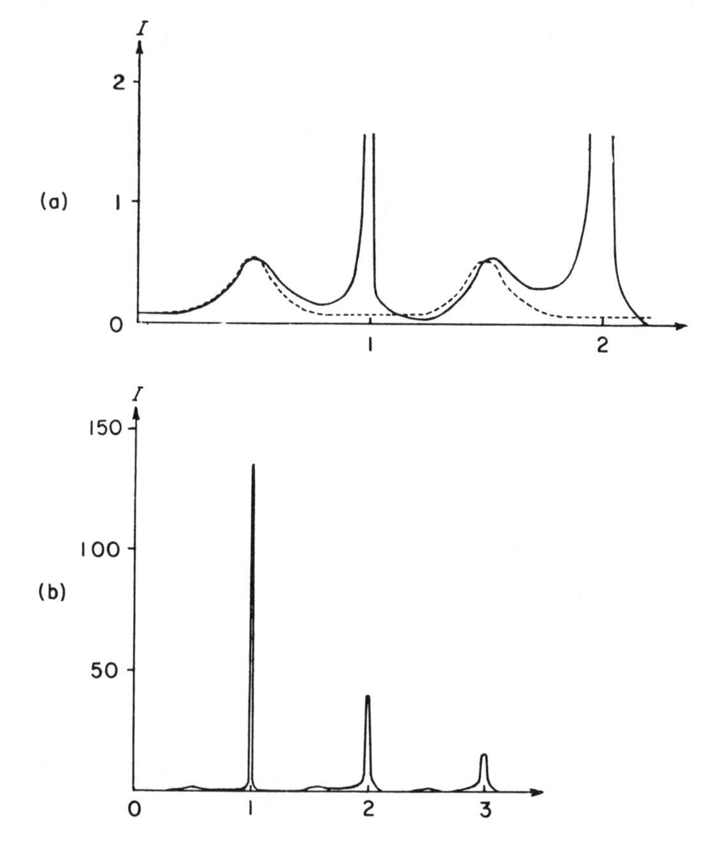

*Near-perfect order.* In this case we have very few similar pairs, so that $1 - p$ is very small. As a first approximation, the superlattice lines are not modified by the disorder and their widths depend only on the degree of order, as in Section 8.2.4. The normal lines are broad, but much less so than in the case of perfect disorder, since their width is multiplied by the very small factor $(1 - p)$. Again as a first approximation, they are symmetrical and not displaced.

*Partial Order.* Figure 8.16 shows the curve calculated for the same atoms as above and for $p = 0.75$. It will be seen that the normal lines are broad, but less so than in the case of perfect disorder, and that the maxima near the superlattice nodes become more and more asymmetrical at points remote from the center of reciprocal space. *In the first unit cell, however, there is little change.*

Thus, for the first unit cells, and especially for high degrees of order, the continuous background is only slightly affected by displacement disorder. We can therefore assume *that it depends only on the degree of order, according to the theory of Section 8.2.2.* Displacement disorder affects the scattering near the normal lines, especially if the solution is disordered. In such a case, however, while the linear model predicts lines which broaden rapidly for increasing node indices, real solid solutions give narrow lines up to large angles. It is therefore certain that position disorder tends to be less important in a three dimensional lattice than in a linear one. This result is easily explained by the fact that in three dimensions a given atom may have neighbors of different types, so that size effects tend to neutralize. In a linear lattice the fluctuations are larger, since only one neighbor enters into consideration.

It is therefore possible to determine the degree of order with a fair accuracy from scattering measurements when the atoms of the solid solution have different volumes, so long as the unit cells which are *closest to the center of the reciprocal space* are the only ones which are used. The local order produces scattering between the nodes, but the scattering in the neighborhood of the normal nodes cannot be explained without displacements of the atoms from their normal positions in the perfect lattice [46].

8.3.3.2. THE STRUCTURE OF THE AGE-HARDENED Al Cu ALLOY. A solid Al Cu solution containing 4% of copper, homogenized at 425°C and then quenched, is supersaturated at room temperature. After aging for a few hours at temperatures varying between 20-100°C, a single crystal gives a pattern as in Figure 8.17 [47]. The diffraction spots correspond closely to those of pure aluminum. The scattering regions are localized

**FIGURE 8.17.** *Scattering pattern for an Al Cu single-crystal during age hardening. Monochromatic Mo Kα radiation. The [001] axis was vertical and the crystal oscillated from − 2° to 20° with the axis [100] normal to the beam at zero degrees. The spots A, A′, and B correspond to the intersection with the Ewald sphere of scattering streaks which are nearly normal to its surface. The indices for the matrix spots are shown (R. Graf).*

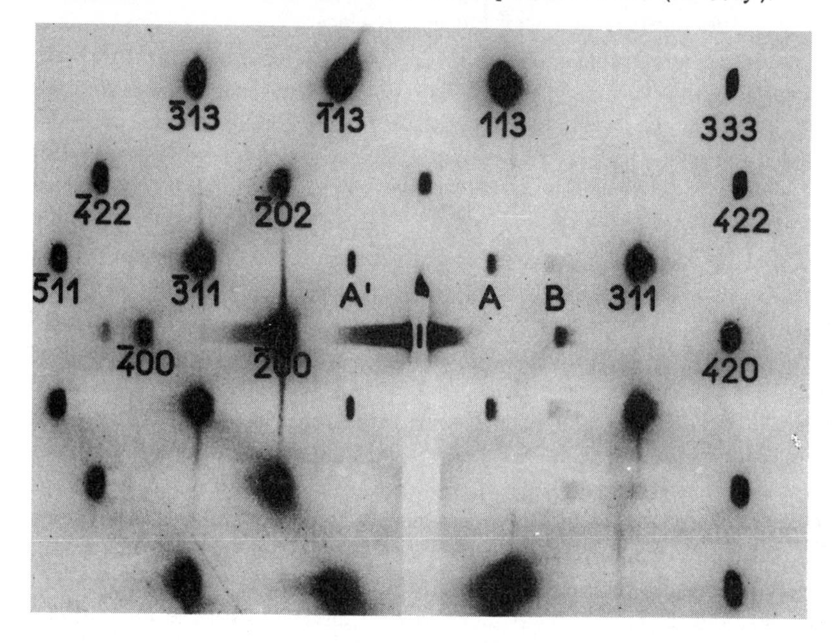


along the ⟨100⟩ rows in the form of streaks; the more intense ones come from the center, while others originate at normal nodes. These streaks are asymmetrical and are always directed in the direction of large angles. They are represented schematically in Figure 8.18.

Let us examine the scattering pattern in the light of the theoretical results of the preceding chapters. First of all, the distribution of the scattering indicates that *the faults occur in the {100} planes.* Since scattering occurs up to the immediate neighborhood of the center, we must have substitution disorder, and since the scattering is maximum at the nodes, including the center of reciprocal space, *the copper atoms have a tendency to segregate* (Section 8.2.4). This could have been expected, since the equilibrium state of the alloy is a mixture of a solid solution containing a very small amount of copper (the matrix) and a precipitate which is rich in copper, $Al_2Cu$. Finally, the scattering regions are

**FIGURE 8.18.** *Reciprocal space of the Al Cu crystal corresponding to the pattern of the previous figure. The widths of the scattering regions are proportional to their intensity. The streaks D are normal to the centers of the faces of the cube.*

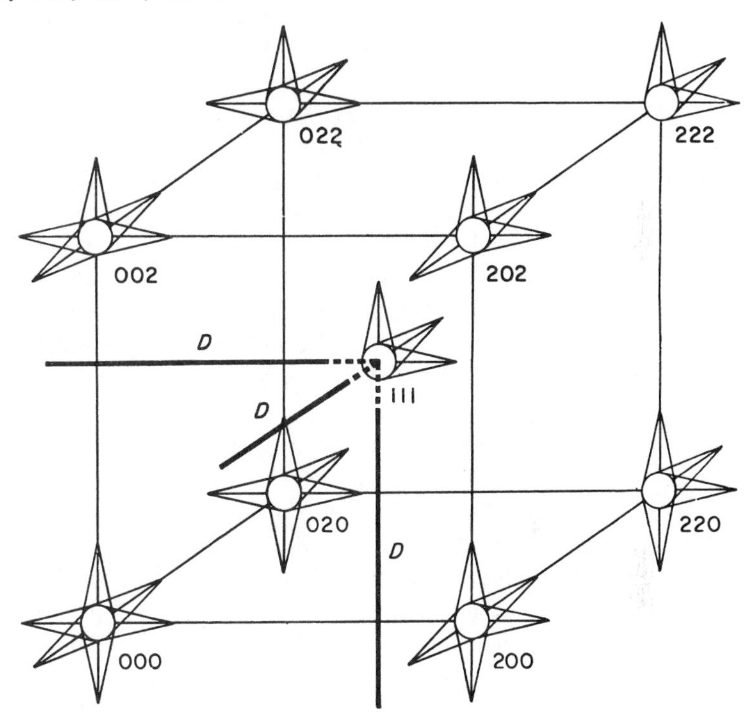


asymmetrical around the nodes, so the substitution disorder is related to a position disorder. *This means that the regions where the copper atoms are assembled must be distorted.* Since the copper atoms have the larger scattering power and are also smaller, the asymmetry is in the direction of the larger angles.

According to the simplest model, the copper atoms are grouped together in the form of small plates on the {100} planes, the copper atoms occupying the positions of aluminum atoms. Such plates have been called Guinier-Preston zones, or G.P. zones, because they were first proposed independently by Guinier and by Preston [48].

If the G.P. zones have large areas, the scattering must be concentrated along the ⟨100⟩ axes and if, on the contrary, they have an area which is not large compared to the unit cell of the plane lattice, the cross-section of the "column" of scattering in the reciprocal lattice must be broad. It is observed in fact that the width of the central streaks depends both on the duration and on the temperature of the aging, and that the average diameter of the zone can be estimated from this width. This diameter is found to be approximately 50 to 100Å.

The simplest hypothesis on the structure of such a zone is that it is composed of a monoatomic layer of copper. Such zones are distributed at random in the matrix and the scattering pattern is that of three plane lattices which are identical to the three {100} planes; it is constituted by the ⟨100⟩ rows of the reciprocal lattice. The intensity along these rows is proportional to the square of the factor $(f_{\mathrm{Cu}} - f_{\mathrm{Al}})$ since the scattering arises from the difference between the scattering powers of the two atoms. One could therefore expect the intensity to decrease regularly with distance from the center of the reciprocal lattice, but that is not what is observed. Other models must therefore be tried. If we now consider copper zones which are several atomic layers thick, the scattering regions concentrate near the nodes of the matrix. These are the domains of reflection corresponding to small crystals having the shape of thin plates. They are *identical* around all the nodes of the matrix, since the lattice within a zone is assumed to be identical to that of the matrix. The length of the central streak should give an approximate value of the thickness of these zones. The length of the scattering regions at half intensity is roughly of the order of $1/na$, where $n$ is the average number of planes in the zone and $a$ is the spacing between (100) planes. The value of $n$ should be different according to the aging treatment, but at the beginning of the age hardening it is very small (2 or 3).

This second model cannot be accepted because, contrary to its prediction, the domains of reflection around the diffraction nodes are *asymmetrical*. It must therefore be assumed that the atoms are not exactly situated at the lattice points of the matrix, the distortions arising from the difference in size between the copper and the aluminum atoms.

Finally, the order of magnitude of the absolute value of the intensity of the central streak at very small angles indicates that a large fraction of the copper atoms in the alloy are segregated.

In order to arrive at a detailed structure of a G.P. zone, the scattered intensity has been calculated for a model in which the unknowns are the copper content and the displacement of a series of (100) planes in an aluminum crystal. A satisfactory model has been deduced from accurate measurements of the scattering along the [100] axis of the reciprocal lattice [48], [49], [50].

## References


1. G. V. Raynor, *Progress in Metal Physics*, Vol. 1, Pergamon Press, London (1949), p. 1.
   W. B. Pearson, *Handbook of Lattice Spacings and Structure of Metals and Alloys*, Pergamon Press, London (1958).
   J. Friedel, *Phil. Mag.*, **46** (1955), 514.
2. H. Lipson, *Progress in Metal Physics*, Vol. 2, Butterworths, London (1950), p. 1.
3. F. W. Jones and C. Sykes, *Proc. Roy. Soc. London, A*, **161** (1937), 440.
4. F. E. Howarth, *Phys. Rev.*, **54** (1938), 693.
   P. Leech and C. Sykes, *Phil. Mag.*, **27** (1939), 742.
5. J. M. Cowley, *Acta Cryst.*, **13** (1960), 1059.
6. O. Midvani and T. Gachechiladze, *Phys. Met. Metallog.*, **8** No. 3 (1959), p. 1.
   A. N. Men, *Phys. Met. Metallog.*, **9** No. 6 (1960), p. 1.
7. Z. W. Wilchinsky, *J. Appl. Phys.*, **15** (1944), 806.
   D. Chipman and B. E. Warren, *J. Appl. Phys.*, **21** (1950), 696.
   J. M. Cowley, *J. Appl. Phys.*, **21** (1950), 24.
8. D. T. Keating and B. E. Warren, *J. Appl. Phys.*, **22** (1951), 286.
9. A. J. Bradley and A. H. Jay, *Proc. Roy. Soc. London, A*, **136** (1932), 210.
10. F. M. Rhines and J. B. Newkirk, *Trans. A.S.M.*, **45** (1953), 1029.
11. V. Daniel and H. Lipson, *Proc. Roy. Soc. London, A*, **181** (1943), 368.
12. M. Hillert, *Acta Met.*, **9** (1961), 525.
13. J. M. Cowley, *J. Appl. Phys.*, **21** (1950), 24.
    N. Norman and B. E. Warren, *J. Appl. Phys.*, **22** (1951), 483.
14. G. Fournet, *J. Phys. Rad.*, **13** (1952), 14A.
15. S. B. Hendricks and E. Teller, *J. Chem. Phys.*, **10** (1942), 147.
16. A. Guinier and R. Griffoul, *Acta Cryst.*, **1** (1948), 188.
17. F. W. Jones and G. Sykes, *Proc. Roy. Soc. London, A*, **166** (1938), 376.
18. A. Guinier and R. Griffoul, *Rev. Met.*, **45** (1948), 387.
19. H. A. Bethe, *Proc. Roy. Soc. London, A*, **150** (1935), 552.
20. C. H. Johannson and J. O. Linde, *Ann Phys.*, **5** (1936), 1.
21. P. Perio and M. Tournarie, *Acta Cryst.*, **12** (1959), 1032 and 1044.
22. D. W. Pashley and A. E. B. Presland, *J. Inst. Met.*, **87** (1958-1959), 419.
23. A. J. C. Wilson, *X-Ray Optics*, Methuen, London (1949), p. 97.
24. A. Guinier and C. B. Walker, *Acta Met.*, **1** (1953), 568.
    B. Belbeoch and A. Guinier, *Acta Met.*, **3** (1955), 370.
25. R. R. Zakharova and N. N. Buinov, *Phys. Met. Metallog.*, **8** (1959) No. 5, 62 and 142.
    P. M. Lerinman and N. N. Buinov, *Phys. Met. Metallog.*, **5** (1957) No. 2, 78.
26. M. B. Webb, *Acta Met.*, **7** (1959), 748.
27. E. J., Freise, A. Kelly and R. B. Nicholson, *Acta Met.* **9** (1961), 250.
28. V. Gerold, *Z. Metalkde*, **46** (1955), 623; *Phys. Status Solidi*, **1** (1961), 37.
29. V. Gerold, *Ergeb. Exact. Naturwiss.*, (1961) Bd33, 105.
30. B. E. Warren, *Acta Cryst.*, **6** (1953), 803.
31. C. B. Walker and D. T. Keating, *Acta Cryst.*, **14** (1961), 1170.
32. G. D. Preston, *Proc. Roy. Soc. London, A*, **167** (1938), 526.
33. B. E. Warren, B. L. Averbach, and B. W. Roberts, *J. Appl. Phys.*, **22** (1951), 1493.

34. P. A. FLINN, B. L. AVERBACH, and P. S. RUDMA, *Acta Cryst.*, **7** (1954), 153.
35. M. A. KRIVOGLAZ, *Sov. Phys. Cryst.*, **4**, No. 6 (1960), 775; **5**, No. 1 (1960), 18; *Phys. Met. Metallog.*, **7**, No. 5 (1959), 11; **8**, No. 4 (1959), 37.
36. W. COCHRAN, *Acta Cryst.*, **9** (1956), 259.
W. COCHRAN and G. KARTHA, *Acta Cryst.*, **9** (1956), 941 and 944.
37. M. A. KRIVOGLAZ, *Phys. Met. Metallog.*, **9**, No. 5 (1960), 1.
38. B. BORIE, *Acta Cryst.*, **10** (1957), 89; **12** (1959), 280.
39. F. H. HERBSTEIN, B. S. BORIE, and B. L. AVERBACH, *Acta Cryst.*, **9** (1956), 466.
C. R. HOUSKA and B. L. AVERBACH, *J. Appl. Phys.*, **30** (1959), 1532.
P. S. RUDMAN and B. L. AVERBACH, *Acta Met.*, **5** (1957), 65.
40. V. HAUK and C. HUMMEL, *Z. Met.*, **47** (1956), 254.
41. A. A. SMIRNOV and E. A. TIKHONOVA, *Sov. Phys. Solid State*, **1** (1960), 1277.
42. B. BORIE, *Acta Cryst.*, **10** (1957), 89.
43. A. GUINIER, *Solid State Physics*, Vol. IX, Academic Press, New York (1958), p. 319.
44. G. V. KURDJUMOV, V. A. ILINA, V. K. KRITSKAYA, and L. I. LYSAK, *Izvest. Akad. Nauk S.S.S.R., Ser. Fiz.*, **17** (1953), 297.
V. I. IVERONOVA and A. P. ZVJAGINA, *Izvest. Akad. Nauk S.S.S.R., Ser. Fiz.*, **20** (1956), 729.
45. A. GUINIER, *Bull. Soc. Franç. Miner. Cryst*, **77** (1954), 680.
46. J. B. NEWKIRK, R. SMOLUCHOWSKI, A. H. GEISLER, and D. L. MARTIN, *Acta Cryst.*, **4** (1951), 507.
47. H. K. HARDY and T. J. HEAL, *Progress in Metal Physics*, Vol. 5, Pergamon Press, London (1954), p. 143.
48. A. GUINIER, *Solid State Physics*, Vol. IX, Academic Press, New York (1958), p. 336.
49. K. TOMAN, *Acta Cryst.*, **8** (1955), 587; **10** (1957), 187.
V. GEROLD, *Acta Cryst.*, **11** (1958), 230.
K. DOI, *Acta Cryst.*, **13** (1960), 45 and 60.
50. Y. A. BAGARYASTSKII, *Sov. Phys. Cryst.*, **4**, No. 3 (1960), 315.

CHAPTER

# 9

# CRYSTAL IMPERFECTIONS DESTROYING LONG-RANGE ORDER

Now that we have completed our study of the first type of imperfection, let us go on to the second type, which was described in Section 6.1 and which is characterized by the fact that the position of an atom with respect to its immediate neighbors fluctuates around an average pattern.

We shall now consider the distribution of the nodes in a distorted lattice. We shall assume that the unit cells are all identical, but it will be easy to generalize the results so that they can be applied to the case where the unit cells fluctuate either in structure or in composition around an average value, provided that these fluctuations bear no relation to the positions of the unit cells. Equation (*2.57*), which we found again as Eq. (*8.29*), shows that every unit cell can be replaced by an average cell if we add the scattering

$$I = \overline{|F|^2} - |\bar{F}|^2,$$

where $\overline{|F|^2}$ and $|\bar{F}|$ are respectively the average of the square of the modulus and the modulus of the average structure factor for the real unit cells. If the nature of the unit cell is related to its position, the general Eqs. (*6.10*) and (*6.3*) still apply, but we have already noted that they are difficult to use.

We shall assume that the object is statistically homogeneous so that the statistics of the distribution of the atoms or atomic groups remain the same, whatever be the point chosen as origin. We can therefore apply directly the general results of Section 2.4, which are valid when there is no long-range order. Let us recall these results briefly. We call $z(\boldsymbol{x})\,dv_x$ the probability of finding a point representing an atom or an atomic group within the volume $dv_x$ at the extremity of the vector $\boldsymbol{x}$, one atom being at the origin of coordinates. The distribution $z(\boldsymbol{x})$ involves a delta function at the origin, $\delta(0)$, and tends to a constant value for large values of $\boldsymbol{x}$. If $v_1$ is the average volume available to each atom or the average volume of the unit cell of the perturbed lattice, $1/v_1$ is the asymptotic value of $z(\boldsymbol{x})$ as well as its average value in all space:

$$\iint\left[z(\boldsymbol{x}) - \frac{1}{v_1}\right] dv_x = 0 .$$

Let us call $Z(\boldsymbol{s})$ the Fourier transform of $z(\boldsymbol{x})$ in reciprocal space. It comprises a delta function at the origin, $(1/v_1)\delta(0)$, and tends to unity for large values of $\boldsymbol{s}$.

From Eq. (*2.35 a*) the scattering power per unit cell is

$$I(\boldsymbol{s}) = \frac{F^2}{V} Z(\boldsymbol{s}) * |\Sigma(\boldsymbol{s})|^2 , \tag{9.1}$$

where $V$ is the volume of the object, $\Sigma(\boldsymbol{s})$ is the transform of its form factor (Section 2.4.2), and $F$ is the structure factor for the contents of the average unit cell.

At the center of the pattern for the perturbed crystal there is a diffraction region corresponding to the central peak of the function $Z(\boldsymbol{s})$ (Section 2.4.2.1), just as at the center of the pattern for any homogeneous substance, whatever its structure,

$$I(\boldsymbol{s})_{\text{center}} = \frac{F^2}{Vv_1} |\Sigma(\boldsymbol{s})|^2 .$$

This region surrounding the center of reciprocal space depends only on the *external shape of the sample* through $\Sigma(\boldsymbol{s})$, and on the *average density* through $F$, which is the structure factor of the unit cell for very small values of $\boldsymbol{s}$.

For very large values of $\boldsymbol{s}$, i.e. for large diffraction angles, the scattering is uniform and equal to $F^2$. In other words, it is the same as if there were no order whatsoever, even at short distances. This is called gas scattering.

Our problem is now to investigate the intermediate regions of reciprocal space by determining $z(\boldsymbol{x})$ as a function of the parameters describing the imperfect crystal. We shall start with a simple one-dimensional problem which will lead to some important applications.

## 9.1. The One-dimensional Distorted Lattice [1]

In this case the points are distributed along a straight line, as illustrated in Figure 9.1. The distance between two successive points varies about an average value $a$, and the distance $A_{n-1}$ to $A_n$ is not related to the next segment, $A_n$ to $A_{n+1}$. That is, the distances between close

**FIGURE 9.1.** (a) *Irregular one-dimensional lattice.* (b) *Distribution function $h(x)$ of the distances between close neighbors.*

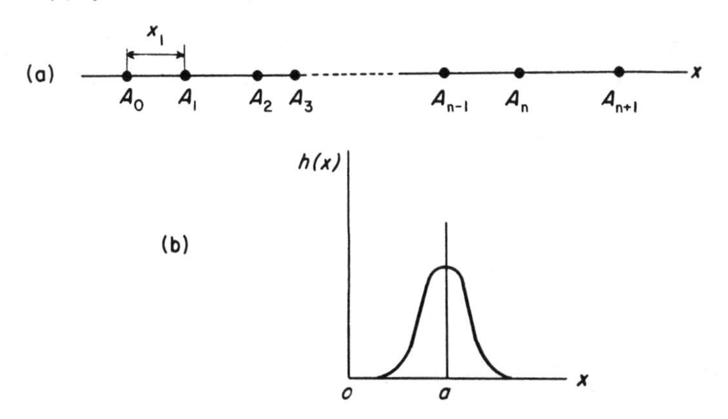


neighbors are independent, and they all follow the same statistical distribution $h(x)$, where $h(x)\,dx$ is the probability that the distance $x_1$ between two close neighbors lies between $x$ and $x + dx$. The quantity $h(x)$ is normalized,

$$\int_0^\infty h(x)\,dx = 1\,, \tag{9.2}$$

and its average value is

$$a = \int_0^\infty x\,h(x)\,dx\,. \tag{9.3}$$

Let us calculate the distribution function $h_2(x)$ for the distance $x_2$ between second neighbors, $x_2$ being the sum of two distance between close neighbors $y$ and $x - y$. The probability for the occurrence of both of these events is the product of the two *individual* probabilities of having a distance $x_1$ equal to $y$ and a distance $x_1$ equal to $x - y$, or $h(y)$ and $h(x - y)$. Since $y$ can have any value, the total probability is

$$h_2(x) = \int_0^\infty h(y)h(x-y)\,dy = h(x) * h(x)\,,$$

which is the faltung of $h(x)$ by itself. Similarly,

$$h_3(x) = h_2(x) * h(x)$$

and, in general,

$$h_n(x) = \underbrace{h(x) * h(x) * h(x) * \cdots * h(x)}_{n \text{ times}} . \tag{9.4}$$

All the functions $h_n(x)$ are normalized:

$$\int_0^\infty h_n(x)\, dx = 1 .$$

The average value of $x_n$ is $na$. Let us show this for the case of second neighbors. We have

$$\int_0^\infty h_2(x)\, dx = \int_0^\infty \int_0^\infty h(y)\, h(x-y)\, dx\, dy$$

and

$$\int_0^\infty x\, h_2(x)\, dx = \int_0^\infty \int_0^\infty x\, h(y)\, h(x-y)\, dx\, dy .$$

Setting $x - y = X$ and $y = Y$, we find that

$$\int_0^\infty h_2(x)\, dx = \int_0^\infty \int_0^\infty h(Y)\, h(X)\, dX dY = \int_0^\infty h(X)\, dX \int_0^\infty h(Y)\, dY = 1$$

and

$$\int_0^\infty x h_2(x)\, dx = \int_0^\infty \int_0^\infty (X + Y)\, h(X)\, h(Y)\, dX\, dY$$

$$= \int_0^\infty \int_0^\infty X h(X)\, h(Y)\, dX dY + \int_0^\infty \int_0^\infty Y h(X)\, h(Y)\, dX dY = 2a .$$

The points are more and more scattered around the average nodes situated at $a$, $2a$, $3a$, $\cdots$. If, for example, the first neighbor can vary by $\pm\,\varepsilon$, the second neighbor by $\pm\, 2\varepsilon$, etc., then at large distances the $n$th region is so diffuse that it overlaps its neighbors $n - 1$ and $n + 1$ and the distribution becomes uniform. On the other hand, $h_{-n}(x) = h_n(-x)$ and the overall distribution is

$$z(x) = \delta(x) + \Sigma_1^\infty h_n(x) + \Sigma_1^\infty h_n(-x) . \tag{9.5}$$

Figure 9.2 shows the general shape of a distribution of points which is characteristic of this type of imperfection: the order gradually decreases and disappears at large distances.

This particular form of $z(x)$ is interesting because its Fourier transform is easily calculated: it is a function of a single variable $s$, which represents the coordinate along the one-dimensional reciprocal space. It is also

**FIGURE 9.2.** *Schematic representation of the distribution of points on the linear lattice of Figure 9-1.*

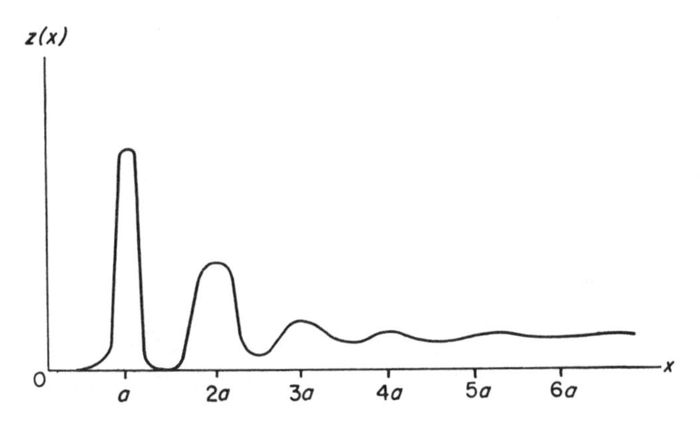


related to an imaginary diffraction experiment on a purely one-dimensional object: $s$ is the projection on the object of the diffraction vector $(\boldsymbol{S} - \boldsymbol{S}_0)/\lambda$ .

The transform of $h(x)$ is as follows:

$$H(s) = \int h(x) \exp(2\pi isx)dx \tag{9.6}$$

and, from the faltung theorem,

$$\text{transf } h_n(x) = \text{transf } [\underbrace{h(x) * \cdots * h(x)}_{n \text{ times}}] = [H(s)]^n . \tag{9.7}$$

But, since $h_{-n}(x) = h_n(-x)$,

$$\text{transf } h_{-n}(x) = [H^*(s)]^n . \tag{9.8}$$

Using Eqs. (*9.6*) and (*9.7*) and also the relation transf $\delta(x) = 1$, the transform of $z(x)$ [Eq. (*9.5*)] is shown to be equal to the series

$$\text{transf } z(x) = Z(s) = 1 + H(s) + H^2(s) + \cdots + H^*(s) + [H^*(s)]^2 + \cdots \tag{9.9}$$

or, in terms of the modulus $\rho$ and of the argument $2\pi u$ of $H(s)$ ,

$$Z(s) = 1 + 2 \textstyle\sum_1^\infty \rho^m \cos(2\pi mu) . \tag{9.10}$$

From Appendix B.2,

$$Z(s) = \frac{1 - \rho^2}{1 + \rho^2 - 2\rho \cos(2\pi a)} . \tag{9.11}$$

If $h(x)$ is negligible everywhere except within a small interval around the average point $x = a$. we can find a simple approximate formula for $H(s)$ by developing the exponential function of Eq. 9.6:

$$H(s) = \exp(2\pi isa)\int h(x)\exp[2\pi is(x-a)]\,dx$$

$$= \exp(2\pi isa)\left[\int h(x)\,dx + 2\pi is\int (x-a)h(x)\,dx - 2\pi^2 s^2\int h(x)(x-a)^2 dx + \cdots\right].$$

Let us set

$$\int h(x)(x-a)^2\,dx = \Delta^2 .$$

Then, using Eqs. (*9.2*) and (*9.3*) and neglecting the higher order terms, we have

$$H(s) \cong \exp(2\pi isa)(1 - 2\pi^2 s^2 \Delta^2) \cong \exp(-2\pi^2 s^2 \Delta^2)\exp(2\pi isa)\,. \qquad (9.12)$$

With this approximation, $\rho$ and $u$ of Eq. (*9.11*) are given by

$$\rho = \exp(-2\pi^2 s^2 \Delta^2)\,,$$
$$u = sa\,,$$

where $\Delta$ is a coefficient which is characteristic of the width of the distribution $h(x)$ or of the degree of order in the lattice. Equation (*9.12*) is exact for any width of the distribution $h(x)$, so long as the distribution is Gaussian. Hosemann [2] has shown that it is an excellent approximation in many other cases. The perturbed lattice is defined by only two parameters, the average period $a$ and an order parameter characterized by $\Delta$.

The function on the right side of Eq. (*9.11*) has a series of maxima which correspond approximately to the minima of the denominator, and therefore to integral values of $u$ or to the nodes of the average lattice $s = (n/a)$. This comes from the fact that the numerator varies slowly with $s$. The minima occur at intermediate values $u = (2n+1)/2$. The maxima and minima of $Z(s)$ are respectively

$$Z_{\max} = \frac{1+\rho}{1-\rho}\,,$$
$$Z_{\min} = \frac{1-\rho}{1+\rho}\,. \qquad (9.13)$$

If $H(s)$ is of the form given in Eq. (*9.12*), and if $\Delta/a$ is not too large, the $n$th maximum is

$$Z_n = \frac{1+\exp[-2\pi^2 n^2(\Delta/a)^2]}{1-\exp[-2\pi^2 n^2(\Delta/a)^2]} \cong \frac{1}{\pi^2 n^2\left(\dfrac{\Delta}{a}\right)^2}\,.$$

The area under the curve of Eq. (*9.11*) over each pseudoperiod $1/a$ is a constant and is equal to $1/a$. This is easily shown by integrating Eq.

**FIGURE 9.3.** *Shape of the transform Z(s) (or of the scattering power for very long objects, corresponding to the lattice of Figure 9.1.*

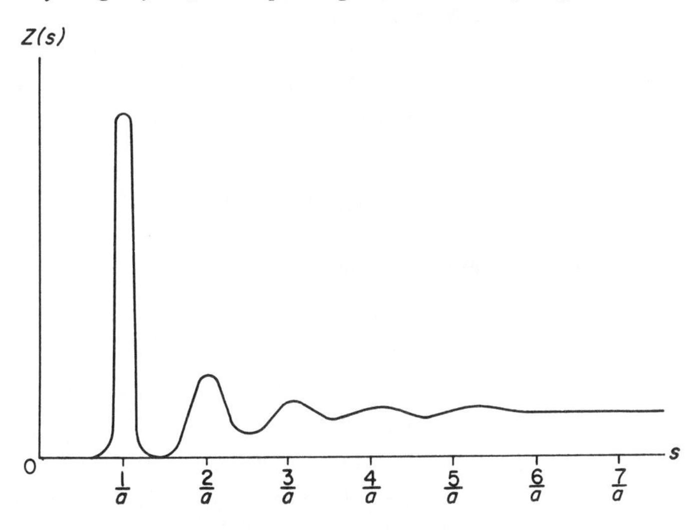


(*9.11*), choosing for $\rho$ a value which is constant and equal to the average value in the interval considered. For small values of $n$, when the minimum between two peaks of $Z(s)$ is low, the integral width of the maximum of order $n$ is

$$\Delta s = \frac{1}{a}\pi^2 n^2 \left(\frac{\Delta}{a}\right)^2 . \qquad (9.14)$$

The transform $Z(s)$ therefore has the general shape shown in Figure 9.3. At the center $s = 0$ and we have an infinitely narrow peak which we were able to foresee from the general properties of $Z(s)$. Its integral width is $1/a$. [See Eq. (*2.93*), replacing $v_1$ by $a$ for a one-dimensional model.] We then have a series of maxima for integral values of $sa$. These maxima decrease in intensity, at first as the square of the order of the node, while their widths increase. For large values of $s$, $\rho$ tends to zero and $Z(s)$ tends to unity. This is also a general result which we have found previously.

We have therefore investigated quantitatively for a particular case the diffraction patterns which are intermediate between those for crystals and those for amorphous bodies, which were described in general terms in Section 3.2.2.

Let us now examine the diffracted intensity. According to Eq. (*9.1*), this is the faltung of $Z(s)$ by the square of the modulus of $\Sigma(s)$, where $\Sigma(s)$ is the transform of the form factor for the object. In the case of a linear object of total length $L = Na$ and containing $N$ atomic groups,

$$|\Sigma(s)|^2 = \left(\frac{\sin \pi Nsa}{\pi s}\right)^2 .$$

The width of $[(\sin \pi Nsa)/(\pi s)]^2$ is $1/Na$. The faltung of Eq. (*9.1*) depends on the ratio of the widths of the peaks for $|\Sigma(s)|^2$ and $Z(s)$, i.e. $1/Na$ and

$$\Delta s = \frac{1}{a}\pi^2 n^2 \left(\frac{\Delta}{a}\right)^2 .$$

The following results were shown in Section 2.4.2.

(a) If $\Delta s$ is negligible compared to $1/Na$, which applies to $n$th order reflections such that

$$n \ll \frac{1}{\pi}\frac{a}{\Delta\sqrt{N}} ,$$

then the intensity is proportional to $|\Sigma(s)|^2$ in the region of the node and the line profile is the same as if the crystal were perfect. This really applies rigorously only to the central peak, but it provides a fair approximation for the first few reflections when the lattice is not much perturbed and when the object is sufficiently small.

(b) If, on the contrary, $1/Na \ll \Delta s$, or if

$$n \gg \frac{1}{\pi}\frac{a}{\Delta\sqrt{N}} ,$$

then the intensity is proportional to $Z(s)$, as in the case of diffraction by liquids and amorphous solids. The width of the peak depends on the degree of disorder in the lattice and not on the size of the object. When the object is large and the disorder appreciable, these conditions can apply even for the first-order reflection.

(c) As the order of reflection $n$ increases, the maxima of $Z$, and therefore of $I(s)$, become less and less distinct until they overlap and become undistinguishable. If we assume, like Hosemann [1], that the maxima are not resolved when the depth of modulation ($Z_{max}/Z_{min}$) is smaller than 4, then the resolved maxima are visible when $(1+\rho)^2/(1-\rho)^2 > 4$, or for $\rho > 0.3$. Since

$$\rho = \exp\left[-2\pi^2 n^2 \left(\frac{\Delta}{a}\right)^2\right],$$

this corresponds to $n\Delta/a < 0.25$.

(d) If $Z$ has maxima which are sufficiently narrow that $\Delta s < 1/Na$, i.e. if there are reflection maxima which satisfy the first condition

$$n \ll \frac{1}{\pi} \frac{a}{\Delta\sqrt{N}},$$

there exists between this first region and the second, where

$$n \gg \frac{1}{\pi} \frac{a}{\Delta\sqrt{N}},$$

an intermediate region where size and disorder effects are superposed. As we have already noted in relation with an analogous problem, the two widths can be added, as a first approximation, and the observed peak can be taken to have a total width of

$$\sqrt{\Delta s^2 + \left(\frac{1}{Na}\right)^2}.$$

To be more accurate, one must calculate the faltung of $Z(s)$ by $(\sin^2 \pi Nsa)/\pi^2 s^2$.

**FIGURE 9.4.** *The interference function corresponding to a file of points forming a perturbed linear lattice. This is a section of reciprocal space along a plane containing the average vector* **a**. *(The lower figures show the corresponding zones of fluctuation for the vectors* **x**). (a) *Fluctuations parallel to* **a**. (b) *Fluctuations isotropic around* **a**. (c) *Fluctuations normal to* **a**: *the distance between neighboring points, projected on* **a**, *is a constant.*

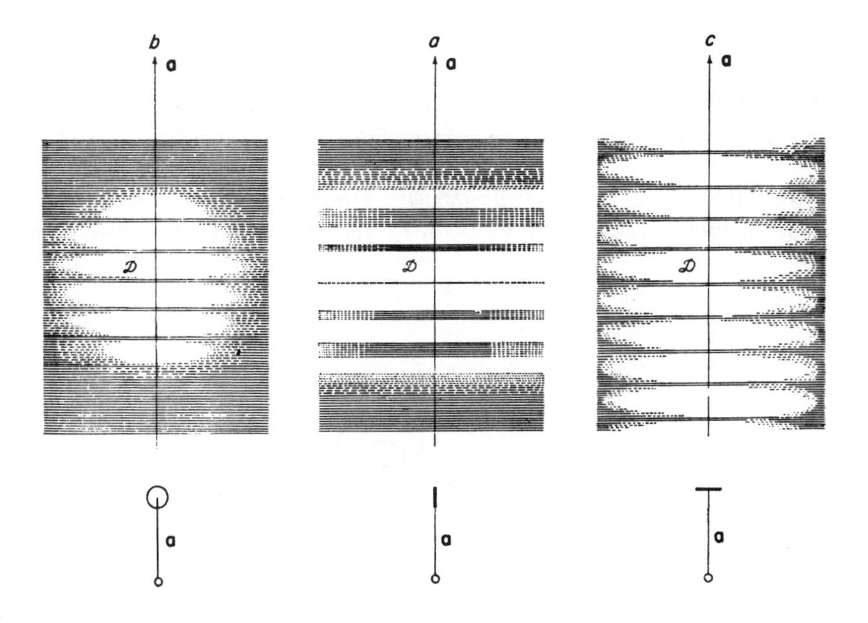

It is also possible to perform a direct calculation, analogous to the one already discussed in Section 7.2.3. The size effect shows up from the fact that, if there are $N$ atomic groups on the line, there are $N-1$ pairs of first neighbors, $N-2$ pairs of second neighbors, etc. The scattering power is then not simply proportional to $Z(s)$ and we must use the following expression, as in Eq. (*7.47*),

$$1 + 2\sum_1^\infty \left(1 - \frac{m}{N}\right)\rho^m \cos(2\pi mu) ,$$

where the summation can be calculated from the formulas given in Appendix B.2.

To summarize, let us recall the characteristics of the diffracting object which we have studied here. A set of $N$ points are placed along a straight line in such a manner that their distances fluctuate irregularly about an average value $a$. We can now establish the distribution of the intensity in reciprocal space, as in Figure 9.4 a. The interference function $I/F^2$ is a constant in any plane which is normal to the direction of the line. The narrowest and highest maximum is situated on the plane passing through the origin, and it is surrounded by symmetrical sets of maxima which gradually become broader and weaker. The maxima are situated at intervals of $1/a$.

## 9.2. Errors of Alignment in a File of Points

Consider now a more general case in which the vectors between close neighbors fluctuate not only in length but also in direction as in Figure 9.5. If these vectors are all drawn with a common origin, their extremities occupy a small volume surrounding an average point. We shall

**FIGURE 9.5.** (a) *File of points with errors of alignment* (b) *Zone of fluctuation of the vectors* $\boldsymbol{x}$ *defined by* $h(\boldsymbol{x})$, *cut by a plane normal to* $\mathbf{s}$.

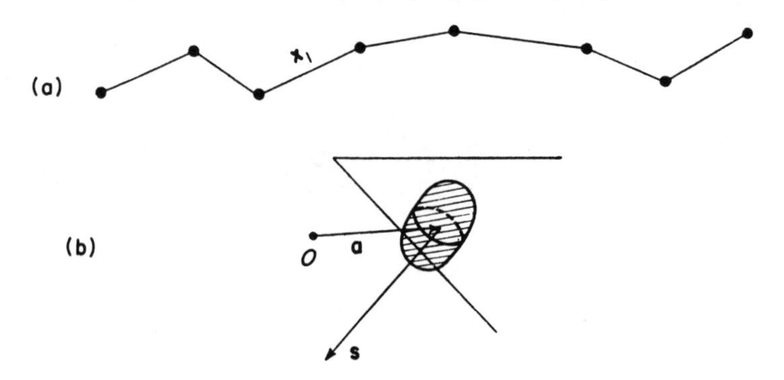

call this volume the zone of fluctuation. The distribution is defined by means of a function $h(\boldsymbol{x})$ such that

$$\int h(\boldsymbol{x})\, dv_x = 1 , \tag{9.15}$$

$$\int \boldsymbol{x} h(\boldsymbol{x})\, dv_x = \boldsymbol{a} . \tag{9.16}$$

Equations (*9.4*), (*9.5*), and (*9.10*) for $h_n(\boldsymbol{x})$, $z(\boldsymbol{x})$, and $Z(\boldsymbol{s})$ remain valid, except that these functions are now defined in space as functions of $\boldsymbol{x}$ or of $\boldsymbol{s}$. The transform $H(\boldsymbol{s})$ can be calculated as in the preceding case. This gives an approximate formula which is analogous to that of Eq. (*9.12*):

$$H(\boldsymbol{s}) = \exp(2\pi i \boldsymbol{s} \cdot \boldsymbol{a}) \exp[-2\pi^2 s^2 \Delta^2(\boldsymbol{s})] ,$$

where $\Delta(\boldsymbol{s})$ is defined by

$$s^2 \Delta^2(\boldsymbol{s}) = \int h(\boldsymbol{x}) [\boldsymbol{s} \cdot (\boldsymbol{x} - \boldsymbol{a})]^2 \, dv_x . \tag{9.17}$$

The function $Z(\boldsymbol{s})$ is given by Eq. (*9.11*), where

$$u = \boldsymbol{s} \cdot \boldsymbol{a} ,$$
$$\rho = \exp[-2\pi^2 s^2 \Delta^2(\boldsymbol{s})] .$$

The function $h(\boldsymbol{x})$ can be considered as the local density inside the zone of fluctuation of the extremity of the vector $\boldsymbol{x}$; therefore Eq. (*9.15*) means that the total mass of the zone is unity, while Eq. (*9.16*) means that its center of gravity is defined by the vector $\boldsymbol{a}$. Also from Eq. (*9.17*), $\Delta^2(\boldsymbol{s})$ is the moment of inertia of the zone with respect to a plane which is normal to $\boldsymbol{s}$ and which passes through the center of the zone, as in Figure 9.5.

Substituting now the above values of $\rho$ and of $u$ in Eq. (*9.11*), we find that so long as the fluctuations are not too strong, i.e. so long as $\rho$ does not vary too rapidly with $\boldsymbol{s}$, the maxima of $Z(\boldsymbol{s})$ are always situated in planes where $u$ is a whole number, or in planes of the reciprocal space such that $\boldsymbol{a} \cdot \boldsymbol{s} = n$.

In the preceding example, $Z(\boldsymbol{s})$ was a constant along these planes, but in the present case $\rho$ varies along the planes, since the moment of inertia of the zone of fluctuation varies with the direction of $\boldsymbol{s}$. It is possible to investigate this variation, using well-known results on moments of inertia. Whatever the shape of the zone of fluctuation or the distribution of the density $h(\boldsymbol{x})$ within it, there exist three mutually perpendicular "principal planes of inertia," and the moments with respect to these planes are the principal moments of inertia $\Delta_1^2$, $\Delta_2^2$, $\Delta_3^2$. If the direction cosines of an arbitrary vector $\boldsymbol{s}$ with respect to the principal axes of inertia are $\alpha$, $\beta$, $\gamma$, the moment of inertia with respect to a plane normal to this line is

$$\Delta^2(\mathbf{s}) = \alpha^2\Delta_1{}^2 + \beta^2\Delta_2{}^2 + \gamma^2\Delta_3{}^2$$

In reciprocal space the locus of points for which $\rho$ has the value $\exp(-2\pi^2 A^2)$ is the surface $s^2\Delta^2(\mathbf{s}) = A^2$, which is the ellipsoid

$$\frac{1}{s^2} = \frac{1}{A^2}(\alpha^2\Delta_1{}^2 + \beta^2\Delta_2{}^2 + \gamma^2\Delta_3{}^2)\,,$$

whose semiaxes are directed along the principal axes of inertia and whose lengths are, respectively, $(A/\Delta_1)$, $(A/\Delta_2)$, and $(A/\Delta_3)$.

Along the planes of the maxima the value of $Z(\mathbf{s})$ decreases as $\rho$ increases and, conversely, it increases on the planes of the minima. The value of $Z(\mathbf{s})$ therefore becomes more and more uniform with increasing distance from the center of reciprocal space, but this effect is anisotropic. To obtain an idea of the general shape of the function $Z(\mathbf{s})$, let us assume, like Hosemann, that the maxima are not resolved for $\rho < 0.3$; this would occur inside an ellipsoid such that $\rho = 0.3$. Figure 9.4 shows schematically the distribution of $Z(\mathbf{s})$ in a section of reciprocal space through the average vector $\boldsymbol{a}$. Figures 9.4 a and 9.4 b show the effect of fluctuations in direction superposed to fluctuations in the distance between close neighbors. The region of reciprocal space $\mathscr{D}$ where there are true scattering maxima is now restricted not only in the direction of the vector $\boldsymbol{a}$, but also in perpendicular directions. The volume of this region decreases in size as the fluctuations increase.

The shape of the volume $\mathscr{D}$ is related to the anisotropy of the zone of fluctuation $h(\boldsymbol{x})$. If the zone exhibits spherical symmetry about its center $\boldsymbol{a}$ as in Figure 9.4b, the region $\mathscr{D}$ is spherical in reciprocal space. In the preceding example of Section 9.1, the vector $\boldsymbol{x}$ has a fixed direction and the zone of fluctuation is therefore a segment of a straight line parallel to $\boldsymbol{a}$. Two of the moments of inertia, $\Delta_2$ and $\Delta_3$, are zero, so the ellipsoid of inertia degenerates into two planes normal to $\boldsymbol{a}$. As shown in Figure 9.4a, the region $\mathscr{D}$ is limited by two such planes. Let us now assume that $\boldsymbol{x}$ has a fixed modulus and that it varies little in direction, so the zone of fluctuation is a small disk normal to $\boldsymbol{a}$. In such a case the moment of inertia $\Delta_1$ is zero and the ellipsoid $\mathscr{D}$ becomes a cylinder whose axis is $\boldsymbol{a}$, as in Figure 9.4 c.

## 9.3. Irregular Stack of Plane Parallel Lattices

Let us examine various applications of these general results. Consider an object constituted by a simple plane lattice of large area with a single atom of scattering factor $f$ per unit cell of area $S_c$. The structure factor $F$ for this object is different from zero only on the rows normal to a plane passing through the node $hk$ of the reciprocal lattice of the plane lattice, and its scattering power atom along these rows is $f^2/S_c$, according to Eq. (*7.35 a*). We now repeat these plane lattices around the points of the linear file $A_1, A_2, A_3, \cdots$, the average interplanar vector $\boldsymbol{a}$ being normal to the plane lattices. In other words, we consider a stack of parallel and identical lattice planes, the displacement between two successive planes being the vectors $\boldsymbol{x}$ which fluctuate independently with a distribution $h(\boldsymbol{x})$.

(a) *The interplanar vectors have a fixed direction* normal to the plane, and a fluctuating modulus. The diffracted intensity is then proportional to the product of the factor $F^2$ and the factor given in Eq. (*9.11*) and in Figure 9.3. The variation of the intensity is the same on all rows $hk$, except for the slow decrease of the atomic scattering factor $f$. In the median plane the diffraction nodes are nearly punctual, the extent of the domain of reflection depending only on the total number of planes. The other nodes $hkl$ are surrounded by a domain which is elongated along the row, especially if the index $l$ of the node is large. The length of these segments in the reciprocal space for a given value of the index $l$ increases as the fluctuations in the plane spacing increase, and the nodes are resolved along each row up to an index of the order of $0.25(a/\Delta)$, according to the calculation of Section 9.1; $\Delta/a$ is the relative fluctuation of the plane spacing, but this is only a very rough approximation. If the fluctuations become large, it is only the nodes of the median plane $l = 0$ which remain distinct, and along the rows the intensity is less and less modulated until it decreases uniformly as $f^2$ (gas scattering).

One particular case which has been investigated because of its applications to the structure of certain clays is a stack of planes spaced by a distance $d_1$, with the abnormal distance $d_2$ appearing at a frequency $\alpha$ [3]. The distribution $h(x)$ is the sum of two delta functions, one at $d_1$ with a probability $(1 - \alpha)$, and another at $d_2$ with a probability $\alpha$; that is,

$$h(x) = (1 - \alpha)\delta(x - d_1) + \alpha\delta(x - d_2) .$$

Then, from the equation of part 2, Table A.1, Appendix A.4, the transform $H(s)$ is given by

$$H(s) = (1 - \alpha)\exp(2\pi i s d_1) + \alpha\exp(2\pi i s d_2) .$$

Figure 9.6 shows an easy method of calculating the modulus $\rho$ and the argument $2\pi u$ of $H(s)$:

$$\rho^2 = \alpha^2 + (1 - \alpha)^2 + 2\alpha(1 - \alpha)\cos[2\pi s(d_2 - d_1)] = 1 - 4\alpha(1 - \alpha)\sin^2[\pi s(d_2 - d_1)],$$

$$2\pi u = 2\pi s d_1 + \arcsin\left\{\frac{\alpha}{\rho}\sin\left[2\pi s(d_2 - d_1)\right]\right\}.$$

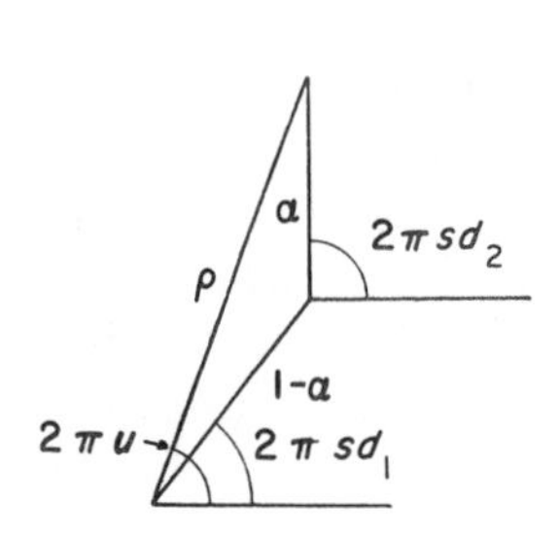


**FIGURE 9.6.** *Calculation of the modulus and of the argument of H(s).*

Substituting in Eq. (*9.11*), we find the scattering power, which is proportional to $Z(s)$ if the number of stacked planes is very large. If $\alpha$ and $d_2 - d_1$ are not too large, we have the approximate relations

$$\rho = 1 - 2\alpha\pi^2 s^2(d_2 - d_1)^2 ,$$
$$u = s[d_1 + \alpha(d_2 - d_1)] .$$

Irregularly distributed stacking faults therefore have the effect of displacing the maxima to positions which would correspond to the average parameter $d_1 + \alpha(d_2 - d_1)$. The maximum of order $l$ for which $s \cong l/d_1$ has a width of $\pi^2\alpha l^2(d_2 - d_1)^2/d_1^2$, which increases as the square of the index $l$ and which is proportional to the probability of the occurrence of the fault.

(b) *The lattice spacing is constant* but the planes are displaced irregularly parallel to themselves. In this case the zone $h(\boldsymbol{x})$ is a flat disk which is normal to the vector $\boldsymbol{a}$ at its extremity. The ellipsoids over which $\rho$ is a constant are cylinders of circular cross-section whose axes are normal to the planes if the displacements of the planes are isotropic, as in Figure 9.4c. All the (00$l$) nodes on the axis normal to the planes have an extent which depends only on the thickness of the object, or on the number of planes. Along a given row $hk$, all the nodes $hkl$ have the same extent along the row, which increases as the distance between the row and the center of reciprocal space increases.

According to the above, we could deduce quantitatively the order of magnitude of the fluctuations of $h(\boldsymbol{x})$ from the radius of the cylinder inside which two successive reflections $hkl$ and $hk(l+1)$ on the same row are quite distinct.

Hosemann [4] has shown that the faults in close-packed stacks of hexagonal planes can be treated in this way, and that in particular one arrives very easily at the results of Paterson, which were stated in Section 7.2.3.

(c) If the interplanar vectors fluctuate both in modulus and in direction, the number of nodes corresponding to distinct maxima is limited. The number of 00$l$ reflections depends on the relative size of the fluctuations in the direction normal to the plane, while the number of $hk$0 reflections depends on the relative size of the fluctuations parallel to the lattice plane. If these are not distributed isotropically, the broadening of the nodes is not isotropic around the [001] axis. An investigation of the rows in terms of the indices $h$ and $k$ is required to determine the shape of the zone of fluctuation in the plane normal to $\boldsymbol{a}$.

## 9.4. Perturbed Three-dimensional Lattice. The Paracrystal

The main characteristic of diffraction by one-dimensional models—*the broadening and the progressive weakening of the reflection spots starting from the center*—can be found in the patterns for many real solids, in particular for macromolecular compounds such as cellulose, proteins, etc. One can easily imagine that the relative positions of such atomic groups can be somewhat indeterminate because the forces between such large molecules are relatively weak. One can therefore expect to find in such three-dimensional structures imperfections which destroy long-range order.

Despite the indefiniteness of the spots, these patterns show a definite analogy with those of crystals. For example, the pattern of a ramie (cellulose) fiber (Figure 9.7) is analogous to that of a rotating crystal. We must locate the center of the spots so as to determine the positions of the nodes, and attribute to them an intensity equal to that integrated over

**FIGURE 9.7.** *Diffraction pattern of a cellulose fiber. Monochromatic Cu Kα radiation. The broadened spots are situated on layer lines as in a rotating crystal pattern.*

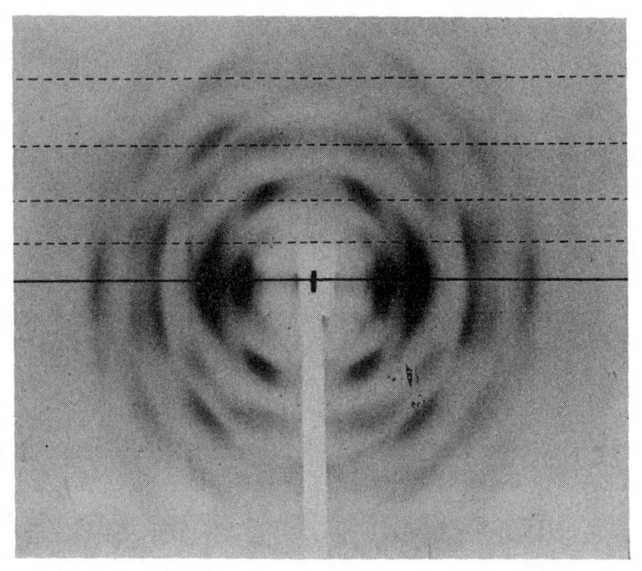

the whole surface of the diffuse spot. From the discussion in Chapter 5 (and also in the present one), it appears possible to derive correctly the average crystal structure of the imperfect crystal from these data. It should therefore be possible to identify the lattice and the atomic structure of the unit cell, whether the broadening arises from the small size of the crystals or from lattice distortions. Unfortunately, because of the small number of measurable spots, it is impossible to find a sufficient number of terms in the Fourier series required to calculate the positions of the atoms in the unit cell, and the structure can therefore be determined only approximately. It can nevertheless be established in special cases—cellulose, for example [5].

On the other hand, the model of the perfect crystal is inadequate to explain the widths of the spots and the continuous background.

Attempts have been made to explain the broadening of the reflection domains solely from the small sizes of the crystals, and the Scherrer formula has been used to calculate their sizes. This interpretation is plausible, however, only if the widths of these domains remain the same for all the nodes of the reciprocal lattice. Otherwise *the results based on a single spot are meaningless*, as can be seen by applying this method to the case discussed in Section 9.1. In fact, it is only the reflection domain surrounding the center of reciprocal space which is determined by the size of the crystal and which therefore permits, at least theoretically, a calculation of the shape and size of the crystal. We shall return to this important subject in the next chapter.

The idea of crystal imperfection is often the most important in accounting for a diffraction pattern, especially if the pattern is characterized by a small number of spots near the center, becoming weaker and broader as their indices increase. In such a case we may assume that the crystal contains imperfections of the second type which destroy long-range order. If the spots of low order are even slightly broadened, and if there is little scattering at the center, the broadening may be attributed *solely* to crystal imperfections. More exactly, this means that the function $I(\boldsymbol{s})/F^2$ can be taken to be proportional to $Z(\boldsymbol{s})$, $F$ being the structure factor of the elementary atomic group. If the distribution of the scattered intensity $I(\boldsymbol{s})$ is known throughout reciprocal space, one can theoretically calculate the statistics of the mutual positions of the elementary groups $z(\boldsymbol{x})$ by inversion of the function $I(\boldsymbol{s})/F^2$.

In the case where size effect is not negligible, the transform of $I(\boldsymbol{s})/F^2$ is not $z(\boldsymbol{x})$, but rather the product $V(\boldsymbol{x})\,z(\boldsymbol{x})$ [Eq. (2.33)]. The function $V(\boldsymbol{x})$ can be found from the scattering at the center; it is the transform of $|\Sigma(\boldsymbol{s})|^2$ [Eq. (2.34)], or $I(\boldsymbol{s})$ in the central spot (Section 2.4.2.1.) on the assumption that it can be measured from the pattern. Therefore

$$z(\boldsymbol{x}) = \frac{\text{transf } I(\boldsymbol{s})/F^2}{(v_1/F^2) \text{ transf } [I(\boldsymbol{s})_{\text{ in central spot}}]}$$

where $v_1$ is the average volume occupied by each atomic group. Theoretically, this equation solves the problem of the determination of the structure of a poorly crystallized single crystal. And there are probably few cases to which it can be applied in this general form.

Our study of amorphous bodies in Chapter 3 was based on this distribution function $z(\boldsymbol{x})$, but the problem was relatively simple because $z(\boldsymbol{x})$ was isotropic. In the case of the perturbed crystal, $z(\boldsymbol{x})$ is essentially anisotropic, since it is more or less related to the crystal lattice which is approximately that of the real imperfect crystal. In fact, $z(\boldsymbol{x})$ is the sum of the functions $h_m(\boldsymbol{x})$, each one of which describes the relative positions of the atomic groups separated in the unperturbed crystal by the lattice vector $\boldsymbol{x}_m$ (Figure 9.8). In the case of the one-dimensional structure, it was possible to deduce all the functions $h_m(x)$ from the first one, $h(x)$, related to close neighbors, but generalization to two and three dimensions is extremely difficult.

We have already met with a similar problem in defining the order

**FIGURE 9.8.** *Schematic illustration of the distribution $z(\boldsymbol{x})$ in a two-dimensional crystal with small perturbations of the second type. The zones of fluctuation gradually increase in size and there is no long-range order.*

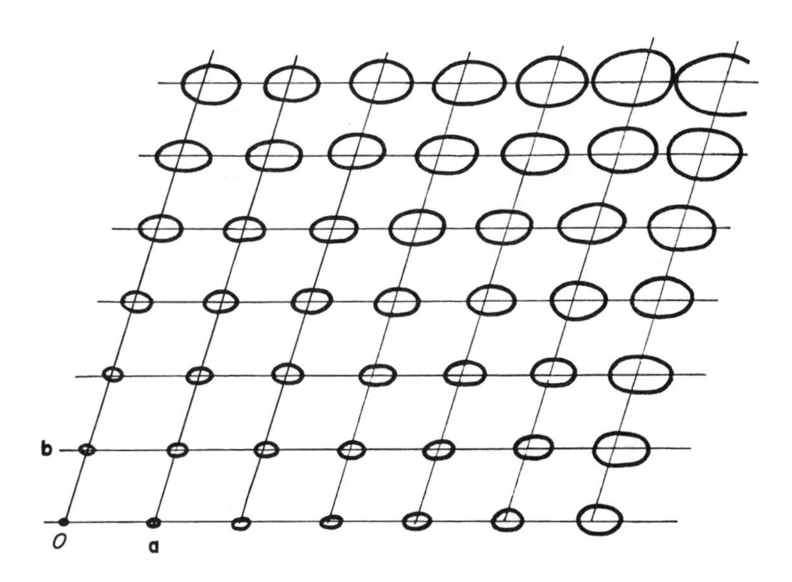

parameters for a mixed crystal with a regular lattice. In the case of the linear lattice, the degree of order for the $n$th neighbors is defined from the parameter for close neighbors but this is not true in general for a three-dimensional lattice. Theoretically the different parameters can be calculated for a given lattice from the energies of interaction between the various types of atoms when the solid solution is in equilibrium at a given temperature. One might also calculate the various distributions $h_m(\boldsymbol{x})$ from the laws of interaction for the atoms as functions of their positions in a given slightly perturbed lattice, but this highly complex calculation has not been made as yet.

Hosemann [1], [5] has found a simple solution for what he has called the "ideal paracrystal." He assumes that the perturbed crystal is built up from three fundamental vectors $\boldsymbol{a}, \boldsymbol{b}, \boldsymbol{c}$ which fluctuate according to three independent distributions $h_a(\boldsymbol{x})$, $h_b(\boldsymbol{x})$, $h_c(\boldsymbol{x})$, the distribution for the vector $\boldsymbol{x}_{m_a m_b m_c}$ being obtained by adding $m_a$ vectors $\boldsymbol{a}$, $m_b$ vectors $\boldsymbol{b}$, and $m_c$ vectors $\boldsymbol{c}$, which *fluctuate independently*.

Under this hypothesis, Eq. (*9.7*) can be generalized; Hosemann writes

$$h_m(x) = \underbrace{h_a(x) * \cdots * h_a(x)}_{m_a \text{ times}} * \underbrace{h_b(x) \cdots * h_b(x)}_{m_b \text{ times}} * \underbrace{h_c(x) \cdots * h_c(x)}_{m_c \text{ times}} .$$

If $H_a(\boldsymbol{s})$, $H_b(\boldsymbol{s})$, and $H_c(\boldsymbol{s})$ are the transforms of the three function $h(\boldsymbol{x})$, we have

$$H_m(\boldsymbol{s}) = [H_a(\boldsymbol{s})]^{m_a}[H_b(\boldsymbol{s})^{m_b}[H_c(\boldsymbol{s})]^{m_c}. \tag{9.18}$$

The quantity $Z(\boldsymbol{s})$ is the product of three geometric sums, analogous to that of Eq. (*9.11*) and, setting $\rho_a$, $\rho_b$, $\rho_c$, and $u_a$, $u_b$, $u_c$, as the moduli and the arguments of the three functions $H(\boldsymbol{s})$, we find that

$$Z(\boldsymbol{s}) = \prod_{i=a,b,c} \frac{1 - \rho_i^2}{1 + \rho_i^2 - 2\rho_i \cos 2\pi u_i} . \tag{9.19}$$

The distribution of $Z(\boldsymbol{s})$ can be obtained by combining three figures analogous to Figure 9.4. Near the center of reciprocal space, $Z(\boldsymbol{s})$ is approximately zero except near the intersection of the planes of the maxima of the three factors appearing in the above equation. These are the nodes of the reciprocal lattice corresponding to the lattice built up on the three average vectors $\boldsymbol{a}, \boldsymbol{b}, \boldsymbol{c}$. As the node indices increase, the domains of reflection broaden until they overlap, approaching a uniform distribution for which $Z = 1$.

A paracrystal therefore gives a diffraction pattern which is somewhat analogous to those observed with many imperfect crystals. It is nevertheless certain that the basic hypothesis cannot be justified in the general case, for it is not possible to have a three-dimensional lattice in which

**FIGURE 9.9.** *Examples of two-dimensional irregular arrangements of points, and the corresponding diffraction patterns obtained by optical means* (*Hosemann* [*6*]).

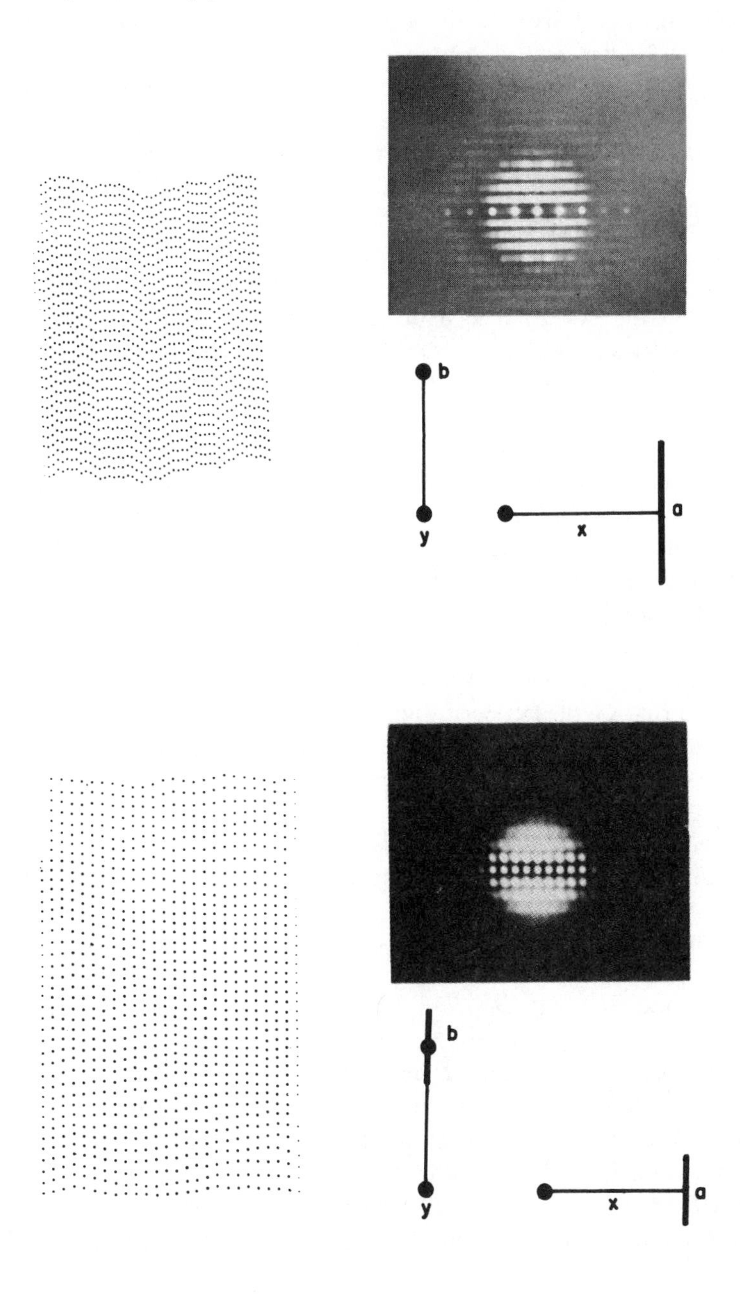

the interatomic vectors fluctuate completely independently. One may even doubt whether this model is at all acceptable, for it can obviously not be applied to close-packed lattices where there are a large number of neighbors; in such cases there is no indication of the manner in which the three fundamental vectors must be selected.

The model of the paracrystal is possibly more acceptable for fibrous crystals, which Hosemann [2] has investigated. Such crystals are in fact built up on a two-dimensional lattice, one of the vectors being parallel to the axis of the fiber, and the other perpendicular. These vectors play a fundamental role, and their fluctuations arise from essentially different sources, namely the interactions along chains and between chains.

Figure 9.9 shows two examples of perturbed plane lattices for which Hosemann [6] has obtained a diffraction pattern by an optical process. This type of analogue calculation has also been used for interpreting the diffraction patterns of alloys which exhibit partial order and stacking faults [7].

These lattices derive from a rectangular lattice built up on the vector $\boldsymbol{a}$ which is horizontal, and on $\boldsymbol{b}$, which is vertical. In the arrangement *a*, the vectors $\boldsymbol{y}$ are all identical, while the vectors $\boldsymbol{x}$ fluctuate within a region which has the shape of a short line normal to $\boldsymbol{a}$. The diffraction pattern is the combination of fine strata which are normal to $\boldsymbol{b}$ and spaced by a distance $1/b$, and of Figure 9.4c along $\boldsymbol{a}$. Only the equatorial spots are clearly visible. For the other strata the spots are diffuse but the scattering is restricted to the stratum.

In the case of the lattice of Figure 9.9b, the fluctuations of $\boldsymbol{x}$ are analogous but not as marked. On the other hand, $\boldsymbol{y}$ fluctuates in length, its direction remaining fixed. The diffraction pattern is then the combination of that of Figure 9.4a along $\boldsymbol{a}$ and of Figure 9.4c along $\boldsymbol{b}$. The extra-equatorial spots are diffuse, especially in regions which are remote from the equator, and there is scattering between the strata.

## 9.5. Fibrous Crystals

In general, the patterns for natural fibers do not yield the reciprocal lattice of the crystal because, instead of single crystals, one has a "two-dimensional powder" resulting from the grouping of crystallites of random orientation along the axis of the fiber. The pattern therefore gives the figure of revolution obtained by rotating the reciprocal lattice for a single crystal around the fiber axis.

Let us consider a set of parallel chains formed of perfect linear lattices, and their intersections with a plane normal to the axis of the fiber. This gives a set of points which form, depending on the regularity of the lateral arrangement of the chains, either a plane lattice or the two-dimensional equivalent of a "paracrystal," a liquid, or a gas. If we consider *all the fibers together*, we can define a distribution $p(x)$ such that $(2\pi x p(x))/s_1 dx$ represents the number of chains situated at a distance lying between $x$ and $x + dx$ of a given chain, where $s_1$ is the average surface available for each point in the plane. At large distances $p(x)$ tends to 1. By analogy with the three-dimensional case, Eq. (*2.22*),

$$z(x) = \delta(x) + \frac{1}{s_1} + \frac{1}{s_1}(p(x) - 1)\,.$$

The transform of $z(x)$ is a function of a single variable, $Z(s)$, which is defined in the plane $(\pi)$ passing through the origin of reciprocal space and normal to the direction of the chains. The variable $s$ represents the distance between the origin and a point of the plane $(\pi)$. For the calculation of the diffracted intensity in one particular experiment, $s$ is the length of the projection $\boldsymbol{s}_\pi$ on the plane $(\pi)$ of the usual vector $(\boldsymbol{S} - \boldsymbol{S}_0)/\lambda$:

$$Z(s) = \frac{1}{s_1}\,\delta(s) + 1 + \frac{1}{s_1}\int [p(x) - 1] \exp(2\pi i \boldsymbol{s}_\pi \cdot \boldsymbol{x})\, ds_x\,.$$

To calculate the above integral, let us call $\alpha$ the angle between $\boldsymbol{s}_\pi$ and $\boldsymbol{x}$. Then

$$\begin{aligned}\int [p(x) - 1] \exp(2\pi i \boldsymbol{s}_\pi \cdot \boldsymbol{x})\, ds &= 2\int_0^\infty\int_0^\pi [p(x) - 1| \exp(2\pi i s x \cos\alpha) x\, dx\, d\alpha \\ &= 2\pi \int_0^\infty [p(x) - 1] J_0(2\pi s x) x\, dx\,,\end{aligned}$$

and therefore

$$Z(s) = \frac{1}{s_1}\,\delta(s) + 1 + \frac{2\pi}{s_1}\int_0^\infty [p(x) - 1] J_0(2\pi s x) x dx\,, \qquad (9.20)$$

where $J_0$ is the Bessel function of order zero [8]. The above equation is analogous to Eq. (*2.47*), except that in the two-dimensional problem $J_0(2\pi s x)$ replaces $\sin 2\pi s x$.

Consider, for example, a set of two points at a distance $a$, the line joining the two points having any given direction in the plane. This is a "diatomic molecule" situated in the plane. Then a calculation analogous to that of Section 2.6 gives

$$\mathscr{I}(s) = 1 + J_0(2\pi a s) \qquad (9.21)$$

for the interference function. This equation should be compared with the Debye formula of Eq. (*2.55a*),

$$\mathscr{I}(s) = 1 + \frac{\sin(2\pi s a)}{2\pi s a}\,, \qquad (9.22)$$

**FIGURE 9.10.** *Interference function for pairs of atoms spaced by a distance a, of random orientation, but restricted to a plane.*

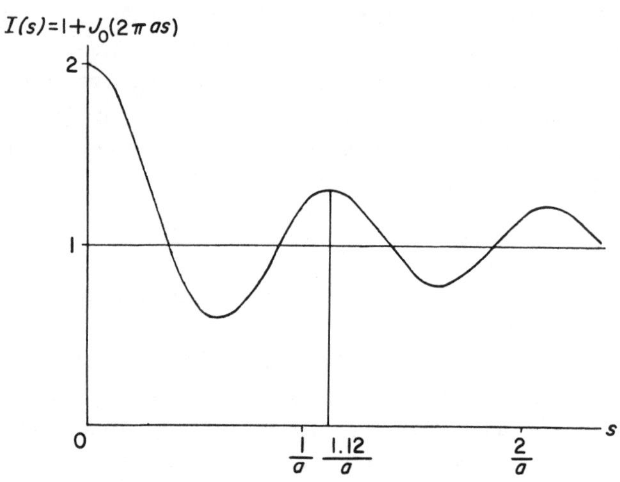


which applies to diatomic molecules in space. The scattering curves shown in Figures 9.10 and 2.4 are analogous and the first maximum occurs at $s = 1.12/a$ [Eq. (*9.21*)] instead of $1.23/a$ [Eq. (*9.22*)].

Let us return to the set of identical parallel chains and call $F$ the structure factor for the chains. The scattering power per chain for the set, neglecting the central peak which depends on the external shape of the object, is given by

$$I(s) = F^2\left\{1 + \frac{2x}{s_1}\int_0^\infty [p(x) - 1)] \times J_0(2\pi sx)x\,dx\right\}. \quad (9.23)$$

When the chain is regular and has a period $c$, $F$ is zero except on planes normal to the axis $\boldsymbol{c}$ and situated at $1/c, 2/c, \cdots$. The diffraction pattern in reciprocal space is then composed of a set of planes which are identical within the variation of the structure factor for the atomic group of the chains, as in Figure 9.11. The variation of the intensity in a plane passing through the origin as a fuction of its distance

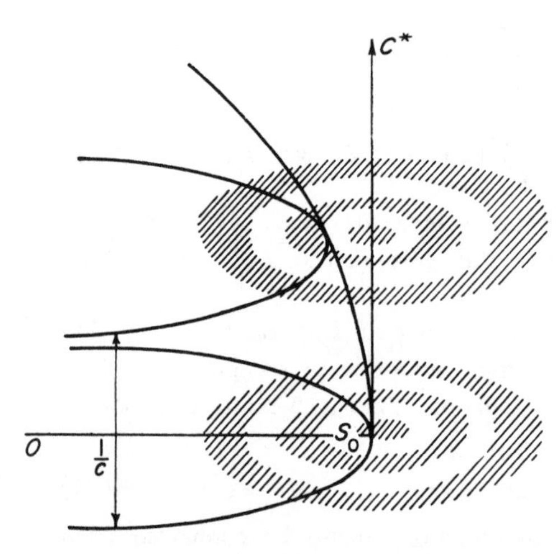


**FIGURE 9.11.** *Reciprocal space for a rigorously periodic fibrous body. Intersection of this space with the reflection sphere for predicting the diffraction pattern.*

to the origin gives the distribution $p(x)$ through a Fourier-Bessel inversion [8]:

$$p(x) - 1 = 2\pi s_1 \int_0^\infty \left[ \frac{I(s)}{F^2} - 1 \right] J_0(2\pi s x) s \, ds \, . \tag{9.24}$$

If the translation vectors from one chain to another are not all normal to the axis of the chain, i.e. if we add to the lateral disorder a lengthwize disorder parallel to the chains, the diffraction is unaffected when the vector $\boldsymbol{s}$ is in the base plane ($\pi$), but the diffraction pattern is not identical on all the planes situated at $1/c$, $2/c$, $\cdots$. From the results of Section 9.3, fluctuations parallel to the axis make the domains of reflection more and more diffuse along the strata as the distance to the equator increases [9].

Finally, if we also assume that the chains have a somewhat irregular structure, the diffraction planes become ill-defined and have a width which increases with the index of the stratum. This broadening is constant for a given stratum if the fluctuations remain parallel to the axis of the chain, and they increase with increasing distance from the central axis if the chain is not exactly straight [10].

## References


1. J. J. Hermans, *Rec. Trav. Chim. Pays-Bas,* **63** (1944), 5.
   R. Hosemann, *Z. Physik,* **128** (1950), 1 and 465.
2. R. Hosemann, *Acta Cryst.,* **4** (1951), 520.
3. G. W. Brindley and J. Mering, *Acta Cryst.* **4** (1951), 441.
   W. H. Zachariasen, *Phys. Rev.,* **71** (1947), 715.
   J. Mering, *Acta Cryst.,* **2** (1949), 371.
   J. Mering, *J. Chimie Phys.,* **51** (1954), 440.
4. R. Hosemann and S. N. Bagchi, *Phys. Rev.,* **94** (1954), 71.
5. R. Hosemann and S. N. Bagchi, *Direct Analysis of Diffraction by Matter,* North Holland, Amsterdam, 1962-
6. R. Hosemann, *Naturwiss.,* **19** (1954), 440.
7. C. A. Taylor, R. M. Hind, and H. Lipson, *Acta Cryst.,* **4** (1951), 261.
8. G. N. Watson, *A Treatise on the Theory of Bessel Functions,* Cambridge Univ. Press, Cambridge (1958).
9. R. W. James, *The Optical Principles of X-Ray Diffraction,* Bell, London (1948), p. 572.
10. R. S. Bear and O. E. A. Bolduan, *Acta Cryst.,* **3** (1950), 230 and 236.

CHAPTER

# 10

# SMALL-ANGLE X-RAY SCATTERING

## 10.1. Characteristics of X-Ray Scattering at Very Small Angles

In Chapter 5 we saw that all the nodes of the reciprocal lattice of a small crystal, including the 000 node, are surrounded by identical reflection domains. Thus, from Eq. (*5.5*), the scattering power per unit cell of a small crystal in the immediate vicinity of the center of the reciprocal space is given by

$$I(\boldsymbol{s}) = \frac{F^2}{V_c V} \mid \Sigma(\boldsymbol{s}) \mid^2 , \tag{10.1}$$

where $\Sigma(\boldsymbol{s})$ is the Fourier transform of the form factor $\sigma(\boldsymbol{x})$ of the crystal, this function being equal to unity inside the crystal and zero outside. The quantity $F$ is the structure factor for the unit cell occupying the volume $V_c$. For very small angles, $F$ is equal to the number of electrons in the unit cell, so $F/V_c$ is equal to the average electron density $\rho$ in the crystal. Thus the total scattering power is

$$I_N(\boldsymbol{s}) = I(s) \frac{V}{V_c} = \rho^2 \Sigma \mid (\boldsymbol{s}) \mid^2 . \tag{10.2}$$

We also saw that for any *homogeneous body*, whatever its internal structure [Eq. (*2.42*)], the diffracted intensity has a central peak corresponding to a scattering power given by an equation similar to Eq. (*10.1*), except that the average volume occupied by the atomic group $v_1$ is substituted for $V_c$.

Equation (*10.2*) is therefore completely general. The central peak in the intensity curve, which depends solely upon the external shape of the object, occurs only in the center in the case of amorphous bodies, while it repeats itself around each node in the case of perfect crystals.

For $s = 0$, the maximum intensity of the peak is $\rho^2 V^2$, since $\Sigma(0) = V$; $V$ is the volume of the small sample, and $\rho V$ is equal to the total number $n$ of electrons in the sample. This result is obvious, since the waves scattered by the $n$ electrons are in phase at small scattering angles. In the case of amorphous bodies it is only at small angles that these waves are in phase, while in the case of crystals they are in phase whenever the conditions for selective reflection are satisfied.

We have neglected the central peak until now because it is so extremely narrow that it cannot be distinguished from the direct beam, even for microscopic samples. The "width" of the central peak (Section 2.4.2) is $1/V$, or $1/d^3$, where $d$ is of the order of magnitude of the size of the sample. The diffraction domain in the reciprocal space extends out to a distance $s$ of the center, which is of the order of $1/d$, or out to a diffraction angle $\theta$ such that $s = (2 \sin \theta)/\lambda \cong 1/d$. Since $\theta$ is small, $s = 2\theta/\lambda = \varepsilon/\lambda$, so the angular width of the central peak is of the order of

$$\varepsilon = \frac{\lambda}{d}\,. \tag{10.3}$$

Assuming that diffraction can be observed down to $\varepsilon \cong 10^{-3}$, the central peak can be observed if the sample is less than $0.1\mu$ in diameter. *The central peak becomes broader as the particle size decreases, and it does not depend on the internal structure of the particles, so long as the particles are homogeneous.*

This phenomenon is similar to a well-known phenomenon in optics, namely the scattering of light by small particles [1]. Droplets of fog, for example, give a diffuse halo around the moon, the intensity of the

**FIGURE 10.1.** *Small-angle scattering experiment.*

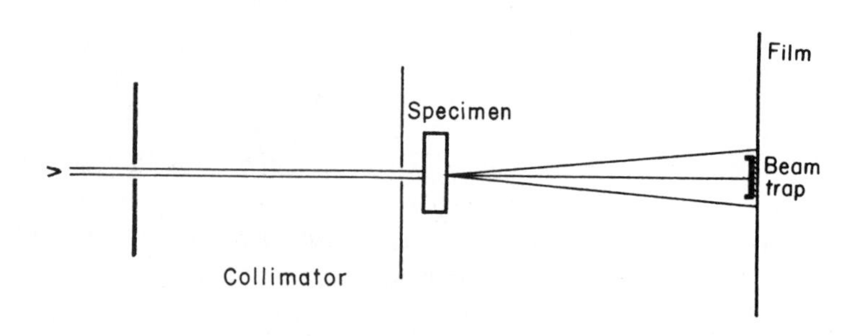

**FIGURE 10.2.** *Examples of small-angle scattering. In each case the scale shown corresponds to the angle* $\varepsilon = 2\theta$. (a) *Amorphous silica in the form of molten quartz.* (b) *Amorphous silica in the form of silica gel.* (c) *Carbon black.* (d) *Hemoglobin in red blood corpuscle.* (e) *Chrysotile fiber. The lines correspond to* $s = 1/130\ \text{Å}^{-1}$. (f) *Al Cu single crystal after age hardening. The incident ray was along the* [100] *axis. Mo* $K\alpha$ *radiation.*

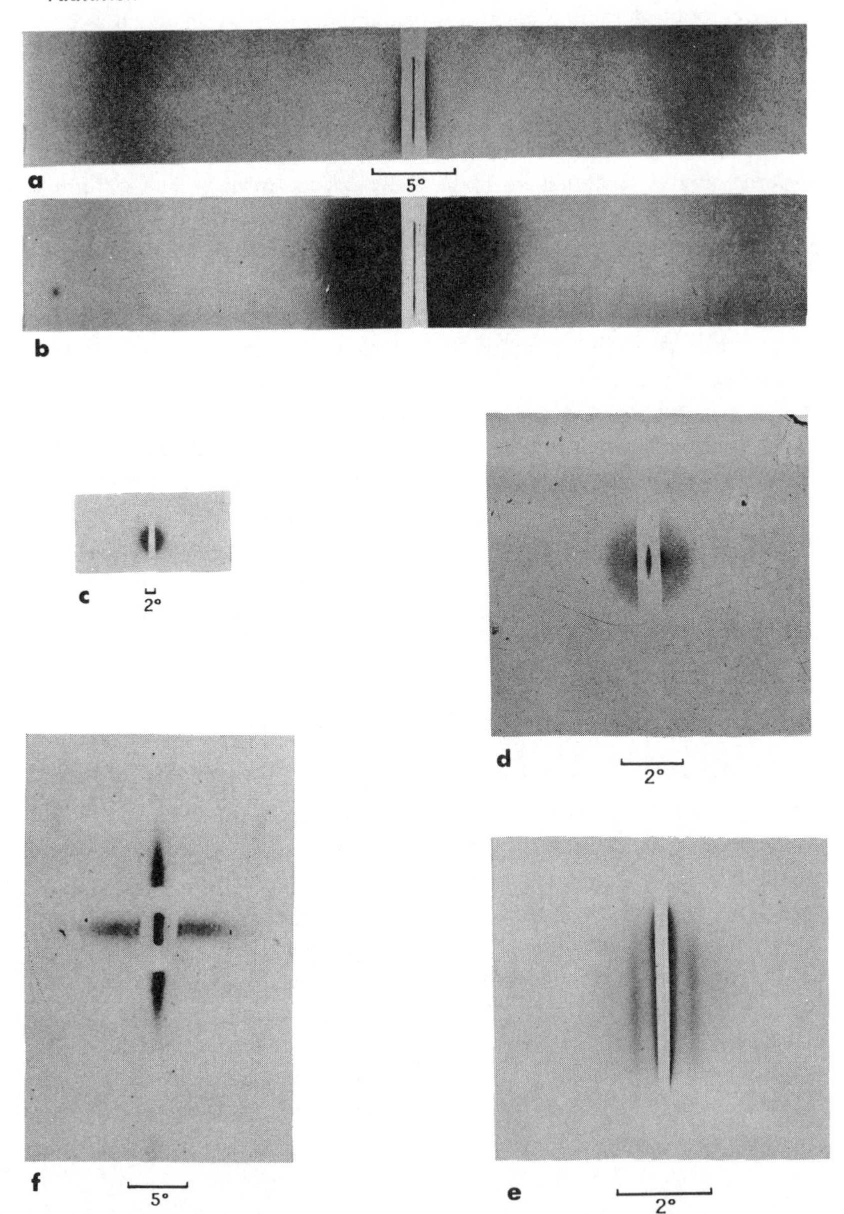

scattered light decreasing with the angle and becoming zero for an angle given approximately by $\lambda/d$, where $d$ is the diameter of the scattering particles. This is the same relation as above, but the particles can be much larger than in the case of X-rays, which have wave lengths more than a thousand times smaller than those of visible light.

Figure 10.1 shows the experimental setup used for observing scattering at very small angles. A very fine, well collimated, monochromatic beam passes through a sample containing submicroscopic grains such as a fine powder or a colloidal solution. A very narrow beam trap prevents the direct beam from reaching the film. Figure 10.2c shows the spot obtained in this manner with carbon black. The central peak is characteristic of the particle sizes. For example, vitreous silica gives the same pattern at large angles, whether it is in the compact form or in the form of silica gel, but *only* the latter, which is finely divided, gives a central peak (Figures 10.2a and 10.2b).

Attempts have been made to determine the sizes of submicroscopic grains and colloidal micelles from the shape of the central peak, just as the broadening of the diffraction spots permits a measurement of the sizes of very small crystals. The central peak has the advantage, however, of being usable even for noncrystalline substances. Furthermore, the broadening of the Debye-Scherrer lines can be due to causes other than crystal size, but *the central peak depends solely on the size of the sample.*

We shall now discuss the theoretical bases for such measurements, the instruments used, and various applications. We shall limit ourselves to generalities, since there exists a monograph by the author which deals with this subject [2], and we shall emphasize the important progress which has been achieved since its publication in 1954.

## 10.2. Theory of Small-angle Scattering

### 10.2.1. Identical Disoriented Particles. Very Low-density System

We assume that the object contains a large number of *identical* particles distributed at random, and that the average volume available for each particle is large compared to the volume of the particle. Under these conditions there is no interference between particles and, as in gas scattering, the intensities diffracted by the various particles simply add. If the particles all have the same orientation, the diffracted intensity per particle is equal to that for an isolated particle; if the particles are oriented at random, the observed intensity is the average intensity diffracted by a single particle for all possible orientations.

In many cases, as in colloidal solutions, for example, the particles are suspended in a homogeneous medium. If $\rho_0$ is the electron density of the continuous medium, the scattering power of the particle is

$$I(\boldsymbol{s}) = (\rho - \rho_0)^2 \,|\, \Sigma(\boldsymbol{s}) \,|^2 \,, \tag{10.4}$$

since the sample can be considered to consist of a medium of uniform density $\rho_0$, plus particles of density $\rho - \rho_0$. As a macroscopic sample of uniform density produces a central peak which is totally unobservable, the observed pattern is therefore correctly given by the above equation.

If the particles are oriented at random, the pattern exhibits symmetry of revolution and the scattering power per particle, as a function of $s = \varepsilon/\lambda$, is given by

$$I(s) = (\rho - \rho_0)^2 \times \begin{Bmatrix} \text{average value of } |\,\Sigma(\boldsymbol{s})\,|^2 \text{ over} \\ \text{the sphere of radius } s = |\,\boldsymbol{s}\,| \end{Bmatrix} \tag{10.5}$$

The calculation of the pattern for the central peak given by homogeneous particles of given shape can therefore be found from Eqs. (*10.4*) and (*10.5*) by a process of integration.

10.2.1.1. THE SCATTERING POWER FOR FOUR DIFFERENT SHAPES OF PARTICLES

(a) *Spheres.* The transform of the form function $\sigma(x)$ is

$$\Sigma(\boldsymbol{s}) = \int_V \exp\,(2\pi i \boldsymbol{s} \cdot \boldsymbol{x})\, dv_x \,,$$

where the integral is evaluated over the volume $V$ of the sphere of radius $a$. From the equation of part 12, Table A.1, Appendix A.4,

$$\Sigma(s) = \frac{4}{3}\pi a^3 \Phi(2\pi s\, a) = \frac{4}{3}\pi a^3 \left[ 3 \frac{\sin 2\pi a s - 2\pi a s \cos 2\pi a s}{(2\pi a s)^3} \right].$$

Thus the scattering power per particle is given by

$$I(s) = \left[ (\rho - \rho_0) \frac{4}{3}\pi a^3 \right]^2 \Phi^2(2\pi a s) \,, \tag{10.6}$$

where the function $\Phi^2(2\pi a s)$ of Figure 10.3 is equal to unity for $s = 0$, is zero for $as$ equal to $0.72, 1.23, 1.73, \cdots, \cong (2k+1)/4, \cdots$, and is maximum for $as$ equal to $0.92, 1.44, 1.95, \cdots, \cong k/2, \cdots$. The intensities of these maxima are respectively $7.4 \times 10^{-3}, 1.28 \times 10^{-3}, 0.4 \times 10^{-3}, \cdots, \cong 0.097/k^4, \cdots$.

The pattern therefore comprises a central spot surrounded by rings of decreasing intensity. Such a pattern has been observed with latex spheres of uniform diameter suspended in water (Figure 10.4). The radius of the spheres, as measured with an electron microscope, was 2,780Å, which agrees very well with the value deduced from the diameter of the rings by Henke and DuMond [3]. One can rarely observe the secondary maxima, which occur only if the particles all have exactly the same radius. If

**FIGURE 10.3.** *Intensity diffracted by a homogeneous sphere of radius a:*

$$\frac{I(s)}{[(\rho-\rho_0)\frac{4}{3}\pi a^3]^2}=\frac{9(\sin 2\pi as-2\pi as\cos 2\pi as)^2}{(2\pi as)^6}.$$

*The scale is multiplied by 100 for* $5/2\pi a < s < 10/2\pi a$ *and by 1000 for* $s > 10/2\pi a$. *The dotted curve shows the asymptotic function* $9/32\pi^4(as)^4$ *of Eq. (10.25).*

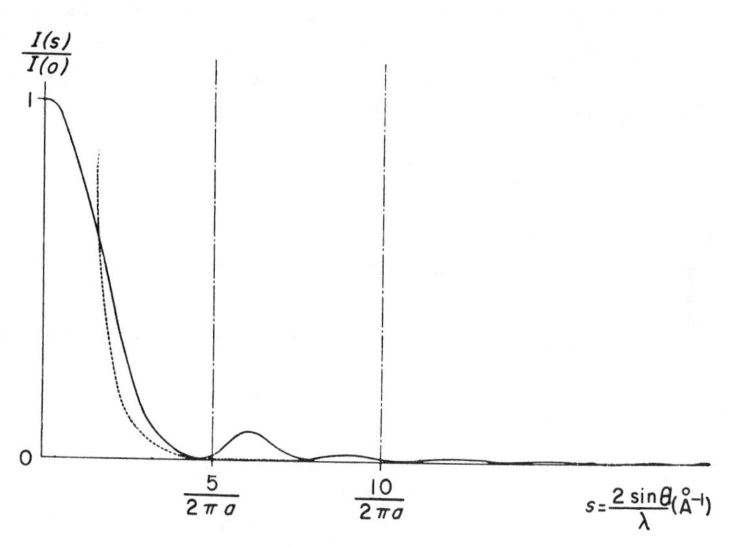


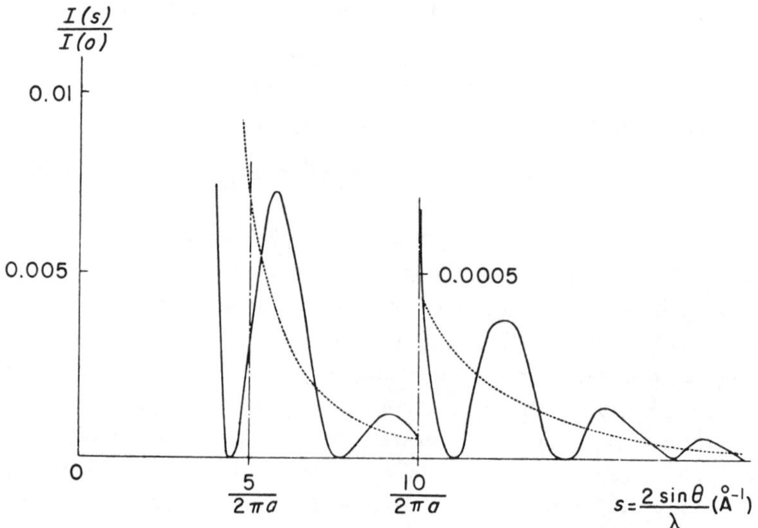

**FIGURE 10.4.** *Diffraction rings for spherical particles 2780 Å in diameter, obtained with a point monochromator. Cu Kα radiation. Sample to film distance 66 cm. Thirteen rings are visible on the original photograph, from the 5th to the 17th* (*Henke and DuMond* [3]).

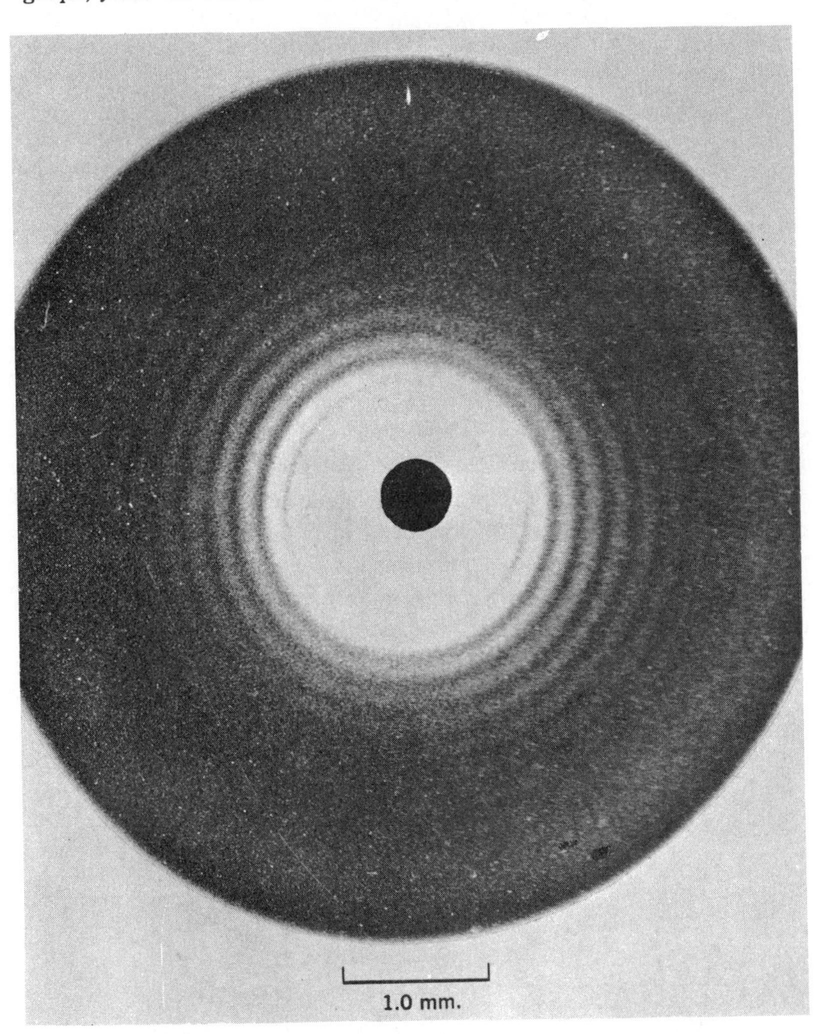


the radii differ only slightly, the rings corresponding to the different radii do not correspond and the resulting curve shows no maxima.

(b) *Very thin cylinders of volume V and length 2H.* In this instance,

$$I(s) = [(\rho - \rho_0)V]^2\left[\frac{\mathrm{Si}(4\pi sH)}{2\pi sH} - \frac{\sin^2 2\pi sH}{(2\pi sH)^2}\right], \qquad (10.7)$$

where $\mathrm{Si}(x) = \int_0^x (\sin t/t)\, dt$.

(c) *Flat disks of radius R and volume V.* Here

$$I(s) = [(\rho - \rho_0) V]^2 \frac{2}{(2\pi sR)^2}\left[1 - \frac{1}{2\pi sR} \mathrm{J}_1(4\pi sR)\right], \tag{10.8}$$

where $\mathrm{J}_1$ is the Bessel function of first order [4].

(d) *Ellipsoids of revolution of volume V and axes a, a, and va:*

$$I(s) = [(\rho - \rho_0) V]^2 \int_0^{\pi/2} \Phi^2\left(\frac{2\pi sav}{\sqrt{\sin^2 \alpha + v^2 \cos^2 \alpha}}\right) \cos\theta\, d\theta, \tag{10.9}$$

where $\Phi$ is as in Eq. (*10.6*) and $\alpha = \arctan(v \tan\theta)$.

### 10.2.2. Approximate Formula Applicable to the Center of the Intensity Curve

The scattering power, which is equal to $n^2 = (\rho - \rho_0)^2 V^2$ for $s = 0$, decreases with $s$ as $n^2(1 - Ks^2)$. We shall now show that the curvature at the center of the curve is related to a simple geometrical parameter of the particle, *whatever the shape of the particle.*

Let $\boldsymbol{S}_0$ be the direction of the rays incident on the particle. For very small scattering angles, $\boldsymbol{s}$ is in the direction $\boldsymbol{D}$ normal to $\boldsymbol{S}_0$ and in the plane of the incident and scattered rays, as in Figure 10.5, and its modulus is $s = 2\theta/\lambda = \varepsilon/\lambda$. If $\boldsymbol{x}$ is any vector in object space, $\boldsymbol{s} \cdot \boldsymbol{x} = s x_D$ where $x_D$ is the projection of $\boldsymbol{x}$ on $\boldsymbol{D}$. To calculate the function $\Sigma(\boldsymbol{s})$, we have

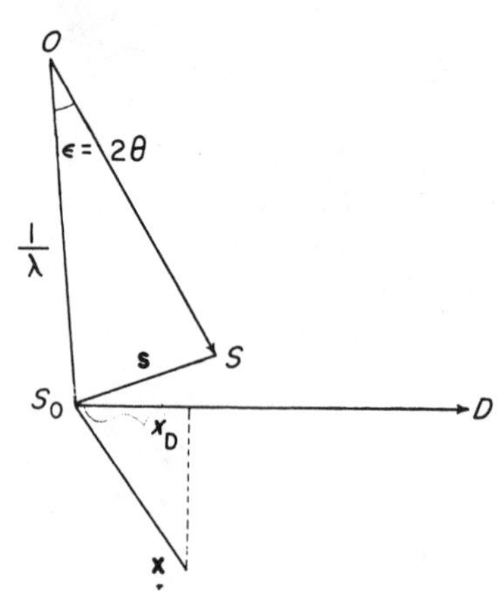


**FIGURE 10.5.** *The vector s in the case of very small-angle scattering.*

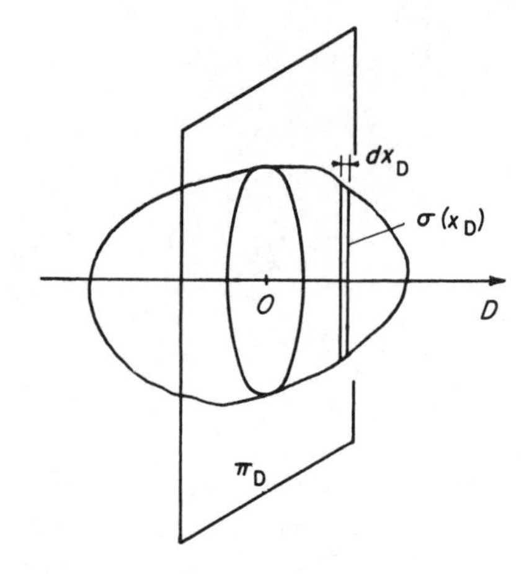


**FIGURE 10.6.** *Calculation of the form function of the particle with respect to its center of gravity O.*

to evaluate the integral

$$\Sigma(\boldsymbol{s}) = \int_V \exp(2\pi i \boldsymbol{s} \cdot \boldsymbol{x})\, dv_x\,,$$

This integral can be written as

$$\int \exp(2\pi i s x_D)\sigma(x_D)\, dx_D\,, \tag{10.10}$$

where $\sigma(x_D)$ is the cross-section of the particle along a plane normal to $\boldsymbol{D}$ at the distance $x_D$ from the origin, as in Figure 10.6.

We select the origin $O$ in object space at the center of gravity of the particle so that

$$\int x_D \sigma(x_D)\, dx_D = 0$$

and, since $s$ remains very small within the central peak, we may expand the exponential function of Eq. (*10.10*) and neglect terms of order higher than $s^2$. Then

$$\Sigma(\boldsymbol{s}) = \int \sigma(x_D)\, dx_D + 2\pi i s \int x_D \sigma(x_D)\, dx_D - 2\pi^2 s^2 \int x_D{}^2 \sigma(x_D)\, dx_D\,. \tag{10.11}$$

The first term is simply the volume $V$ of the particle; the second term is zero because of our choice of origin. Let us now set

$$R_D{}^2 = \frac{1}{V} \int x_D{}^2 \rho(x_D)\, dx_D\,, \tag{10.12}$$

where $R_D$ is the quadratic average of the distances to the plane $(\pi_D)$ normal to $\boldsymbol{D}$ and passing through the center of gravity $O$. This quantity $R_D$ may be called the *average* inertial distance along $\boldsymbol{D}$.

Then Eq. (*10.11*) becomes

$$\Sigma(\boldsymbol{s}) = V - 2\pi^2 s^2 V R_D{}^2.$$

We shall adopt as an approximation for $\Sigma(\boldsymbol{s})$ the exponential form

$$\Sigma(\boldsymbol{s}) = V \exp(-2\pi^2 s^2 R_D{}^2). \tag{10.13}$$

Then the curvature at the origin has the correct value and $\Sigma(\boldsymbol{s})$ tends toward zero as $s$ increases.

From Eqs. (*10.4*) and (*10.13*), we now have an approximate fomula for the scattering power per particle:

$$I(\boldsymbol{s}) = \rho^2 V^2 \exp(-4\pi^2 s^2 R_D{}^2) = n^2 \exp(-4\pi^2 s^2 R_D{}^2)\,. \tag{10.14}$$

For a spherical particle of radius $a$, the inertial distance is independent of the direction $\boldsymbol{D}$ and is equal to $R_D = a/\sqrt{5}$. The approximate expression corresponding to Eq. (*10.6*) is then

$$n^2 \exp\left(-\frac{4}{5}\pi^2 a^2 s^2\right). \qquad (10.15)$$

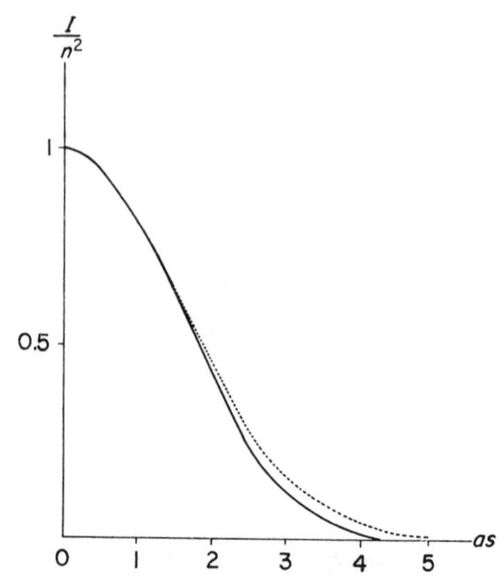


**FIGURE 10.7.** *The exact diffraction curve for a sphere is shown in solid line, while the dotted curve is the exponential approximation.*

As shown in Figure 10.7, this approximation is very satisfactory within the central peak.

In the case of particles of arbitrary shape but having a *common orientation* within the sample, Eq. (*10.14*) gives the shape of the central spot: *its diameter is greatest in the direction in which the diameter of the particle is smallest.* This is illustrated in Figure 10.8. Thus, for ellipsoids of revolution of axes $a$, $a$, and $va$ irradiated perpendicularly to their axes of revolution, $R_D$ is respectively $a/\sqrt{5}$ and $va/\sqrt{5}$ in the directions parallel and normal to these axes. The lines of equal intensity in the central spot are ellipses elongated in the direction of the smaller axes, as in Figure 10.8.

Let us now consider the case where the particles have random orientations. The observed intensity depends only on $s$:

$$I(s) = n^2 \exp(-4\pi^2 s^2 \overline{R_D^2}),$$

$\overline{R_D}^2$ being the average of $R_D^2$ for the various directions of $\boldsymbol{D}$, i.e. of $\boldsymbol{s}$. This average can be calculated by associating with to $\boldsymbol{D}$ two other orthogonal

**FIGURE 10.8.** *Small-angle scattering by an ellipsoid of revolution whose axis is perpendicular to the incident ray.*

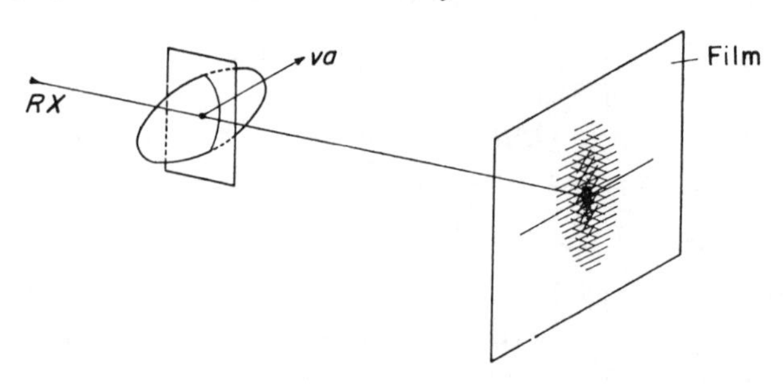

axes, $U$ and $V$, as in Figure 10.9. Then a point inside a particle is given by its coordinates $x_D$, $x_U$, $x_V$, and the square of its distance to the center of gravity is $r^2 = x_D^2 + x_U^2 + x_V^2$.

Taking the quadratic average on both sides,

$$R^2 = R_D^2 + R_U^2 + R_V^2,$$

where the quantity $R$, defined by

$$R^2 = \frac{\int r^2 dv}{V},$$

is the *radius of gyration* of the particle with respect to its center of gravity, and $R_D$, $R_U$, $R_V$ are the inertial distances with respect to the three coordinate planes. Now if these coordinate planes are rotated about the origin, $R^2$ remains constant and the three averages $R_D^2$, $R_U^2$, $R_V^2$ are equal, so $3\overline{R_D^2} = R^2$. Thus the average scattering power particle in the case of random orientations is given by

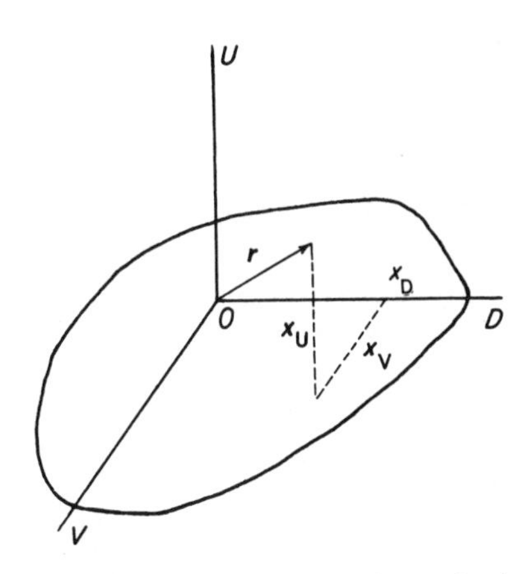


**FIGURE 10.9.** *Calculation of the quadratic average of the inertial distance* $R_D$.

$$I(s) = n^2 \exp\left(-\frac{4\pi^2 s^2 R^2}{3}\right) = n^2 \exp\left(-\frac{4\pi^2 R^2}{3\lambda^2}\varepsilon^2\right), \qquad (10.16)$$

where $\varepsilon$ is the scattering angle.

For a particle of known shape, the radius of gyration is easily calculated. In the case of a sphere of radius $a$, $R = (3/5)^{1/2}a$; for an ellipsoid of revolution with axes $a, a, va$, $R = [(2 + v^2)/5]^{1/2}a$.

10.2.2.1. VALIDITY OF THE EXPONENTIAL APPROXIMATION IN THE CASE OF HETEROGENEOUS PARTICLES. We have assumed above that the particle was homogeneous, first of all because this is often the case, but also because this permitted the use of some general results which we had previously established. The validity of the approximate Eq. (*10.16*) can however be extended [5] to the case of particles composed of an arbitrary group of atoms of atomic numbers $f_k$ at distances $r_k$ from an origin $O$. The particles are all identical but oriented at random. Let us assume that the particles are suspended in a vacuum. Since a particle can have any orientation, its average scattering power is given by the Debye formula [Eq. (*2.54*)], or

$$I_N(s) = \sum\sum f_k f_j \frac{\sin 2\pi s r_{kj}}{2\pi s r_{kj}}$$

Expanding this expression as a power series in $s$,

$$I_N(s) = \sum\sum f_k f_j - \sum\sum f_k f_j \frac{4\pi^2 s^2 r^2{}_{kj}}{6} + \cdots .$$

The contant term is equal to $(\sum f_k)^2 = n^2$, where $n$ is the total number of electrons in the particle. To calculate the second term, we now select the origin $O$ so that $\sum f_k \boldsymbol{r}_k = 0$. This "electronic" center of gravity coincides with the ordinary center of gravity if all the atoms are identical. The two points are always very close to each other since the atomic numbers $f_k$ are closely proportional to the atomic masses. We express $r_{jk}$ in terms of $r_j$ and $r_k$:

$$r_{kj}^2 = (\boldsymbol{r}_j - \boldsymbol{r}_k)^2 = r_k^{\,2} + r_j^{\,2} - 2\boldsymbol{r}_j \cdot \boldsymbol{r}_k ,$$

$$\sum\sum f_k f_j r_{kj}^2 = \sum\sum f_k f_j r_j^{\,2} + \sum\sum f_k f_j r_j^{\,2} - 2 \sum f_k \boldsymbol{r}_k \cdot \sum f_j \boldsymbol{r}_j .$$

The two first terms are equal to

$$\sum f_j \sum f_k r_k^{\,2} = n \sum f_k r_k^{\,2} ,$$

while the third is zero because of the manner in which we have selected our origin ($\sum f_k \boldsymbol{r}_k = 0$). If we now set

$$R^2 = \frac{\sum f_k r_k^{\,2}}{\sum f_k} ,$$

we have again Eq. (*10.16*). The generalized definition of the radius of gyration is of course equivalent to that previously given when applied to homogeneous particles.

10.2.2.2. INTERPRETATION OF THE EXPERIMENTAL CURVES IN TERMS OF THE EXPONENTIAL APPROXIMATION. By taking logarithms to the base 10 of both sides of Eq. (*10.16*),

$$\log I = \log n^2 - \frac{4\pi^2}{3\lambda^2} 0.4343 R^2 \varepsilon^2 . \qquad (10.17)$$

Thus, if $\log I$ is plotted as a function of $\varepsilon^2$, the curve tends to a straight line of slope

$$\alpha = -\frac{4\pi}{3\lambda^2} 0.4343 R^2$$

for small values of $\varepsilon$. This slope gives the radius of gyration

$$R = \sqrt{\frac{3}{4\pi^2 \times 0.4343}} \lambda \sqrt{-\alpha} = 0.416\, \lambda \sqrt{-\alpha} . \qquad (10.18)$$

For example, with Cu K$\alpha$ radiation ($\lambda = 1.54$Å),

$$R = 0.644\sqrt{-\alpha}\ \text{Å}\ .$$

The above calculation of the radius of gyration assumes that the particles are *identical*, that they are *randomly oriented*, and that the sample has a *low density* so that interference between the various particles is negligible. The significant quantity is the slope of the above curve near the origin.

In practice, the curve can be determined only for angles larger than an angle $\varepsilon_0$, which depends on the instrument used. If the radius of gyration of the particles is small enough that $\varepsilon_0$ is considerably less than the angle at which the exponential approximation ceases to be valid, then the curve of $\log I$ *versus* $\varepsilon^2$ has a region of low curvature near the origin which can be used to determine the radius of gyration of the particle from Eq. (*10.18*).

Thus, for a relatively simple setup we may have $\varepsilon_0 = 1/200$ and $\lambda = 1.5$Å. Then the upper limits for the radii of gyration which can be measured

**FIGURE 10.10.** *Intensity curves for small angle scattering.* (a) *Particles with the same radius of gyration. Sphere with* $R = 1$ *and ellipsoid with* $a = 0.41$, $v = 4$. (b) *Particles with the same volume. Sphere with* $R = 1$ *and ellipsoid with* $a = 0.63$, $v = 4$.

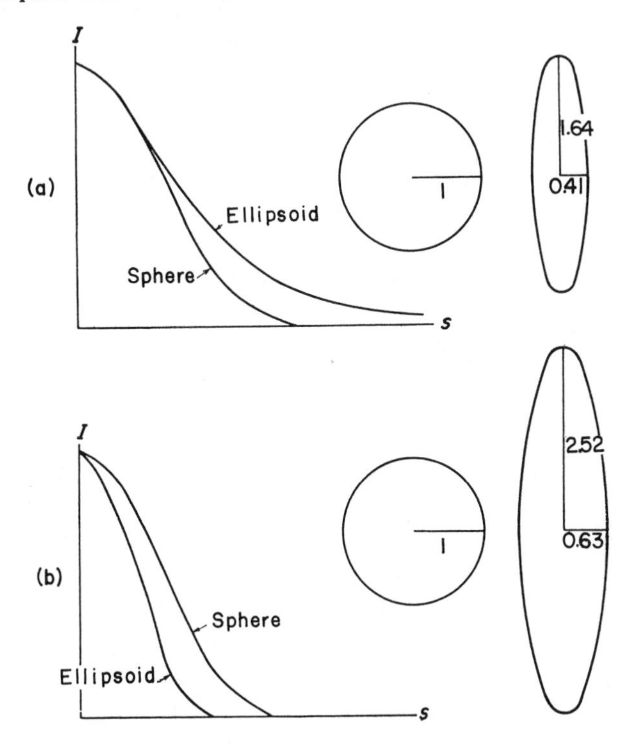

are 60Å for spheres and 35Å for thin cylinders [2].

An infinite number of particle shapes having the same radius of gyration can give identical central peaks. The curves diverge at large angles, but only slightly. *It is found that widely different particle shapes give curves which are only slightly different.* The curves of Figure 10.10a apply to particles with a common radius of gyration. If we now compare particles of a given volume (Figure 10.10b), their radii of gyration, and therefore their central peaks, depend on their shape. Thus in the case of particles of *known volume* the determination of the radius of gyration does give information on the shape. For example, dilute solutions of hemoglobin molecules were investigated by Fournet [6]. For concentrations lower than 10%, the log *I versus* $\varepsilon^2$ curve shown in Figure 10.11 has a long straight portion which yields a radius of gyration of $23 \pm 1$ Å, from Eq. (*10.18*). It is known that the molecular weight is $6.67 \times 10^4$ and that the density is 1.33, so the volume of a molecule is $8.34 \times 10^4$ Å$^3$. Assuming that the molecule has the shape of a cylinder of height $H$ and radius $a$, both the volume $V$ and the radius of gyration $R$ can be calculated in terms of $a$ and $H$. Since $R$ and $V$ are known, we find that $a = 30$ Å and $H = 15$ Å.

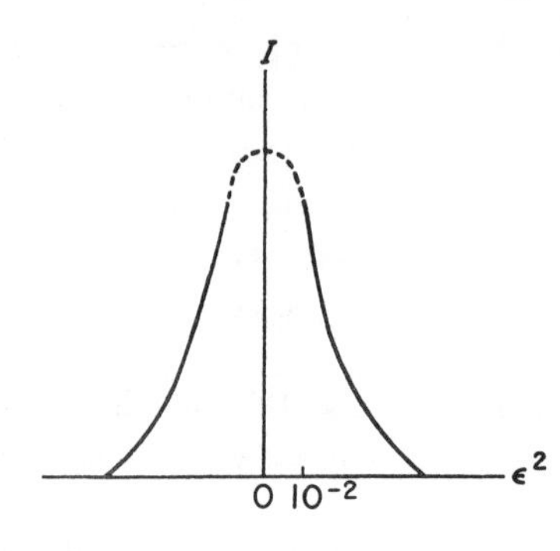


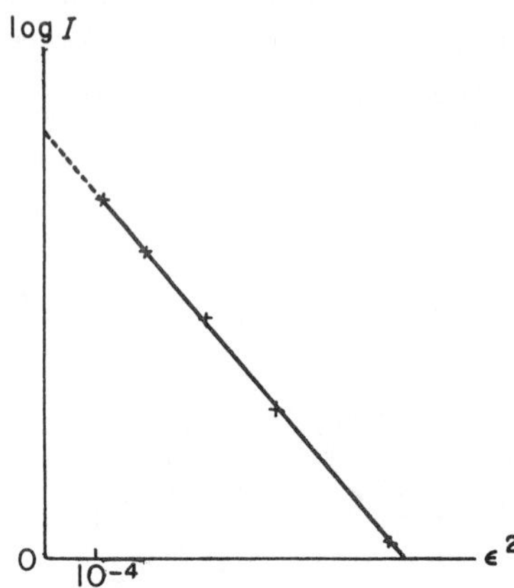


**FIGURE 10.11.** *Curves of* $I$ *vs* $\varepsilon^2$ *and of* log $I$ *vs* $\varepsilon^2$ *for a 10% hemoglobin solution.*

We have so far used only the shape of the curve of diffracted intensity *vs* angle. If the ratio of scattered to incident intensity is known, further information can be obtained, as follows. If the sample contains $N$ particles of scattering power $I(s)$, the flux of energy within the solid angle $d\omega$ of the scattered beam is given by

$$P(\varepsilon) = I_e N I(s)\, d\omega$$
$$= 7.90 \times 10^{-26} I_0\, d\omega N n^2 \exp\left(-\frac{4\pi^2 R^2 \varepsilon^2}{3\lambda^2}\right), \qquad (10.19)$$

where $I_e$, the energy diffracted by a free electron at the small angle $\varepsilon$, is given by the Thomson formula [Eq.(*1.23*)]. The intensity of the direct beam [per $cm^2$ in Eq. (*10.19*)] is $I_0$. If this is measured after the beam has passed through the sample, no absorption correction is necessary. When the log $I(s)$ *versus* $\varepsilon^2$ curve is a straight line, it is easy to extrapolate the measurements up to the zero angle and thus to determine experimentally $P(0)/I_0$. Equation (*10.19*) then gives the *absolute* values of $Nn^2$, where $N$ is the number of particles in the sample and $n$ the number of electrons per particle.

If $N$ is known, $Nn^2$ gives $n$, and also the number of atoms, the mass, and the volume of a particle if the composition and density of a particle are known. On the other hand, if $n$ is known, $Nn^2$ gives $N$, and thus the concentration of the particles in the sample. Luzzati [7] has shown how such measurements can be used in the field of macromolecules. He has also given practical formulas for the case of solutions.

In the absence of absolute measurements and of other information, the complete scattering curve gives only vague indications on the shape of the particles. The method of the radius of gyration has the advantage of providing a quantity which has an exact meaning instead of providing a more or less indefinite average of the dimensions. That is comparable to the modern interpretations of line widths, which lead to parameters which are geometrically well defined but which nevertheless do not entirely determine the shapes of small crystals.

On the other hand, if the particles are oriented, the study of the variation of the intensity with the azimuth angle can provide average inertial distances along interesting directions. For example, Hosemann [8] has been able to determine the length and diameter of cellulose micelles in this way. One can easily find the orientation of the micelles which are known to be elongated. Heyn [9] has also studied small-angle scattering having the shape of a cross in the case of some natural fibers. He found that the micelles are arranged along helices around the fiber axis and he was able to measure the angle between the helix and its axis. We may recall also the diffraction pattern of a single Al Cu crystal during age hardening [10] (Section 8.3.5). When the incident rays are directed along the [001] axis, the pattern involves a cross whose branches are directed along the [100] and [010] axes, as in Figure 10.2*f*. These streaks are produced by copper particles having the form of thin disks parallel to the three planes {100}. It is possible to show that the particles have the form of disks (and not of needles directed along the $<100>$ axes) by rotating the crystal around one of these axes.

10.2.2.3. TOTAL ENERGY SCATTERED AT SMALL ANGLES, AND THE CORRESPONDING SCATTERING COEFFICIENT. Consider a set of identical particles of volume $V$ and radius of gyration $R$. The scattering power *per unit mass* is

$$I_m(\varepsilon) = Nn^2 \exp\left(-\frac{4\pi^2 R^2 \varepsilon^2}{3\lambda^2}\right),$$

$N$ is the number of particles per gram or $1/(V\rho)$ if the specific mass of the particle is $\rho$, and $n$ is the number of electrons per particle. If the particle is composed of atoms of atomic weight $M$ and atomic number $Z$,

$$n = V\rho \frac{Z}{M} 6.06 \times 10^{23}.$$

The ratio $M/Z$ may be set equal to 2—except for hydrogen, which produces negligible scattering. Thus

$$I_m(\varepsilon) = \frac{V\rho}{4}\, 36 \times 10^{46} \exp\left(-\frac{4\pi^2 R^2 \varepsilon^2}{3\lambda^2}\right).$$

By the Thompson formula of Eq. (*1.23*), the power scattered in the solid angle $d\omega$ is

$$i_m(\varepsilon)\, d\omega = 7.3 \times 10^{-3} I_0 V\rho \exp\left(-\frac{4\pi^2 R^2 \varepsilon^2}{3\lambda^2}\right) d\omega,$$

where $I_0$ is the intensity of the incident beam (per cm$^2$), and $V$ is expressed in (Å)$^3$. The total energy in the central peak is then

$$E_m = \int i_m\, d\omega = \int_0^\infty i_m(\varepsilon) 2\pi\varepsilon\, d\varepsilon$$
$$= 1.7 \times 10^{-3} I_0 \lambda^2 \rho \left(\frac{V}{R^2}\right), \qquad (10.20)$$

where $\lambda$ and $R$ are expressed in Å and $V$ in Å$^3$. For particles of similar shapes, $E_m$ increases linearly with their size, since $V/R^2$ is of the order of magnitude of an average diameter. Thus, for a sphere of radius $a$, the ratio $V/R^2 = 7.0a$.

For an incident beam of intensity $I_0$ traversing a sample of mass $dm$ per square centimeter, the energy lost from a beam of 1 cm$^2$ of cross-section by small-angle scattering is

$$dI = E_m\, dm = I_0 \sigma\, dm,$$

where $\sigma$ is a quantity which is analogous to the mass absorption coefficient and is called the mass scattering coefficient at small angles. According to Eq. (*10.20*),

$$\sigma = 1.7 \times 10^{-3}\lambda^2\rho\frac{V}{R^2} \tag{10.21}$$

and, for spherical particles of radius $a$,

$$\sigma = 0.012\lambda^2\rho a\ . \tag{10.22}$$

The attenuation of the beam in passing through the *finely divided sample* is given by the coefficient $\mu + \sigma$, where $\mu$ is the absorption coefficient for the same substance in compact form.

In cases where it is experimentally possible to separate all the central peak from the direct beam, it is possible to use Eq. (*10.22*) to determine the dimensions of the particles (Warren [11]).

It is important to note that $\sigma$ can be large compared to $\mu$ in the case of substances of low atomic weight and for large particles. Thus, for carbon and Cu K$\alpha$ radiation, $\rho = 2$ and $\mu = 4.52$. Then $\sigma$ is 2.85 for $a =$ 50 Å and 28.5 for $a = 500$ Å.

The total scattered energy $E$ is $\sigma p E_1$, where $p$ is the sample thickness in grams per square centimeter, and $E_1$ is the energy of the direct beam *after going through the sample*. For example, a layer of 0.5 mm of carbon has a $p$ of 0.1 g/cm$^2$ and, for $\lambda = 1.5$ Å and spherical particles of radius $a$, $E/E_1 = 5.6 \times 10^{-3}a$. If the radius $a$ is 50 Å, $E/E_1$ is equal to 0.28, and if $a = 500$ Å, $E/E_1$ is 2.8. The transmitted beam is then only a small fraction of the scattered beam.

Under these conditions, if we add a further layer of scattering particles to the sample, their main effect is to scatter the incident beam which has already been scattered by the preceding layers.

Single-scattering phenomena can thus be partly, or even completely masked by multiple scattering. The small-angle scattering theory discussed in this chapter applies only to samples containing a small number of small particles, and thus to scattered radiation of low intensity. On the contrary, multiple scattering occurs in cases where the photoelectric absorption coefficient is small—for then the sample can be thick—and where the scattered intensity is high because the particles are large.

The simplest experiments on multiple scattering are made on carbon black with a very fine monochromatic beam. With a thin sample there is single scattering and a diffuse spot of low intensity compared to that of the undeflected direct beam. As the thickness increases, the probability of multiple scattering increases, and the direct beam gradually decreases in intensity until it disappears entirely. We are then left with a diffuse spot whose size increases as the sample thickness continues to increase [12]. It is a simple matter to measure the intensity as a function of angle in such a case, for there is no need to avoid parasitic scattering due to the direct beam. Luzzati [13] has given the most complete analysis of this multiple-scattering phenomenon, and he has shown how it can be used to obtain information on the size of the scattering particles.

### 10.2.3. Approximation for the Wings of the Curve

The exponential formula [Eq. (*10.16*)], is an approximation of the general Eq. (*10.2*) for the diffracted intensity, which is valid near the center of the curve (small $s$). There is another approximate equation which is valid for the wings of the curve, in the region where the small-angle scattering tends to zero. This *Porod equation* [14] is interesting because it provides a geometrical parameter which is different from the radius of gyration.

Let us consider the intensity diffracted by a sphere. Equation (*10.6*) can be rewritten as

$$I = \frac{(\rho - \rho_0)^2}{8\pi^3}\left[\frac{4\pi a^2}{s^4} + \frac{1}{\pi s^6} - \frac{4a}{s^5}\sin 4\pi as + \left(\frac{4\pi a^2}{s^4} - \frac{1}{\pi s^6}\right)\cos 4\pi as\right]. \tag{10.23}$$

The first two terms decrease uniformly with $s$, the term in $s^{-4}$ being the more important, while the other terms oscillate with a decreasing amplitude and with a pseudoperiod of $s = 1/(2a)$, as in Figure 10.3. Now consider a set of spheres whose radii vary between $a_1$ and $a_2$, the number of spheres of radius $a_k$ per unit mass being $g_k$. The scattering power per unit mass of the sample is then

$$I = \frac{(\rho - \rho_0)^2}{8\pi^3}\left[\sum g_k 4\pi a_k^2 \frac{1}{s^4} + \cdots + \sum \cos 4\pi \alpha_k s[\cdots] + \sum \sin 2\pi a_k s[\cdots]\right]. \tag{10.24}$$

If $(2a_1 s - 2a_2 s)$ is much larger than unity, the last two summations are zero since their positive and negative contributions cancel. It is in fact observed that the oscillations disappear from the curve whenever the sample is not absolutely homogeneous. Therefore, if $s$ is sufficiently large, we are left with the $s^{-4}$ term. Now $\sum g_k 4\pi R_k^2$ is the total surface $S$ of the particles per unit mass, and the asymptotic value of the intensity is given by

$$I_{\text{asympt}} = \frac{(\rho - \rho_0)^2}{8\pi^3}\frac{S}{s^4} \tag{10.25}$$

Porod [14] has shown that this equation *is valid for any shape of particle*, so long as the orientations are random, and so long as none of the particle dimensions are zero ($sL \gg 1$ for every dimension $L$ of the particle). The above asymptotic formula is valid for *dissimilar particles* and also for particles which are quite close to one another, because interference between particles has little effect on the wings of the curve.

This law has been verified experimentally by Van Nordstrand [15] in the case of alumina powders used as catalysts. When $s$ is not too small, the curves of log $I$ *versus* log $\varepsilon$ are straight lines of slope $-4$. He also observed empirically that, for these scattering angles, the scattered intensity is proportional to the total surface of the particles contained in one gram of the powder.

According to Eq. (*10.25*), the ratio of the scattered intensity to the incident intensity gives the absolute value of the specific area of the powder, but analogous samples may be compared by simple relative measurements. Once the validity of the Porod equation has been established, a *single measurement* at a fixed angle is sufficient to compare, for instance, the surfaces of a set of catalyst powders of the same composition. The angle is chosen small enough that the diffracted intensity is sufficiently large but within the region where the $s^{-4}$ law is valid. The various samples are prepared so they all have the same mass per unit area. The reading at the detector is then a relative measurement of the total surface per gram.

For particles having at least one dimension very small, the asymptotic function is different from $s^{-4}$. For instance, Luzzati [16] has shown that for a set of long, thin rods of random orientation, the following equation is valid when $1/s$ is large compared with the diameter of the rod but small compared with its length:

$$I(s) = \frac{1}{2s} Lm^2(1 - 2\pi^2 R_c^2 s^2 + \cdots) , \tag{10.26}$$

where $L$ is the total length of all the rods, $m$ is their electron density per unit length, and $R_c$ is the radius of gyration of the rod relative to its axis.

### 10.2.4. Some Integral Properties of the Scattered Intensity

If the intensity $I(s)$ is integrated over the whole region surrounding the origin, Eq. (*10.4*) gives

$$\int I(\mathbf{s})\, dv_s = (\rho - \rho_0)^2 \int |\Sigma(\mathbf{s})|^2\, dv_s$$

or according to Eq. (*2.38*)

$$\int I(\mathbf{s})\, dv_s = (\rho - \rho_0)^2 V .$$

If the sample contains identical particles of random orientation, the observed intensity $I(s)$ is the average value of $I(\mathbf{s})$ for a given value of $s$. Thus

$$\int I(\mathbf{s})\,dv_s = (\rho - \rho_0)^2 V = \int_0^\infty 4\pi s^2 I(s)\,ds\,. \tag{10.27}$$

Although $I(s)$ is not known experimentally up to $s = 0$, the extrapolation of $s^2 I(s)$ can be made with accuracy. Absolute measurements of $I(s)$ allow the determination of $V$ if the electron density of the particle is known, or conversely of $(\rho - \rho_0)$ if $V$ is known.

Equations (*10.4*) and (*10.27*) give the volume of the particle, in terms of relative measurements only:

$$\frac{I(0)}{\int_0^\infty 4\pi s^2 I(s)\,ds} = \frac{(\rho - \rho_0)^2 V^2}{(\rho - \rho_0)^2 V} = V\,. \tag{10.28}$$

In this case $I(s)$ can be the diffracted intensity in arbitrary units for a sample of undetermined volume.

The integral of Eq. (*10.27*) may be applied to the most general case of a heterogeneous sample defined by its electron density $\rho$, which varies from point to point [17]. If $\rho_0$ is the average density and $I(s)$ the scattering power for the sample of volume $V$,

$$\int 4\pi s^2 I(s)\,\mathrm{ds} = V\overline{(\rho - \rho_0)^2}, \tag{10.29}$$

where $(\rho - \rho_0)$ is the local fluctuation of the electron density.

### 10.2.5. Low-density System of Nonidentical Particles

It is in many cases not permissible to assume that all the particles are identical. The particles may be simply grains of matter or polymers of indefinite molecular weight.

Let us classify them into in groups of identical particles. The $k$th group with $n_k$ electrons per particle and with a radius of gyration $R_k$ is present in the proportion $g_k$. Then the scattering power per particle for the sample is

$$I(\varepsilon) = \sum g_k n_k^2 \exp\left(-\frac{4\pi^2}{3\lambda^2} R_k^2 \varepsilon^2\right).$$

The curve is *not* an exponential any more but, for very small values of $\varepsilon$.

$$I = \sum g_k n_k^2 \left[1 - \frac{4\pi^2}{3\lambda^2}\varepsilon^2 \frac{\sum g_k n_k^2 R_k^2}{\sum g_k n_k^2} + \cdots\right].$$

The tangent to the $\log I$ *versus* $\varepsilon^2$ curve at the origin has a slope of $(4\pi^2/3\lambda^2)R_m^2$. where

$$R_m^2 = \frac{\sum g_k n_k^2 R_k^2}{\sum g_k n_k^2} \tag{10.30}$$

It must be noted, however, that in calculating this average, as above, the large particles are weighted much more heavily than the small ones. Such a measurement is possible only if the observations can extend to scattering angles such that *the exponential approximation is valid for the largest particles of the sample.* If the log $I$ *versus* $\varepsilon^2$ curve (Figure 10.12) has a pronounced curvature throughout the region where it can be determined, it is then impossible to extrapolate it to $\varepsilon = 0$, and the determination of $R_m$ becomes very uncertain.

**FIGURE 10.12.** *Curve of* log $I$ *versus* $\varepsilon^2$ *for a wide range of particle sizes.*

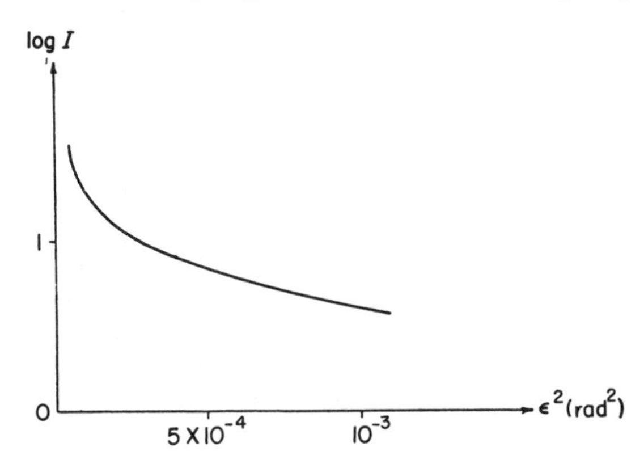


Various authors [18], [19] have attempted to calculate the intensity curve for the case where the particles have the same composition and a known shape, where $n_k$ and $R_k$ are known functions of $k$, and where the distribution function $g_k$ is assumed *a priori* to be a function of the average size of the particles and of the width of the distribution. The intensity curve can then be calculated in terms of these last two quantities. In practice, these are found empirically by fitting theoretical curves to the experimental data. The result depends on the initial hypotheses as to the shape of the particles, for entirely different shapes can be made to fit the data equally well. For example, identical ellipsoidal particles give the same curve as a set of spheres having a proper size distribution.

While the center of the curve gives, in general, only qualitative information, the large-angle approximation (Porod law) is valid for heterodispersed as well as for homodispersed samples.

### 10.2.6. Dense Systems of Identical Particles

If the particles are close to each other, as in the case of a concentrated solution, the observed intensity is not the sum of the intensities scattered by the individual particles, since interference then enters into play. If the particles are identical and distributed uniformly, the problem reduces to that of liquids, which we have already discussed. It is possible to calculate the scattered intensity if one assumes that the orientations are completely random. The distribution of the particles is given by a function $P(x)$ such that $4\pi x^2 P(x)\,dx$ is the number of particles whose centers are situated at a distance lying between $x$ and $x + dx$ of the origin chosen at the center of one of the particles. Let $\overline{F^2(s)}$ and $|\overline{F(s)}|^2$ be the average inetnsity and the square of the average amplitude for an isolated particle whose orientation varies at random. Let $v_1$ be the average volume available to one particle. Then, from Eqs. (*2.47*) and (*2.57*), the scattering power per particle is

$$I(s) = \overline{F^2(s)} + \frac{2}{sv_1} |\overline{F(s)}|^2 \int_0^\infty [P(x) - 1] \sin(2\pi sx) x\, dx\,. \qquad (10.31)$$

If the particles have spherical symmetry, $\bar{F}^2 = |\bar{F}|^2 = F^2$ and

$$I(s) = F^2(s)\left\{1 + \frac{2}{sv_1}\int_0^\infty [P(x) - 1] \sin(2\pi sx) x\, dx\right\}, \qquad (10.32)$$

where $I = F^2(s)$ is the scattering power per particle when there is no interference (gas scattering). As the concentration increases, the second term in the bracket reduces the intensity scattered at the center, since it is negative when $s$ is very small. The curve then has a *hump* which becomes a *maximum* at high concentrations [6] and the central spot is accompanied by a ring, as in case of liquids and amorphous solids (Figure 3.11).

There is no significant interference when $s$ increases; thus the Porod formula [Eq. (*10.25*)] is valid, even for nearly close-packed particles. As we have shown for the case of liquids, the quantity $1/s_m$ gives the order of magnitude of the distance $d_m$ between neighboring particles, where $s_m$ angle corresponding to the maximum or to the hump in the curve. Thus $\lambda \cong \varepsilon_m d_m$. Such formulas, which are analogous to the Bragg formula, $\lambda = 2d \sin\theta$, with or without a correction coefficient, cannot give an accurate measurement of an "average distance" between particles which cannot, in any case, be defined rigorously,

Qualitatively, a scattering curve which does not decrease monotonically indicates the existence of a certain degree of order in the arrangement of the particles, the degree of order being greatest when the curve is

**FIGURE 10.13.** *Scattering curves for the hemoglobin of red cells [20].* (a) *Artificial solution of hemoglobin.* (b) *Normal cells prepared with heparin.* (c) *Abnormal cells.*

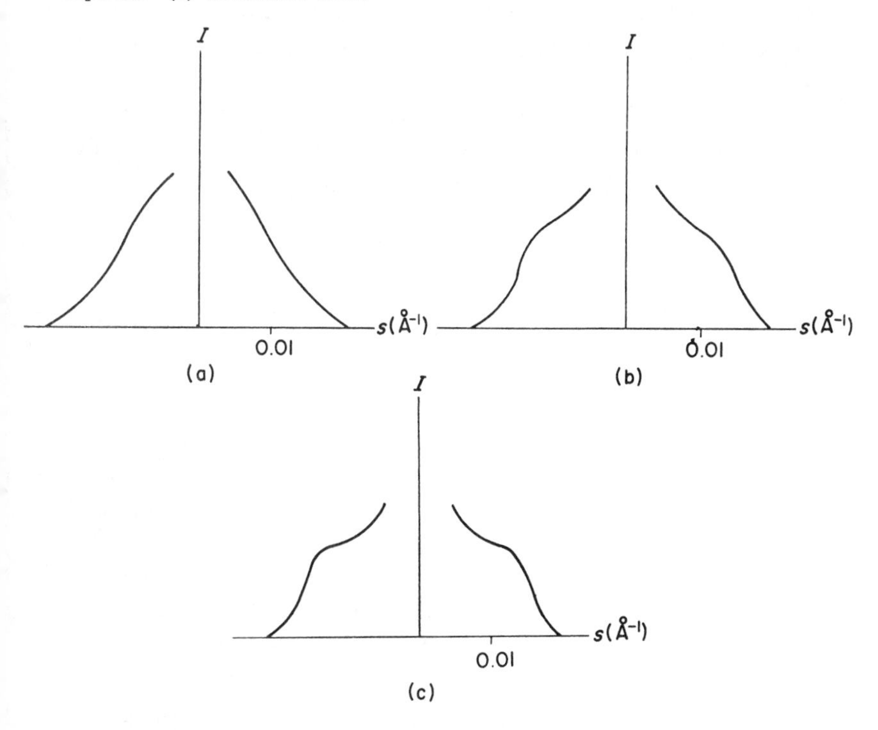


most irregular. For example, red blood cells [20] give a scattering curve like that of Figure 10.13, which corresponds to Figure 10.2d. Dilute solutions of hemoglobin, however, give a smooth curve, as in Figure 10.11. The hemoglobin molecules are therefore arranged according to a certain pattern in the red cells.

If there is a high degree of order in the arrangement of the particles, they can form a highly distorted crystal similar to Hosemann's "para-crystal" of Section 9.4. It even appears that this type of arrangement is rather common for macromolecules. The scattering theory for poorly crystallized substances can thus be important for small-angle scattering; many fibrous substances which form elongated micelles, like nylon and polyethylene, give several complex diffuse spots at small angles [21]. Such patterns can be explained by assuming a more or less irregular arrangement of micelles, and they must be interpreted as in Chapter 9. The scattering theory for low-density systems of particles is then completely inapplicable.

An interesting case is that of patterns which at small angles show a fairly definite ring, whose radius $s=\varepsilon/\lambda$ is of the order of 1/20 or 1/40 Ångstroms$^{-1}$. Such rings have been observed with certain alloys—Al Ag and Al Zn [22]—and with certain types of carbon [23]. In the case of the Al Ag alloy, for example, (Section 8.2.6, Figure 8.10), it may be inferred that the ring arises from the segregation of groups of atoms into particles of electron density different from the rest. From the above discussion, there would appear to exist a fair degree of order between neighboring groups, as in a liquid. A more likely hypothesis is that the particles are far from each other and distributed at random, but that they have a complex structure.

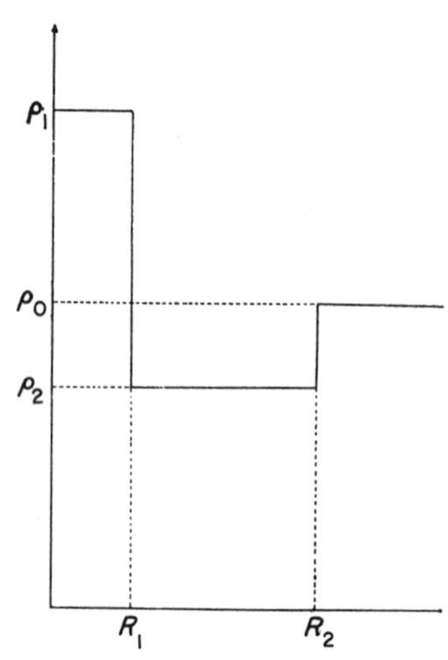


**FIGURE 10.14.** *Schematic representation of the concentration of silver in the clusters of silver atoms in Al Ag alloy as a function of the distance from the center of the zone.*

Let us assume [24] that the silver atoms which are initially distributed uniformly throughout the alloy segregate locally to form nuclei which are then surrounded by regions depleted of silver. This suggests the following model, illustrated in Figure 10.14. In a homogeneous medium of density $\rho_0$ we have a sphere of radius $R_1$ of density $\rho_1 > \rho_0$, surrounded by a spherical shell of radii $R_2$ and $R_1$ and of density $\rho_2 < \rho_0$. Then

$$(\rho_1 - \rho_0)R_1^3 = (\rho_2 - \rho_0)(R_2^3 - R_1^2) .$$

It is as if we had a particle of radius $R_1$ containing $n$ atoms concentric with another particle of radius $R_2$ and also containing $n$ atoms, but which scatter in opposite phase. With the exponential approximation [Eq. (*10.15*)], this composite particle gives rise to a scattered *amplitude* of

$$n \exp\left(-\frac{4\pi^2 s^2 R_1^2}{10}\right) - n \exp\left(-\frac{4\pi^2 s^2 R_2^2}{10}\right)$$

and an intensity of

$$I(s) = n^2\left[\exp\left(-\frac{4\pi^2 s^2 R_1^2}{10}\right) - \exp\left(-\frac{4\pi^2 s^2 R_2^2}{10}\right)\right]^2.$$

This is zero at $s = 0$ and is maximum at

$$s_m = \sqrt{\frac{\ln R_2^2 - \ln R_1^2}{\ln [(4\pi^2/10)(R_2^2 - R_1^2)]}},$$

which explains the observations.

### 10.2.7. Problems Associated with the Theory of Small-angle Scattering

(a) *Scattering by a Large Particle.* Let us consider the scattering by a spherical particle of increasing size. From Eq. (*10.6*), the central peak becomes more and more narrow, and eventually unobservable with the usual instruments when the particle diameter exceeds a few thousand Ångstroms. Simultaneously, the scattering intensity at large angles also increases. From Eq. (*10.25*), which is a simplified form of Eq. (*10.6*), the scattered intensity increases as the square of the particle diameter. Thus, for spheres about $10\mu$ in diameter and for angles of 1° or 2°, these equations lead to scattering intensities which are far larger than the experimental values because such particles give no appreciable scattering at small angles. Equation (*10.6*) can even give a scattering intensity at zero degree which is larger than that of the incident beam!

It is therefore obvious that the above theory cannot be applied to large particles. Warren has shown that it is correct to assume Frauenhofer scattering for small particles, but that for large particles we must assume Fresnel scattering, which yields much smaller intensities [25]. Van de Hulst has developed a general theory that is valid for all particle diameters [26]. Such calculations are interesting in that they explain a paradox, but they are not used in practice since X-ray scattering is never used on large particles.

(b) *Double Bragg Reflection.* Consider a sample composed of small crystallites and a ray reflected from one of these. When the reflected ray meets a second crystallite at the Bragg angle, it is again reflected. If the two crystallites are parallel, the doubly reflected ray emerges parallel to the incident ray, but if they are not quite parallel the ray is slightly deviated. Double reflection can therefore give rise to small-angle scattering. It has been shown quantitatively that the intensity of this scattered radiation is large enough to be observed experimentally and that it decreases with increasing angle. In fact, this scattering can be more intense than the true small-angle scattering resulting from a lack of homogeneity of the electron density.

This phenomenon is now well established and it has been investigated in many crystallites [27]. It is currently common practice to make sure that there is no double reflection when making scattering measurements at small angles. A single crystal, for example, must be oriented in such a way as to avoid Bragg reflections. This requirement can be impossible to satisfy if the crystal is quite distorted, and it now seems extremely difficult to make measurements of small-angle scattering in a crystal which has suffered strong plastic distortions. It has been shown, on the other hand, that double reflection can be neglected in the case of amorphous substances such as carbon.

## 10.3. Experimental Methods

Scattering measurements at very small angles require rather special instruments. It is first necessary to operate by transmission, since reflection would require a very accurate adjustment of the reflecting surface.

Let us consider a beam defined by a pair of slits, as in Figure 10.15, in the path of which the sample is situated. The rays which arrive at a point $P$ on the plane of observation correspond to scattering angles lying within a certain interval $d\varepsilon$ around the average angle $\varepsilon$. The spread $d\varepsilon$ depends on the divergence of the beam, on its cross-section, on the thickness of the sample, etc., and it varies little with $\varepsilon$. The relative error $d\varepsilon/\varepsilon$ is therefore larger for small $\varepsilon$, and the beam must have a much smaller cross-section for small-angle scattering than for the usual type of measurement.

In any case, it is essential to use an extremely fine beam, to permit scattering measurements down to very small angles. Measurements are impossible, not only within the region $MN$ of Figure 10.15, but also within the halo of parasitic radiation which always accompanies the direct beam.

**FIGURE 10.15.** *Effect of beam divergence on the accuracy of the determination of the scattering angle.*

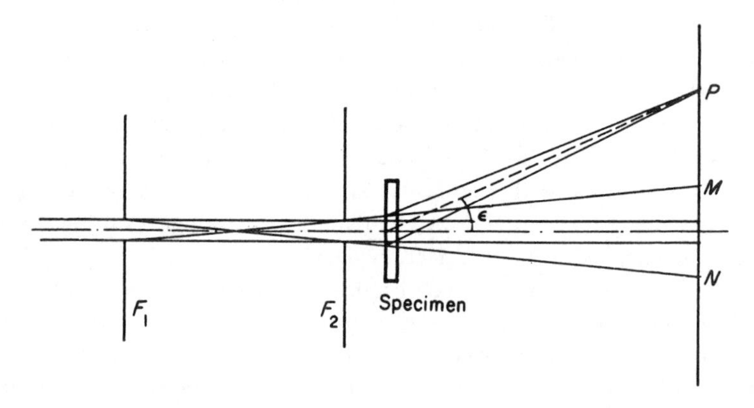

The real difficulty is to produce a sufficiently fine beam with a very small halo, and still have sufficient intensity. Moreover, the radiation must in general be monochromatic. We shall describe two very different experimental methods for solving these difficulties.

### 10.3.1. Measurements with a Geiger Counter [28]

When one does not use a monochromator it is possible to use an ordinary diffractometer with a special slit adjustment and a Ross double filter [Section 6.3.1.]. Three slits define the incident beam. The first one is adjusted for maximum intensity, as close as possible to the tube. The second determines the divergence of the beam, and the third is placed immediately before the sample and adjusted so as to intercept parasitic radiation emerging from the collimator. The third slit may also be placed after the sample, as in Figure 10.16. The counter moves in the plane of observation behind a slit, as shown in the figure. This slit must stop the direct beam completely without producing scattered rays which could be detected by the counter. The two Ross filters are placed successively in front of the counter, and the difference between the two measurements gives the contribution of the monochromatic radiation between the absorption edges of the two filters. The scattering is in general very weak and it is important to operate in a vacuum. If the entire instrument cannot be evacuated, an evacuated tube must be used along the path of the rays between the sample and the counter. The tube can be closed at the ends by means of thin foil. A similar evacuated tube in the collimator serves to reduce absorption.

The distances between the slits as well as their widths are chosen so that (a) there is no appreciable parasitic radiation down to some appropriate angle; (b) the divergence of the scattered beam, and thus the

**FIGURE 10.16.** *Two possible slit arrangements for the measurement of small-angle scattering with a Geiger counter.*

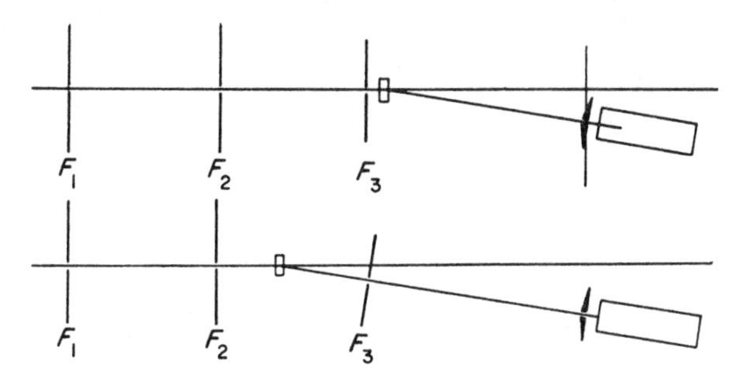

accuracy of the scattering angle, is fixed; (c) the intensity is as large as possible.

An example of such a setup is the following. The X-ray tube has a focus 1 mm wide. At an angle of 1.5° its apparent width, and therefore the width of $F_1$, is 20$\mu$. For Cu K$\alpha$ radiation, it is desirable that parastic radiation be eliminated down to $s = 1/600$ Å$^{-1}$ and that the divergence of the beam incident on the sample be 1/1,000. We then arrive at these conditions:

| | |
|---|---|
| Sample-counter distance | 500 mm |
| Length of collimator | 100 mm |
| Width of second slit | 60 $\mu$ |
| Distance from second to third slit | 80 mm |
| Width of third slit, or effective width of the sample | 120 $\mu$ |

With a Geiger counter it is possible to reach values of $s$ which are of the order of 1/1,000 Å$^{-1}$. One could then resolve lines corresponding to reticular spacings of 1000 Å. For lower resolution it is possible to open up the slits and increase the sensitivity. It is preferable to count the pulses at the counter instead of using an integrator because of the weak intensity involved.

### 10.3.2. Measurements with Monochromators

A crystal-reflected monochromatic beam is generally advantageous and necessary when absolute measurement are required. It is advisable

**FIGURE 10.17.** *Small-angle scattering experiment utilizing a monochromator. The angle $\omega$ is exaggerated for clarity.*

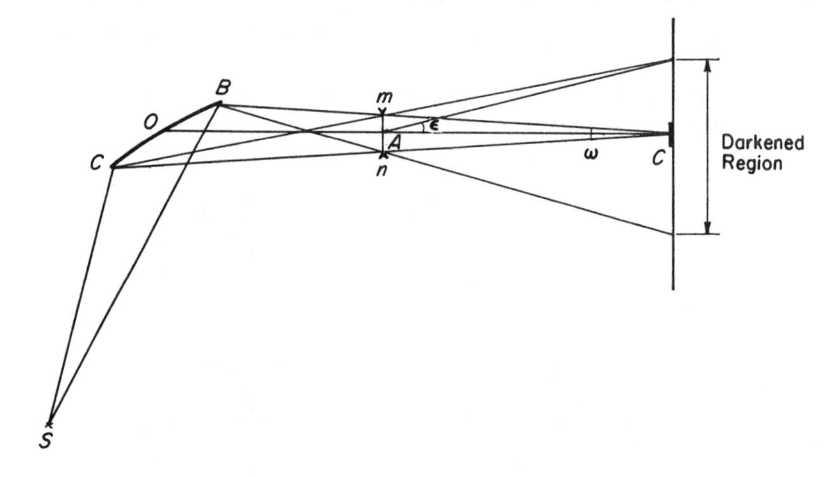

**FIGURE 10.18.** *Small-angle scattering experiment utilizing a double monochromator. $M_1$ and $M_2$, bent-quartz crystals; $F_1$ and $F_2$, slits; ab, sample; B, film (Fournet [6]).*

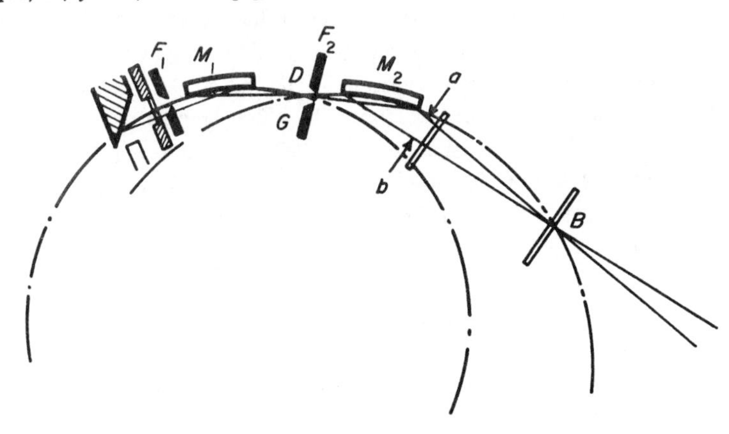


to use a bent-crystal monochromator (Figure 10.17). In this way the reflected beam converges to a fine line so that *the scattering pattern is accurate down to very small angles.* On the other hand, the monochromator crystal is illuminated by the direct beam and acts as a strong source of parasitic radiation which must be eliminated by the slit on the reflected beam. It is thus necessary to reduce the convergence of the rays to an angle which is approximately equal to the smallest angle at which observations are to be made on the pattern. Thus, for an angle of 20′ and for Cu K$\alpha$ radiation, the patterns are limited to $s = \varepsilon/\lambda = (1/250)\ \text{Å}^{-1}$. The sample is placed at *mn* and a narrow copper strip *C* attenuates the direct beam so it is not overexposed. The complete instrument is evacuated.

Parasitic scattering can also be eliminated with a double monochromator [6], as illustrated in Figure 10.18. The beam is reflected successively from two different bent crystals. Since the second crystal receives a beam which is essentially monochromatic, it scatters very little.

As indicated in Chapter 6, a photographic plate is much preferable to a counter for preliminary investigations of unknown patterns, especially in the case of small-angle scattering by macromolecules, for these give complex and unpredictable results. Conversely, a Geiger counter is preferable for a quantitative investigation of an intensity curve when the scattering is continuous and isotropic, as in the case of colloidal solutions.

For absolute measurements the intensity of the direct beam is measured after absorption by a set of calibrated foils.

Whether the beam is defined by a slit collimator or by a monochromator, it is generally flat and not cylindrical. The effect is to distort the pattern, especially at small angles. A small ring, like that of Figure 8.10, can be observed only if the flat beam has a very small height. If the height of the beam were larger than the diameter of the ring, the pattern would reduce to a band parallel to the beam.

It would be preferable to use instruments which give an incident beam converging to a point rather than to a line—X-ray mirrors [3] or point monochromators [29]—but the intensity of such beams is generally very low. Therefore most of the experiments are done with flat beams obtained with a system of slits or with crystal monochromators. The scattering pattern is then distorted.

One measures the intensity $J(s)$ as a function of the distance to the trace of the direct beam, and this is obviously not the function $I(s)$ which appears in the equations of the preceding sections. The quantity $J(s)$ can be used in one of two ways.

(a) It is possible to deduce the true scattering function $I(s)$ from the measured value of $J(s)$. Several techniques of beam height correction have been published, and the calculations are facilitated by the use of analogue computers [30].

(b) The different laws established for $I(s)$ may be adapted to the case of the flat beam.

Some of the transformations are very simple. If $I(s)$ is Gaussian, $J(s)$ is also Gaussian, with the same curvature at the origin [2]. The asymptotic function for the wings of the small-angle scattering corresponding to Eq. (*10.25*) is.

$$s^3 J(s) \rightarrow \frac{S}{16\pi^3}(\rho - \rho_0)^2 .$$

Luzzati [31] has shown that absolute measurements are possible with flat beams.

## 10.4. Applications of Small-angle Scattering

The theory of small-angle scattering is simple in the case of low-density systems of identical particles. This is the case of solutions of large molecules. If the molecular weight is of the order of $10^4$ to $25 \times 10^4$, the molecules have a radius of gyration which permits simple experimental setups. X-Rays are therefore well adapted for the study of the large molecules of biochemistry, and the method has been mostly used for the study of proteins in solution. Measurements of radii of gyration can now be considered to be simple enough to become a routine operation in a biochemical laboratory.

X-Rays have not been used as yet for the determination of the molecular weights of high polymers because, in this case, the particles are not sufficiently similar. Much work has been done on both natural and artificial fibers [32].

Small-angle scattering has also been used for the study of finely divided colloids and catalysts. Although the method cannot provide full information on the particle sizes, since they are very irregular, it can be useful for identification and control in industrial laboratories. The possibility of measuring the total surface of a sample seems particularly interesting.

Finally, the method can be used for both qualitative and quantitative investigations of submicroscopic heterogeneities in solid-state reactions, particularly in metals. In the case of imperfect crystals it is advantageous to use both small-angle scattering and anomalous scattering in order to distinguish between a lattice distortion and a lack of homogeneity. This is because the center of the pattern is affected solely by variations in electron density, while the rest of the pattern depends on both distortion and heterogeneity.

## References


1. J. Strong, *Concepts of Classical Optics*, Freeman, San Francisco (1958).
2. A. Guinier and G. Fournét, *Small-Angle Scattering of X-Rays*, Wiley, New York (1955).
   H. Klug and L. Alexander, *X-Ray Diffraction Procedures*, Wiley, New York (1954), Chapter XII.
3. B. Henke and J. W. M. DuMond, *Phys. Rev.*, **89** (1953), 1300.
   A. Franks, *Proc. Phys. Soc. London, B* **68** (1955), 1054.
4. G. N. Watson, *A Treatise on Bessel Functions*, Cambridge Univ. Press, Cambridge (1958).
5. A. Guinier, *Ann. Phys.*, **12** (1939), 161.
6. G. Fournet, *Bul. Soc. Franç. Minér. Cryst.*, **74** (1951), 39.
7. V. Luzzati, *Acta Cryst.*, **13** (1960) 939.
   V. Luzzati, A. Nicolaieff, and F. Masson, *J. Molec. Biol.*, **3** (1961), 185.
8. R. Hosemann, *Z. Phys.*, **113** (1939), 751; **114** (1939), 133.
9. A. N. J. Heyn, *Text. Res. J.*, **19** (1949), 163.
10. A. Guinier, *Solid State Physics*, Vol. IX, Academic Press, New York (1959). p. 336.
11. B. E. Warren, *J. Appl. Phys.*, **20** (1949), 1782.
12. D. L. Dexter and W. W. Beeman, *Phys. Rev.*, **89** (1953), 1300.
    M. Lambert and A. Guinier, *J. Phys. Rad.*, **17** (1956), 420.
    G. M. Plavnik and B. M. Rovinskii, *Solid State Phys.*, **2** (1960), 994.
13. V. Luzzati, *Acta Cryst.*, **10** (1957), 643.
    R. Baro and V. Luzzati, *Acta Cryst.*, **12** (1959), 144.
14. G. Porod, *Kolloid. Z.*, **124** (1951), 83; **125** (1952), 51 and 109.

15. A. GUINIER and G. FOURNET, *Small-Angle Scattering of X-Rays*, Wiley, New York (1955), p. 193.
16. V. LUZZATI, *Acta Cryst.*, **13** (1960), 939.
17. P. DEBYE and A. M. BUECHE, *J. Appl. Phys.*, **20** (1949), 518.
P. DEBYE, H. R. ANDERSON, JR., and H. BRUMBERGER, *J. Appl. Phys.*, **28** (1957), 679.
18. C. G. SHULL and L. C. ROESS, *J. Appl. Phys.*, **18** (1947), 295 and 308.
19. M. H. JELLINEK, E. SOLOMON, and I. FANKUCHEN, *Ind. Eng. Chem. Anal. Ed.*, **18** (1946), 172.
20. G. FOURNET, D. G. DERVICHIAN, A. GUINIER, and E. PONDER, *Rev. Hematol.*, **7** (1952), 567.
21. B. BELBEOCH and A. GUINIER, *Makromol. Chem.*, **21** (1959), 1.
A. S. POSNER, L. MANDELKERN, C. R. WORTHINGTON, and A. F. PONO, *J. Appl. Phys.*, **31** (1960), 536.
22. A GUINIER, *Solid State Physics*, Vol., IX, Academic Press, New York (1958).
23. D. P. RILEY, *Brit. Coal. Res. Assoc. Conf. London* (1944), p. 232.
24. B. BELBEOCH and A. GUINIER, *Acta Met.*, **3** (1955), 370.
25. B. E. WARREN, *Z. Krist.*, **112** (1959), 255.
26. H. C. VAN DE HULST, *Light Scattering by Small Particles*, Wiley, New York (1957).
27. Proceedings of the Meeting on Small-Angle Scattering, Kansas City, Sept. 1958, *J. Appl. Phys.*, **30** (1959), 601–668.
H. FRICKE and V. GEROLD, *Z. Metal.*, **50** (1959), 136.
M. B. WEBB and W. W. BEEMAN, *Acta Met.*, **7** (1959), 203.
B. E. WARREN, *Acta Cryst.*, **12** (1959), 837.
G. NAGORSEN and B. L. AVERBACH, *J. Appl. Phys.*, **32** (1961), 688.
R. L. WILD, W. T. OGIER, L. M. RICHARDS, and J. C. Nickel, *J. Appl. Phys.*, **32** (1961), 520.
W. T. OGIER, R. L. WILD, and J. C. NICKEL, *J. Appl. Phys.*, **30** (1959), 408.
28. H. N. RITLAND, P. KAESBERG, and W. W. BEEMAN, *J. Chem. Phys.*, **18** (1950), 1237.
29· G. HÄGG and N. KARLSSON, *Acta Cryst.*, **6** (1952), 728.
D. W. BERREMAN, J. W. M. DUMOND, and P. E. MARMIER, *Rev. Sci. Inst.*, **25** (1954), 1219.
30. G. FOURNET and A. GUINIER, *J. Phys. Rad.*, **8** (1947), 345.
K. KRANJC, *Acta Cryst.*, **7** (1954), 709.
V. SYNECEK, *Acta Cryst.*, **13** (1960), 378.
O. KRATKY, G. POROD, and L. KAHOVEC, *Z. Elektroch.*, **55** (1951), 53.
P. W. SCHMIDT and R. HIGHT, *Acta Cryst.*, **13** (1960), 480.
31. V. LUZZATI, *Acta Cryst.*, **10** (1957), 136; **11** (1958), 843.
32. A. N. J. HEYN, *J. Appl. Phys.*, **26** (1955), 519 and 1113.

APPENDIX A

# THE FOURIER TRANSFORMATION

## A.1. Fourier Series for a Function of a Single Variable

Let us consider a periodic function $f(x)$ of period $a$. According to the Fourier theorem, $f(x)$ can be written as a double series of sine and cosine functions whose arguments are of the form $2\pi nx/a$, where $n$ is a positive integer:

$$f(x) = A_0 + 2 \sum_{n=1}^{n=\infty} A_n \cos \frac{2\pi nx}{a} + 2 \sum_{n=1}^{n=\infty} B_n \sin \frac{2\pi nx}{a} . \qquad (A.1)$$

The coefficients of the Fourier series can be calculated from the following identities, in which $m$ is an integer:

$$\int_{-a/2}^{+a/2} \cos \frac{2\pi nx}{a} \cos \frac{2\pi mx}{a} \, dx = \begin{cases} 0 & \text{if } n \neq m, \\ \dfrac{a}{2} & \text{if } n = m, \end{cases}$$

$$\int_{-a/2}^{+a/2} \cos \frac{2\pi nx}{a} \sin \frac{2\pi mx}{a} \, dx = 0 , \qquad (A.2)$$

$$\int_{-a/2}^{+a/2} \sin \frac{2\pi nx}{a} \sin \frac{2\pi mx}{a} \, dx = \begin{cases} 0 & \text{if } n \neq m, \\ \dfrac{a}{2} & \text{if } n = m, \end{cases}$$

To obtain the coefficient $A_m$, we multiply both sides of Eq. ($A.1$) by $\cos(2\pi mx/a)$, and integrate from $-a/2$ to $+a/2$. There remains the single term $\cos(2\pi mx/a)$ on the right:

$$\int_{-a/2}^{+a/2} f(x) \cos \frac{2\pi mx}{a} \, dx = A_m a . \qquad (A.3)$$

Similarly,

$$\int_{-a/2}^{+a/2} f(x) \sin \frac{2\pi mx}{a}\, dx = B_m a\, . \tag{A.4}$$

We also have

$$\int_{-a/2}^{+a/2} f(x)\, dx = A_0 a\, . \tag{A.5}$$

Equation (*A.1*) can also be written in a more symmetrical form:

$$f(x) = \Sigma_{-\infty}^{+\infty} A_n \cos \frac{2\pi nx}{a} + \Sigma_{-\infty}^{+\infty} B_n \sin \frac{2\pi nx}{a}\, , \tag{A.6}$$

where $n$ is an integer which can be either positive, negative, or zero. All the $A_n$ and $B_n$ coefficients can be calculated from Eqs. (*A.3*), (*A.4*), and (*A.5*) and $A_n = A_{-n}$, $B_n = -B_{-n}$, and $B_0 = 0$.

When $f(x)$ is an even function, i.e. when $f(x) = f(-x)$, the $B_n$ coefficients are zero, and when $f(x)$ is odd, i.e. $f(x) = -f(-x)$, the $A_n$ coefficients are zero. If the function is neither even nor odd, it can be written as the sum of an even function plus an odd function:

$$f(x) = f_1(x) + f_2(x) = \frac{f(x) + f(-x)}{2} + \frac{f(x) - f(-x)}{2}\, .$$

The $A_n$ and $B_n$ coefficients correspond, respectively, to series for $f_1(x)$ and $f_2(x)$.

The series for $f(x)$ can be expressed in a more concise form by using imaginary numbers. Let us set

$$C_n = A_n - iB_n\, . \tag{A.7}$$

Then Eq. (*A.6*) can be rewritten as

$$f(x) = \Sigma_{-\infty}^{+\infty} C_n \exp\left(2\pi i \frac{nx}{a}\right), \tag{A.8}$$

for, substituting Eq. (*A.7*) in Eq. (*A.8*), we find that

$$f(x) = \Sigma \left(A_n \cos \frac{2\pi nx}{a} + B_n \sin \frac{2\pi nx}{a}\right) + i\, \Sigma \left(A_n \sin \frac{2\pi nx}{a} - B_n \cos \frac{2\pi nx}{a}\right).$$

Since $A_n = A_{-n}$ and $B_n = -B_{-n}$, the $n$ and $-n$ terms cancel in pairs in the second summation, which is therefore equal to zero, and the right sides of Eqs. (*A.6*) and (*A.8*) are equal. From Eqs. (*A.3*) and (*A.4*), we have

$$C_n = A_n - iB_n = \frac{1}{a} \int_{-a/2}^{a/2} f(x) \exp\left(-\frac{2\pi inx}{a}\right) dx\, . \tag{A.9}$$

It is important to note that the *arguments of the exponential functions of Eqs. (A.8) and (A.9) haoe different signs.* We could, of course, have used different signs in both places.

## A.2. The Fourier Integral

Let us now consider any function $f(x)$ which is continuous in the interval $-a/2$ to $+a/2$, and a periodic function $g(x)$ of period $a$ whose "unit cell" is the function $f(x)$ between $-a/2$ and $+a/2$, the unit cell repeating itself indefinitely on either side.

Let us expand $g(x)$ as a Fourier series, using Eq. (*A.6*). This series gives exactly the function $f(x)$ inside the interval $-a/2$ to $+a/2$, except at $x=-a/2$ and $x=+a/2$ if $f(a/2)$ is different from $f(-a/2)$.

We now introduce the function

$$F(s)=\int_{-a/2}^{+a/2} f(x)\exp(-2\pi isx)\,dx\,, \tag{A.10}$$

which has values $aC_n$, from Eq. (*A.9*), when $s=n/a$, where $n$ is an integer. For some fixed value $x_0$ of $x$, we introduce the following function of $s$:

$$Z(s)=F(s)\exp(2\pi isx_0)\,.$$

For values of $s$ equal to $n/a$,

$$Z\left(\frac{n}{a}\right)=aC_n\exp\left(2\pi i\frac{nx_0}{a}\right).$$

Let us evaluate the summation

$$\Sigma_{n=-\infty}^{n=+\infty}\frac{1}{a}Z\left(\frac{n}{a}\right).$$

From Eq. (*A.8*), this is equal to

$$\frac{1}{a}\Sigma_{-\infty}^{+\infty}\left[aC_n\exp\left(2\pi i\frac{nx_0}{a}\right)\right]=f(x_0)\,.$$

On the other hand, this is an approximate value of the integral

$$\int_{+\infty}^{-\infty}Z(s)\,ds\,,$$

calculated by adding rectangles of width $\Delta=1/a$. Therefore

$$f(x_0)\sim\int_{-\infty}^{+\infty}Z(s)\,ds=\int_{-\infty}^{+\infty}F(s)\exp(2\pi isx_0)\,ds\,. \tag{A.11}$$

We now let $a$ tend to infinity. Then the above approximate equation becomes rigorous, and it is valid for any value $x_0$. Therefore

$$f(x)=\int_{-\infty}^{+\infty}F(s)\exp(2\pi isx)\,ds\,, \tag{A-12}$$

and Eq. (*A.10*) becomes

$$F(s) = \int_{-\infty}^{+\infty} f(x) \exp(-2\pi isx)\, dx\,. \tag{A.13}$$

While this is not a rigorous demonstration, it justifies the relations between the function $f(x)$ and $F(s)$ [1].

*The function $F(s)$ is called the Fourier transform of $f(x)$, and inversely.*

The above two equations lead us to the following two relations. If $f(x)$ is real, $F(s) = F^*(-s)$ and, if we set $F(s) = X(s) - iY(s)$, then $X(s)$ is an even function and $Y(s)$ is an odd function:

$$\begin{aligned} X(s) &= \int_{-\infty}^{+\infty} f(x) \cos(2\pi sx)\, dx\,, \\ Y(s) &= \int_{-\infty}^{+\infty} f(x) \sin(2\pi sx)\, dx\,, \end{aligned} \tag{A.14}$$

and

$$f(x) = \int_{-\infty}^{+\infty} [X(s) \cos 2\pi sx + Y(s) \sin 2\pi sx]\, ds\,. \tag{A.15}$$

If $f(x)$ is both real and even, $F(s)$ is real, so $Y(s)$ is zero; if $f(x)$ is odd, $X(s)$ is zero and $F(s)$ is imaginary.

### A.2.1. The Case of a Real Function Defined for $x > 0$

Let us consider a function $f(x)$ which is defined only for $x > 0$. We set $g(x)$ as a function which is equal to $f(x)$ for $x > 0$, and to $f(-x)$ for $x < 0$. Let us calculate the transform of $g(x)$. Since this function is even,

$$\begin{aligned} G(s) &= \int_{-\infty}^{+\infty} g(x) \cos 2(\pi sx)\, dx \\ &= 2\int_{0}^{+\infty} f(x) \cos(2\pi sx)\, dx\,. \end{aligned}$$

Then, inverting,

$$\begin{aligned} f(x) &= \int_{-\infty}^{+\infty} G(s) \cos(2\pi sx)\, ds \\ &= 2\int_{0}^{\infty} G(s) \cos(2\pi sx)\, ds\,. \end{aligned}$$

Therefore, if we set $F(s)$ as a function defined for $s > 0$ and equal to $G(s)$, the transformation formulas are

$$\begin{aligned} F(s) &= 2\int_{0}^{\infty} f(x) \cos(2\pi sx)\, dx\,, \\ f(x) &= 2\int_{0}^{\infty} F(s) \cos(2\pi sx)\, ds\,. \end{aligned} \tag{A.16}$$

If, instead, $g(x)$ had been chosen equal to $-f(-x)$ for $x < 0$, we would have found the transformation formulas

$$F_1(s) = 2\int_0^\infty f(x)\sin(2\pi sx)\,dx\,,$$
$$f(x) = 2\int_0^\infty F_1(s)\sin(2\pi sx)\,ds\,. \tag{A.17}$$

### A.2.2. Relation Between Object Space and Reciprocal Space

In the case of functions of a single variable, both spaces reduce to straight lines. The variable $x$ is the abscissa in the linear object space, and $s$ is the abscissa on a straight line called the reciprocal space. The Fourier transform thus provides a relation between a given function in object space and a corresponding function in reciprocal space.

Thus, if $f(x)$ has a period $a$, the transform gives the nodes of the reciprocal lattice of period $1/a$ to which are related coefficients which are, in general, complex. If $f(x)$ is not periodic and continuous from $-\infty$ to $+\infty$, the transform is a continuous function of $s$ and is real if $f(x)$ is even, imaginary if $f(x)$ is odd, and, in the general case, complex.

## A.3. Point Functions in Object Space

Let us consider a function of three variables $x_a$, $x_b$, $x_c$ which are the coordinates of a vector $\boldsymbol{x}$ along the three axes $\boldsymbol{a}$, $\boldsymbol{b}$, $\boldsymbol{c}$. The function $f(x_a, x_b, x_c)$ can be written as $f(\boldsymbol{x})$ and is related to the point $M$ of object space defined by the vector $\boldsymbol{x} = \boldsymbol{OM}$.

Let us first consider a triply periodic function in a lattice generated by the vectors $\boldsymbol{a}, \boldsymbol{b}, \boldsymbol{c}$. The function $f(x_a, x_b, x_c)$ is first of all a function of $x_a$, of period $a$, which we can expand as a Fourier series according to Eq. (*A.8*):

$$f(x_a, x_b, x_c) = \sum_{h=-\infty}^{h=+\infty} C_h(x_b, x_c)\exp\left(2\pi i\frac{hx_a}{a}\right),$$

where $h$ is an integer. Each coefficient $C_h$ is a periodic function of $x_b$ with a period $b$. We can therefore write it also as a Fourier series:

$$C_h(x_b, x_c) = \sum_{h=-\infty}^{h=+\infty} C_{hk}(x_c)\exp\left(2\pi i\frac{kx_b}{b}\right),$$

where $C_{hk}(x_c)$ is also a periodic function of $x_c$, or

$$C_{hk}(x_c) = \sum_{l=-\infty}^{l=+\infty} C_{hkl}\exp\left(2\pi i\frac{lx_c}{c}\right).$$

Substituting now in $f(x_a, x_b, x_c)$, we have the triple sum

$$f(\boldsymbol{x}) = f(x_a, x_b, x_c) = \sum_h \sum_k \sum_l C_{hkl} \exp\left[2\pi i\left(\frac{hx_a}{a} + \frac{kx_b}{b} + \frac{lx_b}{c}\right)\right], \qquad (A.18)$$

where the coefficients $C_{hkl}$ are defined by sets of three whole numbers which can be either positive, negative, or zero.

These coefficients can be calculated as in the case of functions of a single variable. Let us calculate the integral

$$\int_{V_c} f(\boldsymbol{x}) \exp\left[-2\pi i\left(h'\frac{x_a}{a} + k'\frac{x_b}{b} + l'\frac{x_c}{c}\right)\right] dv$$

in the unit cell of the lattice, of volume $V_c$, where $h'$, $k'$, $l'$ are three integers. We set $X = x_a/a$, $Y = x_b/b$, $Z = x_c/c$, and observe that the element of volume is given by

$$dv = V_c dX dY dZ\,.$$

Therefore

$$\int_{V_c} f(\boldsymbol{x}) \exp[-2\pi i(h'X + k'Y + l'Z)]\, dv$$
$$= V_c \sum_h \sum_k \sum_l C_{hkl} \int_0^1 \exp[2\pi i(h - h')X]\, dX$$
$$\times \int_0^1 \exp[2\pi i(k - k')Y]\, dY \int_0^1 \exp[2\pi i(l - l')Z]\, dZ\,.$$

But

$$\int_0^1 \exp(2\pi imx)\, dx = \frac{1}{2\pi im}[1 - \exp(2\pi im)]$$

is zero except if $m = 0$, in which case the integral is equal to unity. All the terms of the summation are therefore equal to zero, except the $hkl$ term, and

$$C_{hkl} = \frac{1}{V_c}\int_{V_c} f(\boldsymbol{x}) \exp[-2\pi i(hX + kY + lZ)]\, dv\,. \qquad (A.19)$$

A series of the type shown in Eq. (*A.18*) can be rewritten in an interesting form if we introduce the reciprocal lattice of the periodic lattice in object space. Let $\boldsymbol{a}^*$, $\boldsymbol{b}^*$, $\boldsymbol{c}^*$ be the unit vectors of the reciprocal lattice such that

$$\boldsymbol{a}^*\cdot\boldsymbol{a} = 1\,, \quad \boldsymbol{a}^*\cdot\boldsymbol{b} = 0\,, \quad \boldsymbol{a}^*\cdot\boldsymbol{c} = 0\,, \quad \cdots\,.$$

The vector $\boldsymbol{x}$ is given by $\boldsymbol{x} = X\boldsymbol{a} + Y\boldsymbol{b} + Z\boldsymbol{c}$, and the vector of the reciprocal lattice giving the node $hkl$ is $\boldsymbol{r}^*_{hkl} = h\boldsymbol{a}^* + k\boldsymbol{b}^* + l\boldsymbol{c}^*$. Then

$$\boldsymbol{x}\cdot\boldsymbol{r}^*_{hkl} = hX + kY + lZ\,.$$

The series for $f(x)$ can therefore be written in vector form as

$$f(\boldsymbol{x}) = \sum_h \sum_k \sum_l C_{hkl} \exp(2\pi i\boldsymbol{x}\cdot\boldsymbol{r}^*_{hkl})\,,$$

where $C_{hkl}$ is the *coefficient related to the node hkl of the reciprocal lattice.*

As in the case of a function of a single variable, we consider a function $F(\boldsymbol{s})$ in reciprocal space which is equal to $V_c C_{hkl}$ when $\boldsymbol{s}$ is equal to $\boldsymbol{r}^*_{hkl}$. As $V_c$ tends to infinity, the summations in Eqs. (A.18) and (A.19) become integrals:

$$f(\boldsymbol{x}) = \int F(\boldsymbol{s}) \exp(2\pi i \boldsymbol{s}\cdot\boldsymbol{x})\, dv_s \qquad (A.20)$$

and

$$F(\boldsymbol{s}) = \int f(\boldsymbol{x}) \exp(-2\pi i \boldsymbol{s}\cdot\boldsymbol{x})\, dv_x\,. \qquad (A.21)$$

The two integrals are extended, respectively, to all of object space and to all of reciprocal space.

The following properties are easily demonstrated.

(a) When $f(\boldsymbol{x})$ is real, the transform of $f(-\boldsymbol{x})$ is $F^*(\boldsymbol{s})$ for, if we set $-\boldsymbol{x} = \boldsymbol{u}$, from Eq. (A.21),

$$\int f(-\boldsymbol{x}) \exp(-2\pi i \boldsymbol{s}\cdot\boldsymbol{x})\, dv_x = \int f(\boldsymbol{u}) \exp(2\pi i \boldsymbol{s}\cdot\boldsymbol{u})\, dv_u = F^*(\boldsymbol{s})\,.$$

(b) If $f(\boldsymbol{x})$ is real, from Eq. (A.21) we have

$$F(-\boldsymbol{s}) = F^*(\boldsymbol{s})\,.$$

Then we can set $F(\boldsymbol{s}) = X(\boldsymbol{s}) - iY(\boldsymbol{s})$, where $X(\boldsymbol{s})$ has a center of symmetry and $Y(\boldsymbol{s})$ a center of inversion at the origin.

(c) If $f(\boldsymbol{x})$ is real and centrosymmetric, $F(\boldsymbol{s})$ is also real and centrosymmetric. We have

$$f(\boldsymbol{x}) = \int X(\boldsymbol{s}) \cos(2\pi \boldsymbol{s}\cdot\boldsymbol{x})\, dv_s\,,$$
$$X(\boldsymbol{s}) = \int f(\boldsymbol{x}) \cos(2\pi \boldsymbol{s}\cdot\boldsymbol{x})\, dv_x\,. \qquad (A.22)$$

(d) When the origin is a center of inversion for $f(\boldsymbol{x})$, then $F(\boldsymbol{s}) = iY(\boldsymbol{s})$ is purely imaginary, and $Y(\boldsymbol{s})$ also has a center of inversion at the origin. We then have

$$f(\boldsymbol{x}) = \int Y(\boldsymbol{s}) \sin(2\pi \boldsymbol{s}\cdot\boldsymbol{x})\, dv_s\,,$$
$$Y(\boldsymbol{s}) = \int f(\boldsymbol{x}) \sin(2\pi \boldsymbol{s}\cdot\boldsymbol{x})\, dv_x\,. \qquad (A.23)$$

## A.4. Examples of Fourier Transformations

Table A.1 shows examples of pairs of transforms. Because of the reciprocity of the transformation operation, object space and reciprocal space can be interchanged. The examples can be easily verified by applying the general equations.

TABLE A.1. Functions in Object Space and their Transforms in Reciprocal Space

| Function in object space | Transform in reciprocal space |
|---|---|
| 1. Delta function at the origin:<br>$f(x) = \delta(x)$<br>$\delta(x) = 0$ for $x \neq 0$; $\delta(x) = \infty$ for $x = 0$;<br>$\int \delta(x)\,dx = 1$. | $F(s) = 1$. |
| 2. Delta function at $x = a$:<br>$f(x) = \delta(x - a)$. | $F(s) = \exp(2\pi ias)$<br>$= \cos(2\pi as) + i\sin(2\pi as)$. |
| 3. Two symmetrical delta functions:<br>$f(x) = \delta(x - a) + \delta(x + a)$. | $F(s) = 2\cos(2\pi as)$. |
| 4. Two antisymmetrical delta functions:<br>$f(x) = \delta(x - a) - \delta(x + a)$. | $F(s) = 2i\sin(2\pi as)$. |
| 5. Three-dimensional delta functions:<br>$\delta(\boldsymbol{x}) = 0$ for $\boldsymbol{x} \neq 0$; $\delta(\boldsymbol{x}) = \infty$ for $\boldsymbol{x} = 0$;<br>$\int \delta(\boldsymbol{x})\,dv_x = 1$.<br>Same formulas as in 1, 2, 3, 4; $\boldsymbol{x}$ and $\boldsymbol{a}$ are vectors. | Same formulas with vector $\boldsymbol{s}$ and $as$ replaced by $\boldsymbol{a}\cdot\boldsymbol{s}$. |
| 6. Infinite array of delta functions at the nodes of a linear lattice of period $a$:<br>$f(x) = \sum_{-\infty}^{+\infty} \delta(x - na)$,<br>where $n$ is an integer, positive or negative. | Infinite array of delta functions of value $1/a$ at the nodes of the reciprocal lattice of period $1/a$:<br>$F(s) = \frac{1}{a} \sum_{-\infty}^{+\infty} \delta\left(s - \frac{n}{a}\right)$. |
| 7. Infinite array of delta functions at the nodes of a three-dimensional lattice of unit cell $\boldsymbol{a}, \boldsymbol{b}, \boldsymbol{c}$, volume $V_c$:<br>$z(\boldsymbol{x}) = \sum_u \sum_v \sum_w \delta(\boldsymbol{x} - (u\boldsymbol{a} + v\boldsymbol{b} + w\boldsymbol{c}))$<br>$u$, $v$, and $w$ are integers, positive or negative. | Infinite array of delta functions of value $1/V_c$ at the nodes of the reciprocal lattice $\boldsymbol{a}^*, \boldsymbol{b}^*, \boldsymbol{c}^*$:<br>$Z(\boldsymbol{s}) = 1/V_c \sum_h \sum_k \sum_l \delta[\boldsymbol{s} - (h\boldsymbol{a}^* + k\boldsymbol{b}^* + l\boldsymbol{c}^*)]$,<br>where $h$, $k$, and $l$ are integers, positive or negative. |
| 8. Gaussian function of width $1/K$:<br>$f(x) = K\exp(-\pi K^2 x^2)$.<br>Since $\frac{\int f(x)\,dx}{f(0)} = \frac{1}{K}$,<br>for $K = 1$, $f(x) = \exp(-\pi x^2)$,<br>for $K \to \infty$, $f(x) \to \delta(x)$. | Gaussian function of width $K$:<br>$F(s) = \exp\left(\frac{\pi}{K^2} s^2\right)$.<br>Since $\frac{\int F(s)\,ds}{F(0)} = K$,<br>for $K = 1$, $F(s) = \exp(-\pi s^2)$,<br>for $K \to \infty$, $F(s) \to 1$. |

9. Rectangular function:

$$f(x)\begin{cases}=1 & \text{for } |x| < a/2\,,\\ =0 & \text{for } |x| > a/2\,.\end{cases}$$

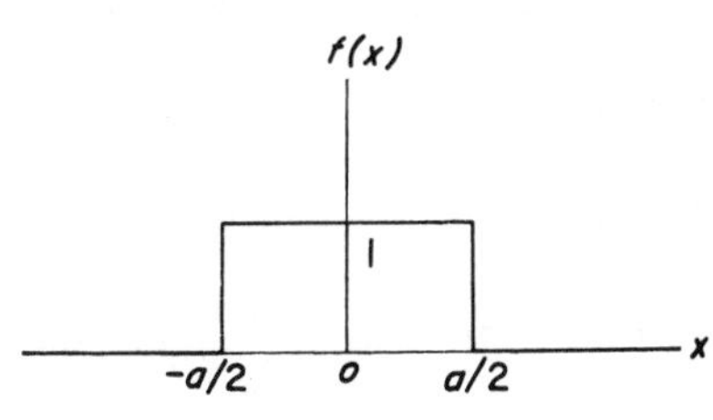


$$F(s) = a\,\frac{\sin \pi sa}{\pi sa}\,.$$

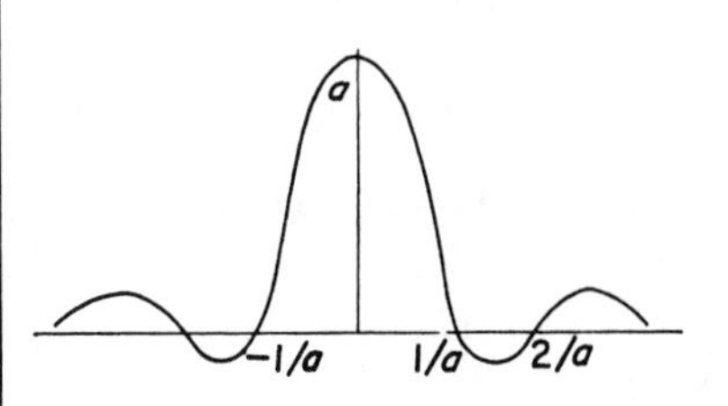


10. Triangular function:

$$f(x)\begin{cases}=a\left(1-\dfrac{|x|}{a}\right) & \text{for } |x| < a\,,\\ =0 & \text{for } |x| > a\,.\end{cases}$$

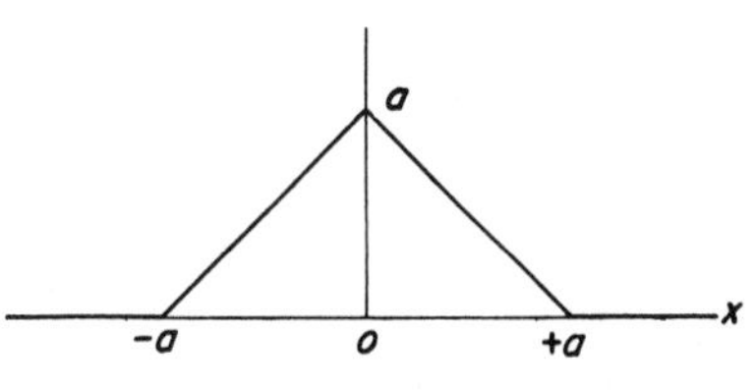


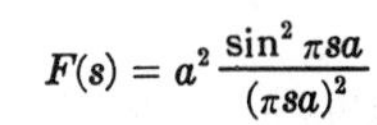

$$F(s) = a^2\,\frac{\sin^2 \pi sa}{(\pi sa)^2}$$

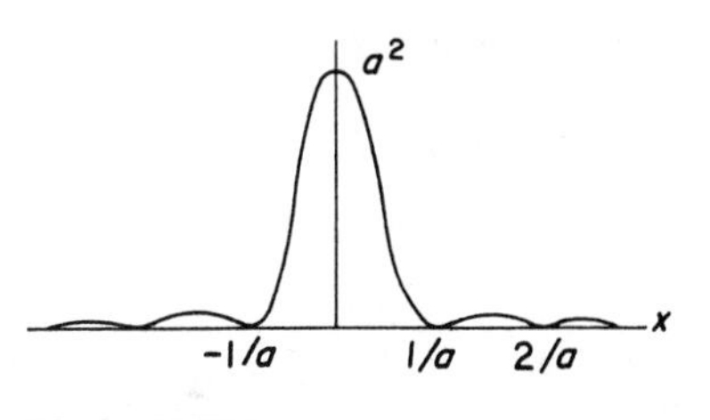


11. Exponential function:

$$f(x) = \exp(-\gamma\,|x|)$$

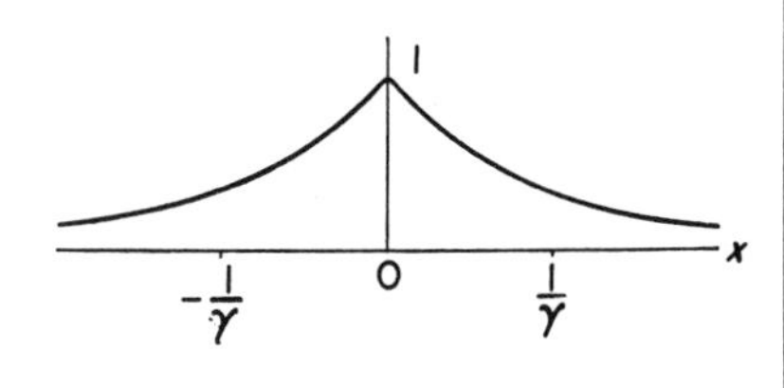


$$F(s) = \frac{2\gamma}{\gamma^2 + (2\pi s)^2} \text{ (Cauchy function)}$$

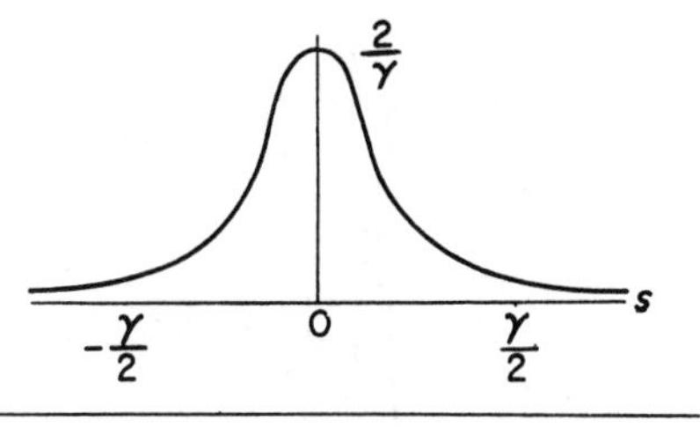


12. Form factor of a sphere:

$$f(\boldsymbol{x})\begin{cases}=1 & \text{for } |\boldsymbol{x}| < R\,,\\ =0 & \text{for } |\boldsymbol{x}| > R\,.\end{cases}$$

$F(\boldsymbol{s})$ has spherical symmetry

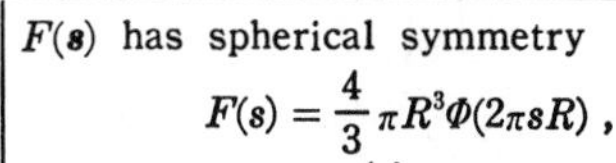

$$F(s) = \frac{4}{3}\pi R^3 \Phi(2\pi s R)\,,$$

$$\text{with } \Phi(u) = 3\,\frac{(\sin u - u\cos u)}{u^3}$$

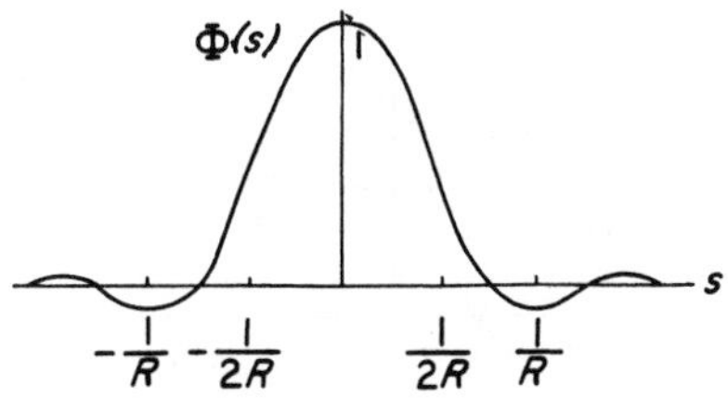

## A.5. The Faltung Product

The faltung or convolution of two functions $f(\boldsymbol{x})$ and $g(\boldsymbol{x})$ is defined as

$$y(\boldsymbol{x}) = \int f(\boldsymbol{u})g(\boldsymbol{x} - \boldsymbol{u})\,dv_u\,, \tag{A.24}$$

where the integral extends to all space. The faltung product is written $y(\boldsymbol{x}) = f(\boldsymbol{x}) * g(\boldsymbol{x})$.

The two functions play completely symmetrical roles, and we can also write

$$y(\boldsymbol{x}) = \int f(\boldsymbol{x} - \boldsymbol{u})g(\boldsymbol{u})\,dv_u\,, \tag{A.25}$$

since the above two equations are equivalent if $\boldsymbol{x} - \boldsymbol{u}$ is substituted for $\boldsymbol{u}$ and $\boldsymbol{u}$ for $\boldsymbol{x} - \boldsymbol{u}$.

**FIGURE A.1.** *Faltung of slowly varying and sharply varying functions.*

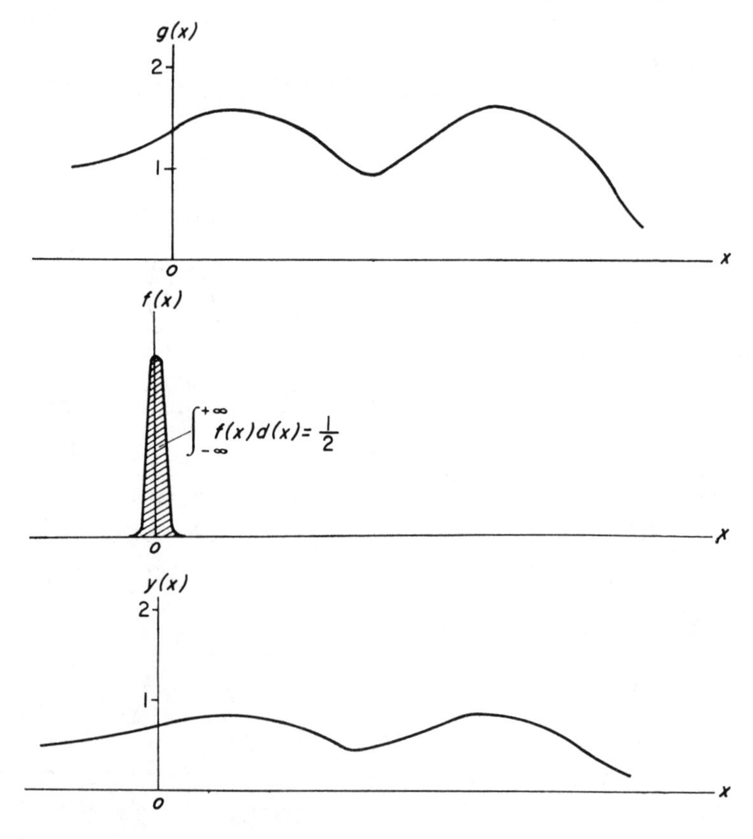

The faltung product becomes very simple if one of the functions, for example $f(\boldsymbol{x})$, reduces to a sharp peak at the origin, while the other function $g(\boldsymbol{x})$ varies slowly with $|\boldsymbol{x}|$ as in Figure A.1. Under these conditions the above integrals cover only a small region near the origin and, for all the values of $\boldsymbol{u}$ where $f(\boldsymbol{u})$ is not zero, $g(\boldsymbol{x}-\boldsymbol{u})$ is approximately equal to its average value $g(\boldsymbol{x})$. Then the function $y(\boldsymbol{x})=f(\boldsymbol{x})*g(\boldsymbol{x})$ becomes

$$y(\boldsymbol{x})=\left[\int f(\boldsymbol{u})\,dv_u\right]g(\boldsymbol{x})$$

and is equal to the product of $g(\boldsymbol{x})$ by a constant factor which is the total area under the curve $f(\boldsymbol{x})$. In particular, if $f(\boldsymbol{x})$ is a Dirac function $\delta(\boldsymbol{x})$, then $\delta(\boldsymbol{x})*g(\boldsymbol{x})=g(\boldsymbol{x})$, since

$$\int \delta(\boldsymbol{x})\,dv_x=1\,.$$

### A.5.1. Fourier Transform of a Faltung

Let us calculate the transform of $y(\boldsymbol{x})=f(\boldsymbol{x})*g(\boldsymbol{x})$:

$$\begin{aligned}Y(\boldsymbol{s})&=\int y(\boldsymbol{x})\exp(2\pi i\boldsymbol{s}\cdot\boldsymbol{x})\,dv_x\\&=\iint f(\boldsymbol{u})g(\boldsymbol{x}-\boldsymbol{u})\exp(2\pi i\boldsymbol{s}\cdot\boldsymbol{x})\,dv_x\,dv_u\,.\end{aligned}$$

We set $\boldsymbol{x}-\boldsymbol{u}=\boldsymbol{W}$ and express $Y(\boldsymbol{s})$ in terms of $\boldsymbol{U}=\boldsymbol{u}$ and $\boldsymbol{W}=\boldsymbol{x}-\boldsymbol{u}$. The differentials $dv_x dv_u$ and $dv_u dv_w$ are equal and

$$\begin{aligned}Y(\boldsymbol{s})&=\iint f(\boldsymbol{U})g(\boldsymbol{W})\exp[2\pi i\boldsymbol{s}\cdot(\boldsymbol{U}+\boldsymbol{W})]\,dv_U\,dv_W\\&=\int f(\boldsymbol{U})\exp(2\pi i\boldsymbol{s}\cdot\boldsymbol{U})\,dv_U\int g(\boldsymbol{W})\exp(2\pi i\boldsymbol{s}\cdot\boldsymbol{W})\,dv_W\\&=F(\boldsymbol{s})G(\boldsymbol{s})\,.\end{aligned}\tag{A.26}$$

*The transform of a faltung is therefore the product of the transforms.* Inversely, the transform of the product of two functions is the faltung of the transforms of the functions.

The faltung of a function $f(\boldsymbol{x})$ by $f(-\boldsymbol{x})$ is

$$z(\boldsymbol{x})=\int f(\boldsymbol{u})f(\boldsymbol{u}-\boldsymbol{x})\,dv_u$$

or, setting $\boldsymbol{w}$ as the vector $\boldsymbol{u}-\boldsymbol{x}$,

$$z(\boldsymbol{x})=\int f(\boldsymbol{x}+\boldsymbol{w})f(\boldsymbol{x})\,dv_w\,,$$

*where* $z(\boldsymbol{x})$ *is sometimes called the autocorrelation function of* $f(\boldsymbol{x})$. The transform of $z(\boldsymbol{x})$ is the product of $F(\boldsymbol{s})$ by the transform of $f(-\boldsymbol{x})$, or $F^*(\boldsymbol{s})$. Therefore

$$\text{transf } z(\boldsymbol{x}) = |F(\boldsymbol{s})|^2 . \tag{A.27}$$

One can show that function 10 of Table A.1 is the autocorrelation function of function 9: its transform is the square of the transform of function 9.

## A.6. Calculation of Fourier Transforms

Two examples of Fourier transforms used in this book are the transformation of the diffracted intensity curve by an amorphous substance, and the transformation of the line profile of a powder pattern.

### A.6.1. Calculation of the Function $P(x)$ from the Intensity Curve

We have seen in Section 3.2.5 that experimental measurements give a function $f(s) = si(s)$, from which the distribution function $P(x)$ is derived through Eq. (*3.18*):

$$P(x) - 1 = \frac{v_1}{x} F(x) ,$$

where $F(x)$ is the transform of $f(s)$, according to Eq. (*A.17*), or

$$F(x) = 2 \int_0^\infty f(s) \sin (2\pi sx) ds . \tag{A.28}$$

Experimentally, one obtains $i(s)$ for values of $s$ ranging from $s_1$, which is small but not zero, and a maximum value $s_0$, which depends both on the geometry of the experiment and on the wavelength used. For example, if diffraction is observed up to $\theta = 90°$ and if $\lambda = 1.54$ Å, then $s_0 = (2 \sin \theta)/\lambda = 1.3$ Å$^{-1}$. It is a simple matter to extrapolate $f(s)$ to $s = 0$ because it is known that $si(s)$ tends to zero. At large angles, on the contrary, the value of $f(s)$ is rather uncertain because $s$ increases while $i(s)$ tends to zero. We shall nevertheless assume that $f(s)$ can be set equal to zero when $s$ is larger than a certain value $s_0$. We shall give two methods of calculating the integral of Eq. (*A.28*).

A.6.1.1. HARMONIC ANALYZER There exist various types of harmonic analyzers (Coradi, Mader-Ott) one of which is illustrated in Figure A.2, which can perform graphically the following operations. We first draw the curve $f(s)$ as a function of $s$ between zero and $p$, and then follow the curve with a pointer as when using a planimeter, returning to the starting point along the axis. The instrument gives directly the value of

**FIGURE A.2.** *Harmonic analyzer* (*Ott*) *with four planimeters giving simultaneously the sine and cosine terms for two harmonics.*

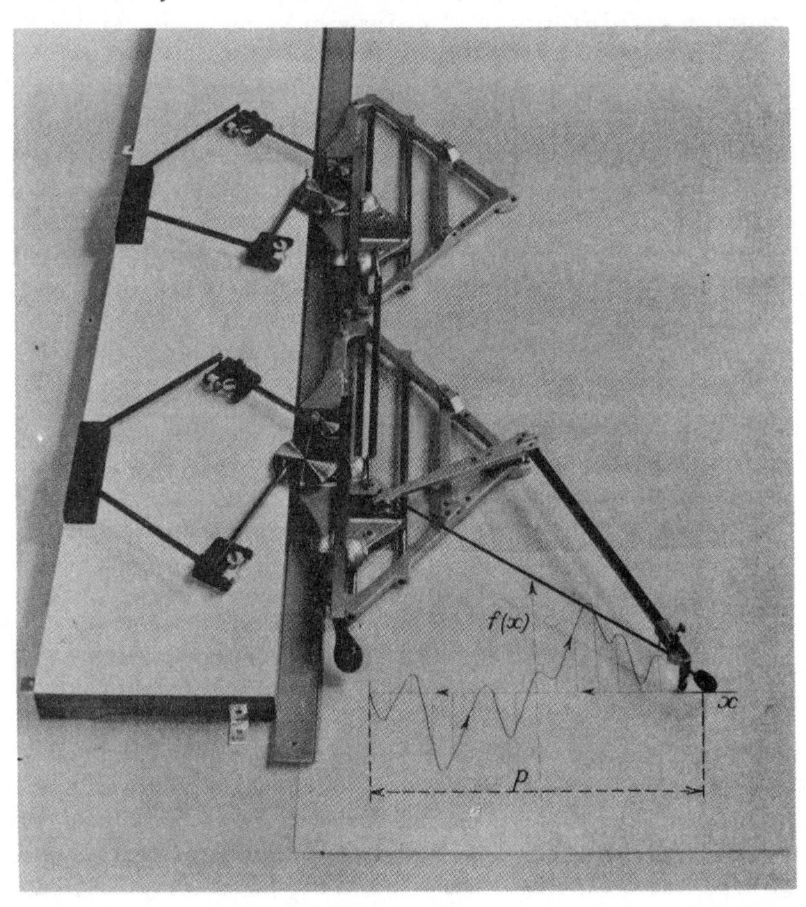


$$K \int_0^p f(s) \sin\left(2\pi \frac{sn}{p}\right) ds \,,$$

where $K$ is a multiplying factor which is determined by the setting of the instrument. The rank $n$ of the harmonic is an integer that can vary between 1 and $N$, according to the setting of the instrument. For usual instruments, $N$ is of the order of 20 to 25. For each value of $n$, we can also calculate the integral with a cosine function instead of a sine. The range $p$ can vary from a few centimeters to approximately 40 cm.

We draw the curve $f(s)$ in such a manner that the amplitude $p$ represents a value of $s$ which is larger than $s_0$, with $f(s)$ being zero from then on. To each harmonic $n$ there corresponds a value of the transform $F(x)$ for $x = n/p$. In $N$ operations the analyzer gives $N$ points on the transform, spaced by distances $1/p$ and extending over the range $0<x<N/p$. For example, let us assume that $s_0$ is equal to 0.9 $Å^{-1}$. We draw the curve in such a way that the amplitude $p$ of the instrument corresponds to a slightly larger value, say 1 $Å^{-1}$. If $N = 25$, $F(x)$ is determined for $x = 1, 2, 3, \cdots, 25$ Å. In most cases $F(x)$ is known to be equal to zero, so $P(x)$ is equal to unity for values of $x$ which are much less than 25 Å. On the other hand, it is useful to have closely spaced points at the beginning of the curve for $P(x)$. It is therefore useful to have $p$ represent a larger value of $s$, say 2, in which case the points found would be 0.5, 1, 1.5, 2, $\cdots$, 12.5 Å. Some analyzers can be used only with a given value of $p$; it is then necessary to redraw the curve of $f(s)$ when the interval of integration is changed.

A.6.1.2. NUMERICAL INTEGRATION [2]. To perform a numerical integration, we replace the integral of Eq. (*A.28*) by a summation, setting $\Delta s = 1/(2x_0)$:

$$F_1(x) = 2 \sum_0^\infty f\left(\frac{n}{2x_0}\right) \sin\left(\pi \frac{nx}{x_0}\right) \frac{1}{2x_0} \,. \tag{A-29}$$

It can be seen that the function $F_1(x)$ has a period of $2x_0$ and that it is odd; it represents the required function $F(x)$ in the interval 0 to $x_0$, as in Figure A.3. The quantity $x_0$ must therefore be chosen considerably larger than the abscissa beyond which $F(x)$ is zero, or $P(x)$ is equal to unity. Since $f(s)$ is nearly zero for $s>s_0$, we must sum a finite number of terms. The interval $\Delta s = 1/(2x_0)$ must be small enough that the fine details of the curve can be taken into account. For the diffraction curves obtained with amorphous substances, one can use $\Delta s = 0.02$ $Å^{-1}$, which gives $x_0 = 25$ Å, and this is amply sufficient to represent $P(x)$. The number of terms in the summation of Eq. (*A.29*) must be such as to give a value of $n/(2x_0)$ which is at least equal to $s_0$, with $f(s)$ being zero from

**FIGURE A.3.** *Calculation of the Fourier transform of a function defined in the interval* 0 *to* $x_0$.

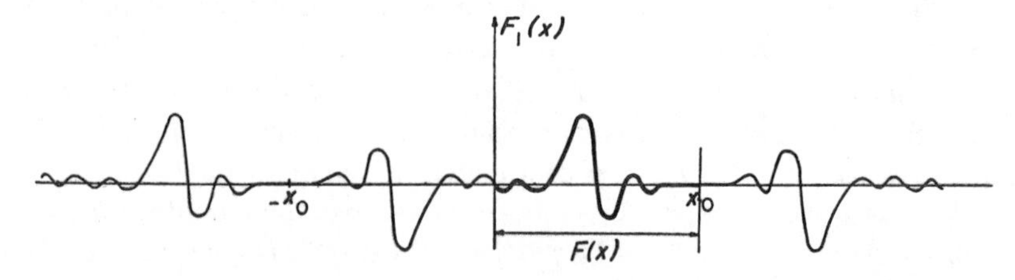

then on. Thus $n/2x_0 < s_0$, or $n < s_0/0.02$, and, in the preceding example, $s_0 = 0.9$ Å$^{-1}$ and we must use 45 terms. Most laboratories now use electronic computers. An integral of the Fourier type is very easily and rapidly calculated even with a medium-sized machine.

### A.6.2. Calculation of the Transform Corresponding to a Line of a Powder Pattern

The line profile is first drawn as a function of $s = (2 \sin \theta)/\lambda$, as far as possible from the center. Let the interval $-s_0$ to $+s_0$ be the region inside which $i(s)$ is different from zero. With a harmonic analyzer we use the interval $2s_0$ as the amplitude of the function to be analyzed, and the instrument gives directly

$$X(t) = \int_{-s_0}^{+s_0} f(s) \cos \frac{\pi n s}{s_0}\, ds\,,$$
$$Y(t) = \int_{-s_0}^{+s_0} f(s) \sin \frac{\pi n s}{s_0}\, ds\,, \qquad (A\text{-}30)$$

where $t = n/2s_0$ and where the rank of the harmonic $n$ varies from 1 to $N = 25$. We therefore determine the values of the transforms for values of $t$ spaced by $1/(2s_0)$ from 0 to $N/(2s_0)$. With a larger interval $2s_0$, we obtain points which are more closely spaced over a smaller interval. For the analysis of diffraction pattern lines, $s_0$ is always quite small and the variable $t$ extends up to several hundreds of Ångströms. It is important to note that the first term of order zero of the cosine series is equal to the integral $\int f(s)\, ds$. If it is possible to determine the area under the curve with the planimeter of the analyzer, it is easy to normalize the transform by using $X(t)/X(0)$ and $Y(t)/Y(0)$ as coefficients.

The numerical integration of Eq. ($A.30$) is performed by an electronic computer. The calculation becomes especially simple when the line profile has an axis of symmetry. This axis is then chosen as origin and we are left only with the cosine terms in the transform.

Let us find the relation between the transforms for the two functions $i(s)$ and $i_1(s) = i(s - s_0)$, the difference between these two functions being due to a change in the origin of $s$. We have

$$I_1(t) = \int_{-\infty}^{+\infty} i_1(s) \exp(2\pi i s t)\, ds = \int_{-\infty}^{+\infty} i(s - s_0) \exp[2\pi i(s - s_0)t] \exp(2\pi i s_0 t)\, ds$$
$$= \exp(2\pi i s_0 t) I(t)\,.$$

Changing the origin in object space is equivalent to multiplying the transform by a factor whose modulus is unity and whose argument is $2\pi s_0 t$. If, for example, the function $i(s)$ is symmetrical, its transform $X(t)$ is real, while the transform of $i_1(s)$ is complex:

**FIGURE A.4.** (a) *Separation of a function into two symmetrical functions $i_1$ and $i_2$.* (b) *Transforms of $i_1$ and $i_2$.* (c) *Transform of the function i. Real part X(t) and imaginary part Y(t).*

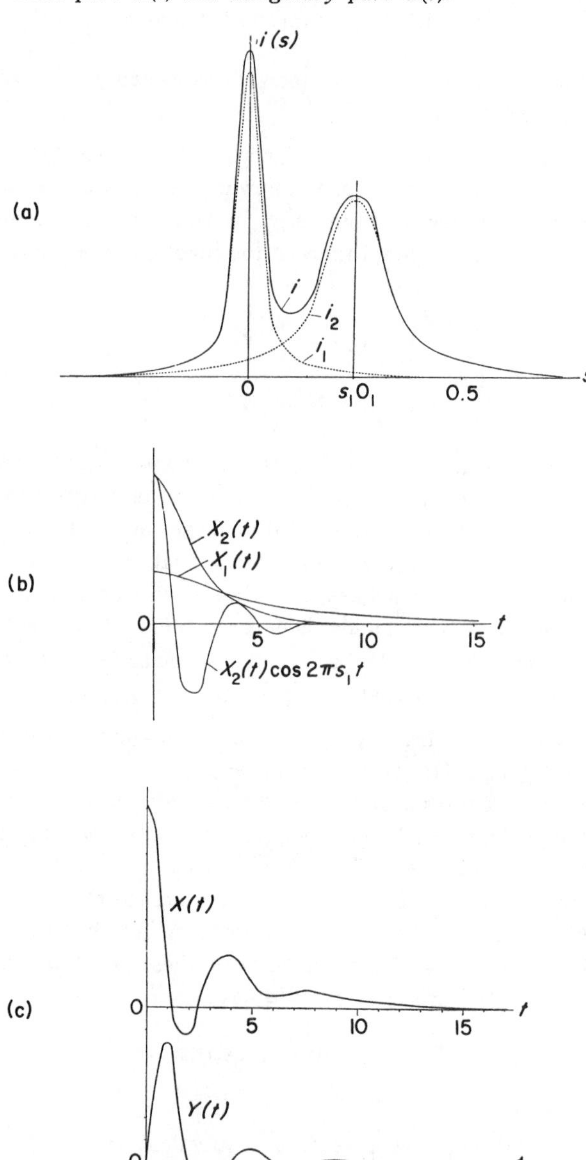

$$I_1(t) = X(t)\cos 2\pi s_0 t + iX(t)\sin 2\pi s_0 t \ .$$

The function $I_1(t)$ oscillates more rapidly as $s_0$ increases. This makes it difficult to calculate the transforms of highly asymmetrical functions because $I(t)$ is in general determined only at points situated a considerable distance apart, and interpolation becomes impossible if the functions oscillate rapidly.

Let us consider a profile $i(s)$ representing the very common case of a doublet, as illustrated in Figure A.4. One simple and accurate way of calculating the transform is to decompose the function into the sum of two different lines. One of these, $i_2(s)$ centered at $s_1$, is chosen to be rigorously symmetrical. Then $i_1(s)$ is very approximately symmetrical with respect to the origin $O$. We then calculate the transforms $X_1(t)$ of $i_1(s)$ and $X_2(t)$ of $i_2(s - s_1)$. These are simple and easy to draw. We can then find the transform of the function $i(s)$:

$$\begin{aligned} X(t) &= X_1(t) + X_2 \cos 2\pi s_1 t \ , \\ Y(t) &= X_2 \sin 2\pi s_1 t \ . \end{aligned} \tag{A.31}$$

In general, a given experimental profile has no fixed origin unless there is an axis of symmetry. Let us examine how this origin must be chosen so that the Fourier transform can be as simple as possible. We expand the transform as a series near the origin:

$$\begin{aligned} I(t) &= \int i(s)\exp(2\pi ist)\,ds \\ &= \int_{-\infty}^{+\infty} i(s)\,ds + 2\pi it \int_{-\infty}^{+\infty} si(s)\,ds - 2\pi^2 t^2 \int_{-\infty}^{+\infty} s^2 i(s)\,ds + \cdots \ . \end{aligned} \tag{A.32}$$

The imaginary part of the transform is therefore tangent to the $t$ axis if the origin is such that

$$\int_{-\infty}^{+\infty} si(s)\,ds = 0 \ .$$

If the origin is chosen at the center of gravity of the line (first moment equal to zero), *the imaginary part is small and the real part hardly oscillates.* This is the optimum position.

According to Eq. (*A.32*), the real part has a horizontal tangent at the origin and its curvature is proportional to the second moment of the function. Example 10 of Table A.1 has a break at the origin because the second moment of the transform,

$$\int_{-\infty}^{+\infty} \frac{\sin^2(\pi sa)\,ds}{\pi^2 a^2} \ ,$$

is infinite, and the radius of curvature is zero. If we calculated numerically the transform of $(\sin^2 \pi sa)/\pi^2 s^2 a^2$, we would necessarily have to use a finite curve and the transform would be somewhat different from the theoretical straight line. This error arises from the finite extent of the function whose transform is required. It is important to realize the existence of this error, which arises in the transformation of lines broadened by the size effect. The transform should theoretically start with a straight portion, as in Section 5.3.2 [3], but the curves obtained experimentally never show this feature and we must therefore never use the section near the origin.

## A.7. Application of the Stokes Method for the Correction of Line Profiles [4]

The experiments give the profiles $i_1(s)$ and $i_0(s)$ of the broadened and of the theoretically infinitely narrow lines, as in Figure A.5a. The problem is to find the true profile of the line $i(s)$, which would be given by an ideal instrument.

The method consists in taking the transforms $I_1(t)$ and $I_0(t)$ of the two experimental functions $i_1$ and $i_0$, as explained in the preceding paragraph, either with a harmonic analyzer or by numerical integration. In general, the two functions are not symmetrical and their transforms are complex, as illustrated in Figure A.5b. We have shown in Section 5.5 that the transform of $I(t)$ is the ratio of $I_1(t)$ divided by $I_0(t)$. Thus

$$I(t) = \frac{I_1(t)}{I_0(t)} = \frac{X_1(t) - iY_1(t)}{X_0(t) - iY_0(t)},$$
$$I(t) = X(t) - i\,Y(t) = \frac{X_0X_1 + Y_1Y_0}{X_0^2 + Y_0^2} - i\,\frac{Y_1X_0 - X_1Y_0}{X_0^2 + Y_0^2}. \qquad (A\text{-}33)$$

The curves of Figure A.5 can therefore give the two functions $X(t)$ and $Y(t)$, using Eq. (*A.33*). Then $i(s)$ is the transform of $I(t)$ and, from Eq. (*A.15*),

$$i(s) = \int_{-\infty}^{+\infty} [X(t)\cos 2\pi tx + Y(t)\sin 2\pi tx]\,dt\,.$$

The same type of calculation is required as for the inverse transformation, and the result gives the curve $i(s)$.

It is important to note that the origins chosen for the two curves $i_1(s)$ and $i_0(s)$ can be arbitrary, for both transforms are multiplied by a complex number of modulus unity. In other words, displacing the origin gives functions $X(t)$ and $Y(t)$, which are different but which correspond to the same function $i(s)$ with a different origin; this has no effect on the shape of the profile.

**FIGURE A.5.** *The Stokes method of correction for a line profile* (*Stokes* [4]). (a) $i_1(s)$, *observed profile*; $i_0(s)$, *profile of standard line*; $i(s)$, *true profile*. (b) *Corresponding transforms* $I_1(t)$, $I_0(t)$, *and* $I(t)$. *The real parts are shown as solid lines and the imaginary parts as dotted lines.*

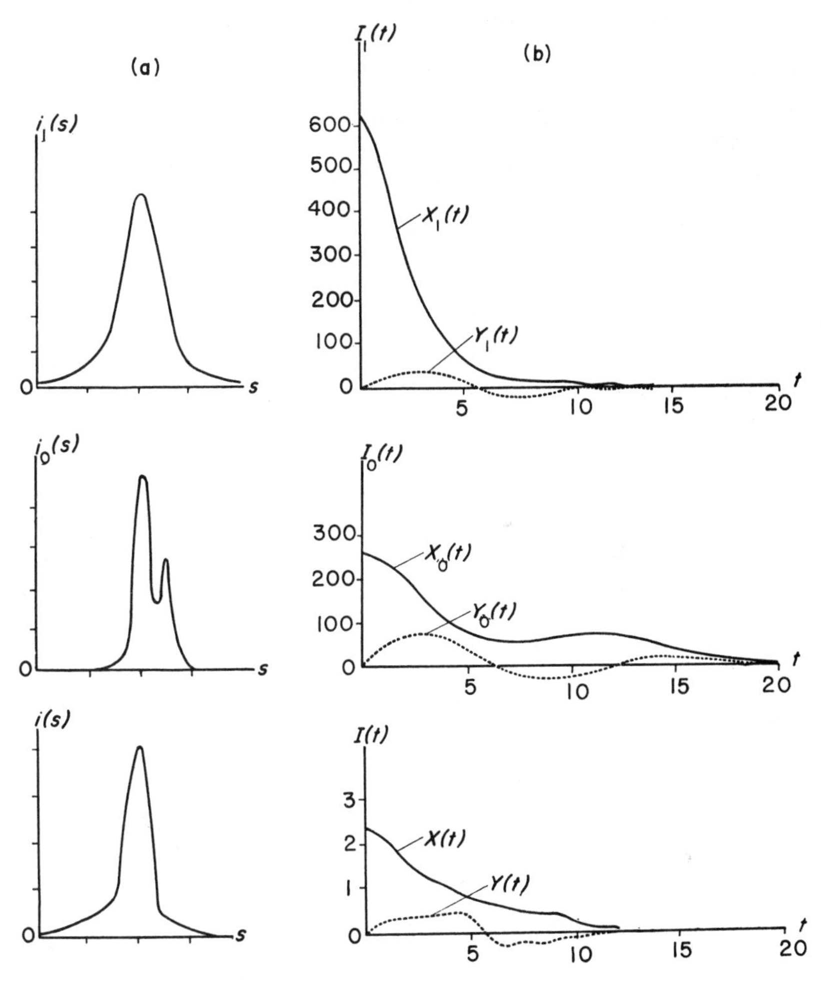


The transforms cannot be evaluated accurately when they are small because they then result from the partial cancellation of positive and negative terms which are both large and which both involve errors proportional to their absolute values. If $I(t)$ is equal to the ratio of two functions which are both small, then it is known only roughly and the inverse transformation gives a function $i(s)$ which is unreliable. In practice, the Stokes method can be used only if the function $I_0(t)$ has an appreciable value throughout the interval where $I_1(t)$ is different from zero. In other words, $I_0(t)$ must be very broad with respect to $I_1(t)$, which means that $i_0(s)$ must be narrow compared to $i_1(s)$. *This method of correction is therefore reliable only if the line broadening due to the apparatus is small compared to the required line width.*

## References

1. E. C. TITCHMARSH, *Theory of Fourier Integrals*, Clarendon Press, Oxford (1937).
2. E. J. W. WHITTAKER, *Acta Cryst.*, **1** (1948), 165.
3. E. F. BERTAUT, *Acta Cryst.*, **5** (1952), 117.
4. A. R. STOKES, *Proc. Phys. Soc. London*, **61** (1948), 382.

APPENDIX B

# CALCULATION OF $\sum \cos nx$, $\sum a^n \cos nx$, $\sum n \cos nx$, and $\sum na^n \cos nx$

Calculations related to X-ray diffraction by crystals often involve the summation of trigonometric terms in which the argument increases according to an arithmetical progression.

## B.1. Calculation of $\sum \cos nx$ and of $\sum \sin nx$

Let us consider the two series

$$A_n = 1 + \cos x + \cos 2x + \cdots + \cos (n-1)x,$$

$$A'_n = \sin x + \sin 2x + \cdots + \sin (n-1)x .$$

Their sums can be evaluated from $A_n + iA'_n$, which is a power series in $\exp(ix)$. This is easily evaluated, and $A_n$ and $A'_n$ can be found by separating the real and imaginary parts.

This result can also be found geometrically, using the Fresnel construction. Vectors of unit length forming angles $x$, $2x$, $\cdots$ with the $x$-axis, as in Figure B.1,

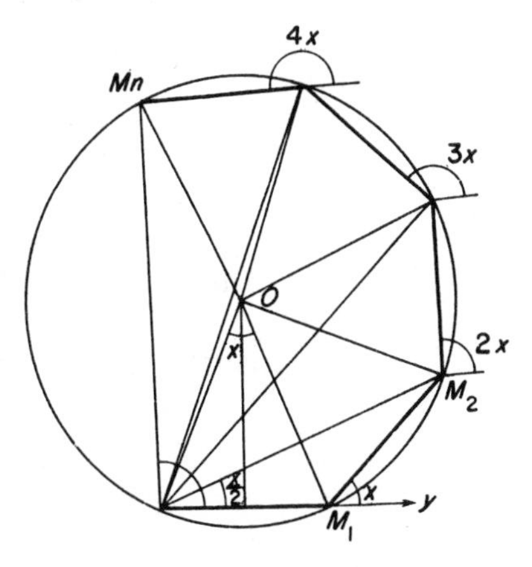


**FIGURE B.1.** *Fresnel construction.*

are added together to give a polygon inscribed inside a circle of radius $1/[2\sin(x/2)]$; the angle subtended by the chord $OM_n$ is $nx$. Thus

$$OM_n = 2\frac{1}{2\sin(x/2)}\sin\frac{nx}{2}.$$

The projections of $OM_n$ on the coordinate axes give $A_n$ and $A'_n$. From the figure, the angle between $OM_n$ and the $x$-axis is $(n-1)x/2$. Therefore

$$\begin{aligned} A_n &= 1 + \cos x + \cdots + \cos(n-1)x = \frac{\sin(nx/2)}{\sin(x/2)}\cos\left[\frac{(n-1)x}{2}\right], \\ A'_n &= \quad\ \sin x + \cdots + \sin(n-1)x = \frac{\sin(nx/2)}{\sin(x/2)}\sin\left[\frac{(n-1)x}{2}\right]. \end{aligned} \tag{B.1}$$

## B.2. Calculation of $\sum a^n \cos nx$ and of $\sum a^n \sin nx$

Let us now set

$$\begin{aligned} B_n &= 1 + a\cos x + a^2\cos 2x + \cdots + a^{n-1}\cos(n-1)x, \\ B'_n &= \quad a\sin x + a^2\sin 2x + \cdots + a^{n-1}\sin(n-1)x. \end{aligned}$$

Then

$$B_n + iB'_n = \frac{1-u^n}{1-u}, \tag{B.2}$$

where $u = a\exp(ix)$. Separating real and imaginary parts,

$$\begin{aligned} B_n &= \frac{1 - a\cos x - a^n\cos nx + a^{n+1}\cos(n-1)x}{1 + a^2 - 2a\cos x}, \\ B'_n &= \frac{a\sin x - a^n\sin nx + a^{n+1}\sin(n-1)x}{1 + a^2 - 2a\cos x}. \end{aligned} \tag{B.3}$$

If $a \leqq 1$, the limiting values for $n \to \infty$ are

$$\begin{aligned} B_\infty &= \frac{1 - a\cos x}{1 + a^2 - 2a\cos x}, \\ B'_\infty &= \frac{a\sin x}{1 + a^2 - 2a\cos x}. \end{aligned} \tag{B.4}$$

For $x \to 0$, we have the approximate formulas

$$\begin{aligned} B_\infty &= \frac{1}{1-a} - \frac{ax^2(1+a)}{2(1-a)^3}, \\ B'_\infty &= \frac{ax}{(1-a)^2}. \end{aligned} \tag{B.5}$$

## B.3. Calculation of $\sum n \cos nx$ and of $\sum n \sin nx$

The quantity

$$C_n = \cos x + 2 \cos 2x + \cdots + (n-1) \cos (n-1)x$$

is *the derivative with respect to x of the series $A'_n$ given in Eq.* ($B.1$):

$$C_n = \frac{d}{dx}\left\{\frac{\sin (nx/2) \sin [(n-1)x/2]}{\sin (x/2)}\right\}.$$

Similarly,

$$C'_n = \sin x + 2 \sin 2x + \cdots + (n-1) \sin(n-1)x = -\frac{dA_n}{dx}$$

$$= -\frac{d}{dx}\left\{\frac{\sin (nx/2) \cos [(n-1(x/2)]}{\sin (x/2)}\right\}$$

and, finally,

$$C_n = \frac{1}{2}\,\frac{n \sin \{[n-(1/2)]x\} \sin (x/2) - \sin^2 (nx/2)}{\sin^2 (x/2)},$$

$$C'_n = \frac{1}{2}\,\frac{n \cos \{[n-(1/2)]x\} \sin (x/2) - \sin (nx/2) \cos (nx/2)}{\sin^2 (x/2)}. \qquad (B.6)$$

From Eqs. ($B.1$) and ($B.6$), we find that

$$\frac{\sin^2 (nx/2)}{\sin^2 (x/2)} = n(2A_n - 1) - 2C_n,$$

whence

$$\frac{\sin^2 (nx/2)}{\sin^2 (x/2)} = n + 2(n-1) \cos x + 2(n-2) \cos 2x + \cdots + 2 \cos (n-1)x. \qquad (B.7)$$

This expansion is a Fourier series which is analogous to the transformation formula 10 of Table A.1.

Another useful expression can be immediately deduced from Eq. ($B.1$):

$$\begin{aligned}\Sigma_{-n/2}^{+n/2} \exp (i\,px) &= 1 + 2\left(\cos x + \cos 2x + \cdots + \cos \frac{n}{2}x\right)\\ &= 2A_{(n/2)+1} - 1 \qquad (B.8)\\ &= \frac{\sin [(n+1)(x/2)]}{\sin (x/2)}.\end{aligned}$$

## B.4. Calculation of $\sum na^n \cos nx$ and of $\sum na^n \sin nx$

These sums can be calculated by using the fact that they are equal to the $x$ derivatives of $B_n$ and $B_n'$. The exact formulas are very complicated and will not be given here.

# INDEX

A CATALOG OF SELECTED

# DOVER BOOKS

IN SCIENCE AND MATHEMATICS

A CATALOG OF SELECTED

# DOVER BOOKS

IN SCIENCE AND MATHEMATICS

## *Astronomy*

BURNHAM'S CELESTIAL HANDBOOK, Robert Burnham, Jr. Thorough guide to the stars beyond our solar system. Exhaustive treatment. Alphabetical by constellation: Andromeda to Cetus in Vol. 1; Chamaeleon to Orion in Vol. 2; and Pavo to Vulpecula in Vol. 3. Hundreds of illustrations. Index in Vol. 3. 2,000pp. 6⅛ x 9¼.
23567-X, 23568-8, 23673-0 Three-vol. set

THE EXTRATERRESTRIAL LIFE DEBATE, 1750–1900, Michael J. Crowe. First detailed, scholarly study in English of the many ideas that developed from 1750 to 1900 regarding the existence of intelligent extraterrestrial life. Examines ideas of Kant, Herschel, Voltaire, Percival Lowell, many other scientists and thinkers. 16 illustrations. 704pp. 5⅜ x 8½. 40675-X

A HISTORY OF ASTRONOMY, A. Pannekoek. Well-balanced, carefully reasoned study covers such topics as Ptolemaic theory, work of Copernicus, Kepler, Newton, Eddington's work on stars, much more. Illustrated. References. 521pp. 5⅜ x 8½.
65994-1

AMATEUR ASTRONOMER'S HANDBOOK, J. B. Sidgwick. Timeless, comprehensive coverage of telescopes, mirrors, lenses, mountings, telescope drives, micrometers, spectroscopes, more. 189 illustrations. 576pp. 5⅜ x 8¼. (Available in U.S. only.)
24034-7

STARS AND RELATIVITY, Ya. B. Zel'dovich and I. D. Novikov. Vol. 1 of *Relativistic Astrophysics* by famed Russian scientists. General relativity, properties of matter under astrophysical conditions, stars, and stellar systems. Deep physical insights, clear presentation. 1971 edition. References. 544pp. 5⅜ x 8¼. 69424-0

## *Chemistry*

CHEMICAL MAGIC, Leonard A. Ford. Second Edition, Revised by E. Winston Grundmeier. Over 100 unusual stunts demonstrating cold fire, dust explosions, much more. Text explains scientific principles and stresses safety precautions. 128pp. 5⅜ x 8½. 67628-5

THE DEVELOPMENT OF MODERN CHEMISTRY, Aaron J. Ihde. Authoritative history of chemistry from ancient Greek theory to 20th-century innovation. Covers major chemists and their discoveries. 209 illustrations. 14 tables. Bibliographies. Indices. Appendices. 851pp. 5⅜ x 8½. 64235-6

CATALYSIS IN CHEMISTRY AND ENZYMOLOGY, William P. Jencks. Exceptionally clear coverage of mechanisms for catalysis, forces in aqueous solution, carbonyl- and acyl-group reactions, practical kinetics, more. 864pp. 5⅜ x 8½.
65460-5

THE HISTORICAL BACKGROUND OF CHEMISTRY, Henry M. Leicester. Evolution of ideas, not individual biography. Concentrates on formulation of a coherent set of chemical laws. 260pp. 5⅜ x 8½. 61053-5

A SHORT HISTORY OF CHEMISTRY, J. R. Partington. Classic exposition explores origins of chemistry, alchemy, early medical chemistry, nature of atmosphere, theory of valency, laws and structure of atomic theory, much more. 428pp. 5⅜ x 8½. (Available in U.S. only.) 65977-1

GENERAL CHEMISTRY, Linus Pauling. Revised 3rd edition of classic first-year text by Nobel laureate. Atomic and molecular structure, quantum mechanics, statistical mechanics, thermodynamics correlated with descriptive chemistry. Problems. 992pp. 5⅜ x 8½. 65622-5

## Engineering

DE RE METALLICA, Georgius Agricola. The famous Hoover translation of greatest treatise on technological chemistry, engineering, geology, mining of early modern times (1556). All 289 original woodcuts. 638pp. 6¾ x 11. 60006-8

FUNDAMENTALS OF ASTRODYNAMICS, Roger Bate et al. Modern approach developed by U.S. Air Force Academy. Designed as a first course. Problems, exercises. Numerous illustrations. 455pp. 5⅜ x 8½. 60061-0

DYNAMICS OF FLUIDS IN POROUS MEDIA, Jacob Bear. For advanced students of ground water hydrology, soil mechanics and physics, drainage and irrigation engineering and more. 335 illustrations. Exercises, with answers. 784pp. 6⅛ x 9¼. 65675-6

ANALYTICAL MECHANICS OF GEARS, Earle Buckingham. Indispensable reference for modern gear manufacture covers conjugate gear-tooth action, gear-tooth profiles of various gears, many other topics. 263 figures. 102 tables. 546pp. 5⅜ x 8½. 65712-4

MECHANICS, J. P. Den Hartog. A classic introductory text or refresher. Hundreds of applications and design problems illuminate fundamentals of trusses, loaded beams and cables, etc. 334 answered problems. 462pp. 5⅜ x 8½. 60754-2

MECHANICAL VIBRATIONS, J. P. Den Hartog. Classic textbook offers lucid explanations and illustrative models, applying theories of vibrations to a variety of practical industrial engineering problems. Numerous figures. 233 problems, solutions. Appendix. Index. Preface. 436pp. 5⅜ x 8½. 64785-4

STRENGTH OF MATERIALS, J. P. Den Hartog. Full, clear treatment of basic material (tension, torsion, bending, etc.) plus advanced material on engineering methods, applications. 350 answered problems. 323pp. 5⅜ x 8½. 60755-0

A HISTORY OF MECHANICS, René Dugas. Monumental study of mechanical principles from antiquity to quantum mechanics. Contributions of ancient Greeks, Galileo, Leonardo, Kepler, Lagrange, many others. 671pp. 5⅜ x 8½. 65632-2

METAL FATIGUE, N. E. Frost, K. J. Marsh, and L. P. Pook. Definitive, clearly written, and well-illustrated volume addresses all aspects of the subject, from the historical development of understanding metal fatigue to vital concepts of the cyclic stress that causes a crack to grow. Includes 7 appendixes. 544pp. 5⅜ x 8½. 40927-9

STATISTICAL MECHANICS: Principles and Applications, Terrell L. Hill. Standard text covers fundamentals of statistical mechanics, applications to fluctuation theory, imperfect gases, distribution functions, more. 448pp. 5⅜ x 8½. 65390-0

THE VARIATIONAL PRINCIPLES OF MECHANICS, Cornelius Lanczos. Graduate level coverage of calculus of variations, equations of motion, relativistic mechanics, more. First inexpensive paperbound edition of classic treatise. Index. Bibliography. 418pp. 5⅜ x 8½. 65067-7

THE VARIOUS AND INGENIOUS MACHINES OF AGOSTINO RAMELLI: A Classic Sixteenth-Century Illustrated Treatise on Technology, Agostino Ramelli. One of the most widely known and copied works on machinery in the 16th century. 194 detailed plates of water pumps, grain mills, cranes, more. 608pp. 9 x 12. 28180-9

ORDINARY DIFFERENTIAL EQUATIONS AND STABILITY THEORY: An Introduction, David A. Sánchez. Brief, modern treatment. Linear equation, stability theory for autonomous and nonautonomous systems, etc. 164pp. 5⅝ x 8¼. 63828-6

ROTARY WING AERODYNAMICS, W. Z. Stepniewski. Clear, concise text covers aerodynamic phenomena of the rotor and offers guidelines for helicopter performance evaluation. Originally prepared for NASA. 537 figures. 640pp. 6⅛ x 9¼. 64647-5

INTRODUCTION TO SPACE DYNAMICS, William Tyrrell Thomson. Comprehensive, classic introduction to space-flight engineering for advanced undergraduate and graduate students. Includes vector algebra, kinematics, transformation of coordinates. Bibliography. Index. 352pp. 5⅜ x 8½. 65113-4

HISTORY OF STRENGTH OF MATERIALS, Stephen P. Timoshenko. Excellent historical survey of the strength of materials with many references to the theories of elasticity and structure. 245 figures. 452pp. 5⅜ x 8½. 61187-6

ANALYTICAL FRACTURE MECHANICS, David J. Unger. Self-contained text supplements standard fracture mechanics texts by focusing on analytical methods for determining crack-tip stress and strain fields. 336pp. 6⅛ x 9¼. 41737-9

## *Mathematics*

HANDBOOK OF MATHEMATICAL FUNCTIONS WITH FORMULAS, GRAPHS, AND MATHEMATICAL TABLES, edited by Milton Abramowitz and Irene A. Stegun. Vast compendium: 29 sets of tables, some to as high as 20 places. 1,046pp. 8 x 10½. 61272-4

FUNCTIONAL ANALYSIS (Second Corrected Edition), George Bachman and Lawrence Narici. Excellent treatment of subject geared toward students with background in linear algebra, advanced calculus, physics and engineering. Text covers introduction to inner-product spaces, normed, metric spaces, and topological spaces; complete orthonormal sets, the Hahn-Banach Theorem and its consequences, and many other related subjects. 1966 ed. 544pp. 6⅛ x 9¼. 40251-7

ASYMPTOTIC EXPANSIONS OF INTEGRALS, Norman Bleistein & Richard A. Handelsman. Best introduction to important field with applications in a variety of scientific disciplines. New preface. Problems. Diagrams. Tables. Bibliography. Index. 448pp. 5⅜ x 8½. 65082-0

FAMOUS PROBLEMS OF GEOMETRY AND HOW TO SOLVE THEM, Benjamin Bold. Squaring the circle, trisecting the angle, duplicating the cube: learn their history, why they are impossible to solve, then solve them yourself. 128pp. 5⅜ x 8½. 24297-8

VECTOR AND TENSOR ANALYSIS WITH APPLICATIONS, A. I. Borisenko and I. E. Tarapov. Concise introduction. Worked-out problems, solutions, exercises. 257pp. 5⅝ x 8¼. 63833-2

THE ABSOLUTE DIFFERENTIAL CALCULUS (CALCULUS OF TENSORS), Tullio Levi-Civita. Great 20th-century mathematician's classic work on material necessary for mathematical grasp of theory of relativity. 452pp. 5⅝ x 8¼. 63401-9

AN INTRODUCTION TO ORDINARY DIFFERENTIAL EQUATIONS, Earl A. Coddington. A thorough and systematic first course in elementary differential equations for undergraduates in mathematics and science, with many exercises and problems (with answers). Index. 304pp. 5⅜ x 8½. 65942-9

FOURIER SERIES AND ORTHOGONAL FUNCTIONS, Harry F. Davis. An incisive text combining theory and practical example to introduce Fourier series, orthogonal functions and applications of the Fourier method to boundary-value problems. 570 exercises. Answers and notes. 416pp. 5⅜ x 8½. 65973-9

COMPUTABILITY AND UNSOLVABILITY, Martin Davis. Classic graduate-level introduction to theory of computability, usually referred to as theory of recurrent functions. New preface and appendix. 288pp. 5⅜ x 8½. 61471-9

ASYMPTOTIC METHODS IN ANALYSIS, N. G. de Bruijn. An inexpensive, comprehensive guide to asymptotic methods–the pioneering work that teaches by explaining worked examples in detail. Index. 224pp. 5⅜ x 8½ 64221-6

ESSAYS ON THE THEORY OF NUMBERS, Richard Dedekind. Two classic essays by great German mathematician: on the theory of irrational numbers; and on transfinite numbers and properties of natural numbers. 115pp. 5⅜ x 8½. 21010-3

# Physics

OPTICAL RESONANCE AND TWO-LEVEL ATOMS, L. Allen and J. H. Eberly. Clear, comprehensive introduction to basic principles behind all quantum optical resonance phenomena. 53 illustrations. Preface. Index. 256pp. 5⅜ x 8½. 65533-4

ULTRASONIC ABSORPTION: An Introduction to the Theory of Sound Absorption and Dispersion in Gases, Liquids and Solids, A. B. Bhatia. Standard reference in the field provides a clear, systematically organized introductory review of fundamental concepts for advanced graduate students, research workers. Numerous diagrams. Bibliography. 440pp. 5⅜ x 8½. 64917-2

QUANTUM THEORY, David Bohm. This advanced undergraduate-level text presents the quantum theory in terms of qualitative and imaginative concepts, followed by specific applications worked out in mathematical detail. Preface. Index. 655pp. 5⅜ x 8½. 65969-0

ATOMIC PHYSICS (8th edition), Max Born. Nobel laureate's lucid treatment of kinetic theory of gases, elementary particles, nuclear atom, wave-corpuscles, atomic structure and spectral lines, much more. Over 40 appendices, bibliography. 495pp. 5⅜ x 8½. 65984-4

AN INTRODUCTION TO HAMILTONIAN OPTICS, H. A. Buchdahl. Detailed account of the Hamiltonian treatment of aberration theory in geometrical optics. Many classes of optical systems defined in terms of the symmetries they possess. Problems with detailed solutions. 1970 edition. xv + 360pp. 5⅜ x 8½. 67597-1

THIRTY YEARS THAT SHOOK PHYSICS: The Story of Quantum Theory, George Gamow. Lucid, accessible introduction to influential theory of energy and matter. Careful explanations of Dirac's anti-particles, Bohr's model of the atom, much more. 12 plates. Numerous drawings. 240pp. 5⅜ x 8½. 24895-X

ELECTRONIC STRUCTURE AND THE PROPERTIES OF SOLIDS: The Physics of the Chemical Bond, Walter A. Harrison. Innovative text offers basic understanding of the electronic structure of covalent and ionic solids, simple metals, transition metals and their compounds. Problems. 1980 edition. 582pp. 6⅛ x 9¼. 66021-4

HYDRODYNAMIC AND HYDROMAGNETIC STABILITY, S. Chandrasekhar. Lucid examination of the Rayleigh-Benard problem; clear coverage of the theory of instabilities causing convection. 704pp. 5⅜ x 8¼. 64071-X

INVESTIGATIONS ON THE THEORY OF THE BROWNIAN MOVEMENT, Albert Einstein. Five papers (1905–8) investigating dynamics of Brownian motion and evolving elementary theory. Notes by R. Fürth. 122pp. 5⅜ x 8½. 60304-0

THE PHYSICS OF WAVES, William C. Elmore and Mark A. Heald. Unique overview of classical wave theory. Acoustics, optics, electromagnetic radiation, more. Ideal as classroom text or for self-study. Problems. 477pp. 5⅜ x 8½. 64926-1

PHYSICAL PRINCIPLES OF THE QUANTUM THEORY, Werner Heisenberg. Nobel Laureate discusses quantum theory, uncertainty, wave mechanics, work of Dirac, Schroedinger, Compton, Wilson, Einstein, etc. 184pp. 5⅜ x 8½. 60113-7

ATOMIC SPECTRA AND ATOMIC STRUCTURE, Gerhard Herzberg. One of best introductions; especially for specialist in other fields. Treatment is physical rather than mathematical. 80 illustrations. 257pp. 5⅜ x 8½. 60115-3

AN INTRODUCTION TO STATISTICAL THERMODYNAMICS, Terrell L. Hill. Excellent basic text offers wide-ranging coverage of quantum statistical mechanics, systems of interacting molecules, quantum statistics, more. 523pp. 5⅜ x 8½. 65242-4

THEORETICAL PHYSICS, Georg Joos, with Ira M. Freeman. Classic overview covers essential math, mechanics, electromagnetic theory, thermodynamics, quantum mechanics, nuclear physics, other topics. First paperback edition. xxiii + 885pp. 5⅜ x 8½. 65227-0

PROBLEMS AND SOLUTIONS IN QUANTUM CHEMISTRY AND PHYSICS, Charles S. Johnson, Jr. and Lee G. Pedersen. Unusually varied problems, detailed solutions in coverage of quantum mechanics, wave mechanics, angular momentum, molecular spectroscopy, more. 280 problems plus 139 supplementary exercises. 430pp. 6½ x 9¼. 65236-X

THEORETICAL SOLID STATE PHYSICS, Vol. 1: Perfect Lattices in Equilibrium; Vol. II: Non-Equilibrium and Disorder, William Jones and Norman H. March. Monumental reference work covers fundamental theory of equilibrium properties of perfect crystalline solids, non-equilibrium properties, defects and disordered systems. Appendices. Problems. Preface. Diagrams. Index. Bibliography. Total of 1,301pp. 5⅜ x 8½. Two volumes. Vol. I: 65015-4 Vol. II: 65016-2

A TREATISE ON ELECTRICITY AND MAGNETISM, James Clerk Maxwell. Important foundation work of modern physics. Brings to final form Maxwell's theory of electromagnetism and rigorously derives his general equations of field theory. 1,084pp. 5⅜ x 8½. Two-vol. set. Vol. I: 60636-8 Vol. II: 60637-6

OPTICKS, Sir Isaac Newton. Newton's own experiments with spectroscopy, colors, lenses, reflection, refraction, etc., in language the layman can follow. Foreword by Albert Einstein. 532pp. 5⅜ x 8½. 60205-2

THEORY OF ELECTROMAGNETIC WAVE PROPAGATION, Charles Herach Papas. Graduate-level study discusses the Maxwell field equations, radiation from wire antennas, the Doppler effect and more. xiii + 244pp. 5⅜ x 8½. 65678-5

INTRODUCTION TO QUANTUM MECHANICS With Applications to Chemistry, Linus Pauling & E. Bright Wilson, Jr. Classic undergraduate text by Nobel Prize winner applies quantum mechanics to chemical and physical problems. Numerous tables and figures enhance the text. Chapter bibliographies. Appendices. Index. 468pp. 5⅜ x 8½. 64871-0